VIRAL INFECTIONS AND CELL BIOLOGY

VIRAL INFECTIONS AND CELL BIOLOGY

VOLUME 2

Simin Liu

Kruger Brentt
Publishers

2025

Kruger Brentt Publishers UK. LTD.
Company Number 9728962

Regd. Office: 68 St Margarets Road, Edgware, Middlesex HA8 9UU

ISBN: 978-1-78715-435-3 (Vol. 2)

For information on all our publications visit our website at http://krugerbrentt.com/

PREFACE

"Viral Infections and Cell Biology" offers an in-depth exploration into the intricate interactions between viruses and host cells, providing insights into the molecular mechanisms underlying viral infections and their impact on cellular processes. This book serves as an essential resource for researchers, clinicians, students, and educators interested in understanding the complex interplay between viruses and the host cellular machinery.

Viral infections pose significant threats to human health, agriculture, and ecosystems, causing a wide range of diseases, including respiratory infections, gastroenteritis, hepatitis, and cancer. Understanding the molecular mechanisms of viral infection is crucial for developing effective antiviral therapies, vaccines, and diagnostic tools.

As editors of this volume, we have endeavored to compile a diverse array of perspectives, methodologies, and case studies from leading experts and practitioners in the field of viral infections and cell biology. Through a blend of theoretical foundations, experimental approaches, and clinical insights, this book covers a wide range of topics, including viral entry and replication mechanisms, host-virus interactions, immune responses to viral infections, and the development of antiviral strategies.

Our overarching goal in presenting this book is to provide readers with a comprehensive understanding of the molecular and cellular basis of viral infections. By offering insights into the latest advancements, best practices, and emerging trends in viral pathogenesis and host response, we aim to facilitate the development of innovative approaches for the prevention, diagnosis, and treatment of viral diseases.

We extend our sincere gratitude to all the contributors who have generously shared their expertise, experiences, and insights in the field of viral infections and cell biology. It is our hope that this book will serve as a valuable reference and guide for researchers, clinicians, students, and educators as they work towards combating viral diseases and improving global health outcomes.

Simin Liu

CONTENTS

13 CELL DEATH HAS A ROLE IN THE INFECTION OF MAMMALS AND MOSQUITOES BY ALPHAVIRUSES

1. INTRODUCTION TO ALPHAVIRUSES

Viruses of the Alphavirus genus are classified as arboviruses, which refers to viruses transmitted by arthropods.[1] These viruses are spread by a blood-sucking arthropod to a vertebrate host while feeding on blood. Mosquitoes are the main carriers of alphaviruses [2].

Alphaviruses are diminutive, enclosed viruses with a diameter of roughly 70 nm. The RNA molecule is single-stranded and has a positive-sense. It comprises two open reading frames (ORFs) which code for four non-structural proteins (nsp1–4) and five structural proteins (capsid, E3, E2, 6K, and E1) [3]. Alphaviruses consist of over 30 members, and infection leads to a wide spectrum of clinical symptoms, varying from moderate to severe [3]. Traditionally, alphaviruses were classified as New World and Old World alphaviruses based on their worldwide distribution, evolutionary history, ability to cause disease, and their preference for specific tissues and cells or interactions with their respective hosts. Old World alphaviruses, such as chikungunya virus (CHIKV), Sindbis virus (SINV), Semliki Forest virus (SFV), and Ross River virus (RRV), were primarily distributed in Asia, Africa, and Europe. On the other hand, New World alphaviruses, including Eastern, Western, Venezuelan, and Equine Encephalitis Viruses (EEVs), as well as Mayaro virus (MAYV), were predominantly found in North and South America. Nevertheless, due to the worldwide dissemination of these viruses and their carriers, the distinction between the New and Old World has become outdated. Alphaviruses are currently classified into three distinct categories: aquatic viruses, arthritogenic viruses, and encephalitic viruses [4,5].

Human infections caused by arthritogenic viruses are distinguished by the presence of rashes, fever, joint and muscular discomfort, and in some cases, encephalitis (such as SINV and SFV). Arthralgia and myalgia can persist for an extended period of time, ranging from months to years, following infection with some viruses such as CHIKV, RRV, and MAYV. Encephalitic virus infections are marked by severe fever illness and inflammation of the brain and spinal cord, which can result in mortality in certain instances (such as EEEV and VEEV) [3,6]. When a mosquito infected with an arbovirus, like an alphavirus, bites a mammal, the first cells to be affected are the skin cells. The specific cell types that serve as the initial line of defense against alphaviruses are not well-defined for each virus. However, potential candidates include dermal fibroblasts [7], dermal dendritic cells, enterocytes, or keratinocytes [8,9]. Subsequently, viruses will spread to other organs, including joints, muscles [10], and the brain, where they will initiate disease by causing cell death in the acute phase or prolonged inflammation in the chronic phase [6].

The arboviral infection in mosquito vectors is chronic and endures throughout the insect's whole lifespan. When comparing the impact on the vertebrate host, alphaviruses do not appear to produce considerable damage in the mosquito vector. While certain arboviruses have been found to have fitness drawbacks [11,12,13], numerous studies have indicated that arboviral infection is predominantly asymptomatic and that mosquito vectors exhibit a high level of tolerance against arboviruses [14,15]. Alphaviruses reproduce in arthropod cells and must overcome antiviral pathways from the moment they acquire a viremic bloodmeal orally to when they transmit to a new uninfected vertebrate host. In mosquitoes, the initial obstacle to overcome is the gut epithelium, also known as the midgut barrier. This is where the virus reproduces before entering the hemocoel, which is the bodily cavity containing blood. Once inside the hemocoel, the virus can then spread throughout the entire body. In order to be transmitted again through blood feeding, the virus needs to cross the basal lamina of the salivary glands, known as the salivary gland barrier, and then enter the acinar cells where it can multiply [16].

Remarkably, out of the approximately 112 mosquito genera, it appears that the Aedes and Culex genera, including species like Aedes Albopictus, Aedes Aegypti, Culex Quinquefasciatus, and Culex Pipiens, are the primary vectors capable of transmitting viruses to humans [17]. Indeed, certain mosquitoes may exhibit a preference for biting animals or may lack the ability to transmit viruses. The competence of a mosquito in transmitting a virus is determined by its ability to overcome various tissue barriers. This ability is influenced by the innate immune response and the regulation of cell death in infected cells [15,17,18].

2. CELL DEATH IN MAMMALS

Cell death pathways can be divided into two opposite processes: accidental cell death (ACD) and regulated cell death (RCD). If ACD is a consequence of a severe and rapid injury (osmotic forces, pH variations, lytic viral replication), RCD is based on dedicated molecular machinery, implying that it can be modulated by pharmacological, genetic or infectious interventions [19]. Regulated cell death occurs under two different circumstances. The first one is programmed cell death (PCD) [20], which occurs during embryonic development or in the event of tissue homeostasis. The second one regroups different RCD pathways that occur following an external or internal, prolonged, and intense stress event. This contributes to tissue homeostasis and protection by eliminating useless or potentially dangerous cells (i.e., malignant or infected cells).

Dying cells present different and well described morphological features that have been used so far to classify cell death processes into three main types [19]: apoptosis, autophagy-dependent cell death, and necrosis. Apoptosis is characterized by chromatin condensation, nuclear fragmentation, cytoplasmic shrinkage, membrane blebbing, and the formation of "apoptotic bodies", which are subsequently destroyed by professional or non-surrounding phagocytes. Autophagy-dependent cell death is essentially defined by its distinctive features of extensive cytoplasm vacuolization, leading to phagocytosis and degradation by lysosomes. Necrosis is mainly characterized by swelling, plasma membrane disruption, and cytoplasmic content efflux in the extracellular environment, without evident phagocytosis or lysosomal degradation by the neighboring cells. Intuitively, necrosis is associated to ACD, but it is nowadays clear that some RCD can also lead to non-apoptotic cell death as recently described in necroptosis [21], pyroptosis [22], and ferroptosis [23]. We will focus here on the cell death pathways that have been shown to be involved in antiviral response so far, without considering subtypes such as attachment dependence (i.e., anoïkis) and entotic cell death, parthanos, etc. For complete reviews see References [19,24].

2.1. Intrinsic Apoptosis

"Intrinsic apoptosis is a form of RCD initiated by perturbations of the intracellular or extracellular microenvironment, demarcated by mitochondrial outer membrane permeabilization (MOMP) and precipitated by executioner caspases, mainly caspase 3 (CASP3)" (Nomenclature Committee on Cell Death (NCCD) [19]). It can be induced by numerous dysregulations including DNA damage, endoplasmic reticulum (ER) stress, reactive oxygen species (ROS) overload or infection. The main characteristic is that cells still present plasma membrane integrity and metabolic activity, leading, in vivo, to the removal of apoptotic bodies by surrounding phagocytic cells that recognize phosphatidylserine (PS) at the cell surface. In vitro, unless the cultured cells present phagocytic capacities, apoptosis usually ends by a "secondary necrosis", exposing damaged plasma membrane [25].

The decisive step of intrinsic apoptosis is the irreversible and extensive MOMP, leading to the release of numerous pro-apoptotic factors contained in the intermembrane space [26]. Mitochondrial outer membrane permeabilization is controlled by a family of 20 pro- or anti-apoptotic proteins: the B cells lymphoma 2 (Bcl-2) family proteins, which share one to four Bcl-2 homology domains (BH1 to BH4) [27]. All of them are finely regulated at the transcriptional and/or post-translational level (degradation, phosphorylation, localization, oligomerization, etc.) in order to integrate the extracellular or intracellular signals, which will potentially lead to apoptosis.

In mammals, only three of them (Bax, Bak, and Bok) have been described as able to form pores in the mitochondrial outer membrane (MOM) and other cellular membranes after oligomerization. These proteins are thus considered as "effectors" that can be activated, transcriptionally or post-translationally, after a cellular stress to induce MOMP [28,29]. Moreover, a pool of BH3-only proteins, described as "activators" promotes MOMP induction by interacting with Bax and Bak, allowing the conformational changes necessary for pore formation. These proteins can be post-translationally modified (e.g., Bid, cleaved in the pro-apoptotic truncated form "t-Bid") [30,31] or transcriptionally activated (e.g., PUMA, Noxa and Bim). In particular, the transcription factor p53 represents one of the links between DNA damage or oxidative stress and intrinsic apoptosis. Indeed, after a stress signal, post-translational modifications induce p53 stabilization and translocation in the nucleus. p53 will induce pro-apoptotic Bcl-2 family proteins transcription (i.e., Bax, Bak, PUMA, and Noxa) [32,33]. In the absence of stress conditions, other members of the Bcl-2 family (Bcl-2, Bcl-Xl, Bcl-W, Mcl-1, and Bfl-1) constantly block MOMP [34]. They contain all four BH domains and are inserted in the MOM or ER membrane, interacting with and inhibiting the effectors members (Bax, Bak, and Bok) or the BH3-only activators (PUMA, Noxa, Bim, and tBid) [35,36]. In addition, these anti-apoptotic proteins have been shown to regulate Ca2+ homeostasis in the ER [37,38] and cellular redox equilibrium [39,40]. Finally, it has been shown that some BH3-only proteins (Bad, Bmf, and Hrk) carry out their pro-apoptotic effect without interacting with the effector proteins but by inhibiting the pro-survival ones [41].

Interestingly, it is now clear that mitochondria and ER are physically connected, forming platforms called mitochondria associated (ER) membranes (MAMs) [42]. Mitochondria associated membranes regulate numerous cellular processes such as calcium (Ca2+) homeostasis, autophagy, lipid metabolism, apoptosis, and the rapid exchange of biological molecules [43]. They are involved in inflammasome formation and activation and participate in the antiviral response through the mitochondrial antiviral protein (MAV)/RNA sensors (retinoic acid-inducible gene I (RIG-I) or melanoma differentiation-associated protein 5 (MDA5)) complex activation [44].

Mitochondrial outer membrane permeabilization induces the release of intermembrane space elements, among which cytochrome C (CYC), Omi, and DIABLO (also called SMAC) [45,46,47]. Furthermore, following MOMP, the mitochondrial transmembrane potential ($\Delta\Psi m$) is usually lost, mostly due to CYC release in the cytosol and the consequent stop of the respiratory chain [48].

In the cytosol, the association of CYC with Apaf1 and pro-caspase 9 (CASP9) forms a complex called apoptosome that will activate CASP9 in an ATP-dependent process [45]. In turn, the activated CASP9 will then activate the executioner caspases (i.e., mainly CASP3 and -7) that are involved in the final cellular destruction: poly (ADP-ribose) polymerase (PARP) cleavage, DNA fragmentation, PS exposure, apoptotic bodies formation [49,50,51,52]. Omi and DIABLO enhance cell death by inhibiting the inhibitor of apoptosis protein (IAP) family, which includes XIAP, c-IAP1, and c-IAP2. XIAP is constitutively bound to executioner CASP3 and -7 and, thus, blocks their activity [53,54]. c-IAP1 and c-IAP2, for their part, are two E3 ubiquitin ligases. They upregulate the CASP8 inhibitor c-Flip, induce caspases' degradation through ubiquitination, and promote NF-κB pro-survival pathway through receptor interacting serine/threonine kinase 1 (RIPK1) ubiquitination. These functions have mostly been maintained throughout evolution, from insects to mammals [55,56,57] (Figure 1).

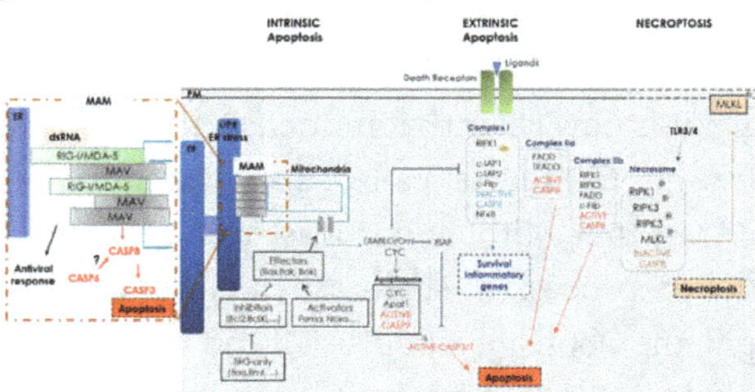

Figure 1. Apoptosis and necroptosis in mammals. PM: plasma membrane, ER: endoplasmic reticulum, MAM: mitochondria-associated membranes, CYC: cytochrome c, IAP: inhibitors of apoptosis proteins, RIPK1/3: receptor

interacting serine/threonine kinase 1/3, Ub: uiquitin, p: phosphorylation, MLKL: mixed-lineage kinase domain-Like, FADD: Fas-associated protein with death domain, TRADD: tumor necrosis factor receptor super family (TNFR1)-associated death domain protein, TLR3/4: Toll-like receptor.

Finally, executioner caspases can positively or negatively regulate the emission of multiple damage-associated molecular patterns (DAMPs) by dying cells, including immunostimulatory [58] as well as immunosuppressive [59] factors.

2.2. Extrinsic Apoptosis

"Extrinsic apoptosis is a type of RCD initiated by perturbations of the extracellular microenvironment that are detected by plasma membrane receptors, propagated by CASP8 (with the optional involvement of MOMP), and precipitated by executioner caspases, mainly CASP3" (NCCD [19]). Extrinsic apoptosis is mainly carried out by the activation of two main receptors types: death receptors and dependence receptors.

Dependence receptors are a functional family of around 20 receptors, characterized by the induction of a positive signal when bound by their ligand (survival, proliferation, differentiation, etc.) while they activate RCD in the absence of the ligand. Among them can be found netrin-1 receptors (deleted in colorectal carcinoma (DCC) [60], uncoordinated 5 homologs (UNC5Hs, UNC5H1,2,3,4 also called UNC5A,B,C,D) [61], the neogenin receptor [62], the low affinity neurotrophin receptor, p75 neurotrophin receptor (p75NTR) [63], and receptors with tyrosine kinase activity (e.g., rearranged during transfection (RET) [64], tropomyosin receptor kinase A and C (TrkA and TrkC) [65], and c-kit (CD117) [66]). Their physiological role is mainly cell guidance and they are mostly involved in tumor progression when dysregulated [67]. Even if Netrin-1 plays a role in inflammation regulation [68,69,70,71], dependence receptors have not been involved, so far, in antiviral response [67]. As such, they will not be described in this review.

Death receptors include Fas (CD95, APO-1), the tumor necrosis factor receptor super family TNFRSF1A (TNFR1), 10a (TNF-related apoptosis-inducing ligand receptor TRAILR1, DR4), and 10b (TRAILR2, DR5) [72,73]. The general mechanism is that ligand binding induces the receptors' oligomerization and the subsequent recruitment through their death domains (DDs), of adapter proteins in the intra-cellular side, to form a "death-inducing signaling complex" (DISC). Fas ligand or TRAIL binding drives the oligomerization of their receptors, the recruitment of Fas-associated protein with death domain (FADD) through the DD and the subsequent formation of the DISC through interaction with CASP8 via death effector domain (DED) and different isoforms of cFlip [74,75]. Bound TNFR1 interacts with TNFR1-dssociated death domain protein (TRADD), through its DD, which enables the formation of "Complex I". The subsequent formations of "Complex II" ("IIa" or "IIb") operate as molecular platforms to regulate the activation and functions of CASP8 (or CASP10, in some cases) [76,77]. CASP8 activation leads to RCD following two different pathways. In Type I cells, CASP8 directly activates CASP3 and -7 thus inducing the execution of the apoptotic pathway. In Type II cells, where CASP3 and -7 are sequestrated by XIAP, extrinsic apoptosis occurs through the cleavage by CASP8 of Bid, a BH3-only protein, and the release of its truncated form, t-Bid. t-Bid acts as an activator on Bax and Bak to provoke MOMP and the subsequent CASP9-dependent RCD, described above in intrinsic apoptosis [78].

CASP8 activation is the key process of extrinsic apoptosis; its regulation is complex and also involved in inflammation and antiviral response [77]. cFlip is one of the key components that promotes or inhibits CASP8 oligomerization and ensuing activation by autoproteolytic cleavage. As cFlip is transcriptionally regulated by NF-κB, it can also participate in a pro-survival pathway induced by TNFR1 in some conditions [79]. It is indeed increasingly clear that the activation of death receptors by their ligands does not necessarily lead to RCD but can also activate pro-survival signals. Specifically, the TNFR1-induced pathway depends on the RIPK1 ubiquitination level, which directly influences the formation of pro-survival versus pro-death complexes.

Briefly, in Complex I, RIPK1 polyubiquitination by cIAP1 and cIAP2 leads to NF-κB activation, pro-survival, and inflammatory genes transcription, where a high level of cFlip is correlated to survival. Subsequently, deubiquitinated RIPK1 is released from Complex I and forms Complex IIa in the cytosol with FADD, TRADD, cFlip, and CASP8. If cFLIP concentration is low, this complex leads to the degradation of RIPK1 and RIPK3, allowing CASP8 dimerization and activation and the subsequent apoptotic cell death through CASP3. In a context of high cFlip concentration, CASP8/cFlip heterodimers are formed and apoptosis is blocked [56,57,78]. Moreover, in the absence of cIAP (after MOMP and IAP inhibitors release for instance), phosphorylated RIPK1 leads to non-canonical NF-κB activation and subsequent association with RIPK3, FADD, and cFlip to activate CASP8 (Complex IIb) [73].

Finally, another possible pathway is induced when CASP8 is inhibited by chemical caspase inhibitors or virally encoded proteins. In this case, deubiquitinated RIPK1 and RIPK3 bind in microfilaments, "amyloid-like" complexes called necrosomes (most likely trimers or tetramers) [80,81]. The auto- and transphosphorylation of RIPK1 and RIPK3 and the recruitment of mixed lineage kinase domain-Like (MLKL) to the plasma membrane, triggering membrane permeabilization, initiate what is called necroptosis [82]. It is of interest to note that MLKL oligomers also lead to PS exposure, a feature usually considered as a hallmark of apoptosis [80].

If CASP8 is inhibited, RIPK3 phosphorylation can be triggered by some activated PRR (pathogen recognition receptors), such as TL3 and TL4 [83], nucleic acid sensors, such as RIG-I and MDA5, and some adhesion receptors [84,85]. In addition, IFNα and β receptor subunit 1 (IFNAR1) and IFNγ receptor 1 (IFNGR1) are also able to trigger necroptosis through TRIF and ISGF3 activation [86] (Figure 1).

2.3. Inflammasome Activation and Pyroptosis

"Pyroptosis is a form of inflammatory RCD that critically depends on the formation of plasma membrane pores by members of the gasdermin (GSDM) protein family, often as a consequence of inflammatory caspase (CASP1, 4 or 5) activation" (NCCD [19]). Pyroptotic cells present PS exposure, chromatin condensation, TUNEL staining but no DNA laddering, and a slight MOMP. Final GSDM-dependent membrane permeabilization allows the release of pro-inflammatory cytokines (IL1β and IL18, both

NF-κB-target genes), maturated by interleukin-1β-converting enzyme (ICE/CASP1)-dependent cleavage. Other factors are also released, thus participating in the defense against pathogens through inflammation and the induction of an adaptive response [19,87,88]. In fact, pyroptosis seems to be mainly involved in the innate immunity against intracellular pathogens [88]. Pyroptosis was first thought to be restricted to monocyte/ macrophage lineage, but it has been observed in other cells [89]. The inflammasome is activated by different DAMPs or pathogen-associated molecular pattern (PAMPs). It is a multiprotein complex that, like the apoptosome (intrinsic apoptosis) or the DISC (extrinsic apoptosis), acts as a caspase-activating platform. It is formed by a receptor (NOD-like receptor (NLR) family or non-NLR (AIM2)), an adapter protein (ASC, apoptosis-associated speck-like protein containing a CARD) and CASP1 that cleaves pro-IL18, pro-IL1β and GSDM. However, it is now clear that pyroptosis can also be activated by other caspases such as CASP3 [90] (Figure 2).

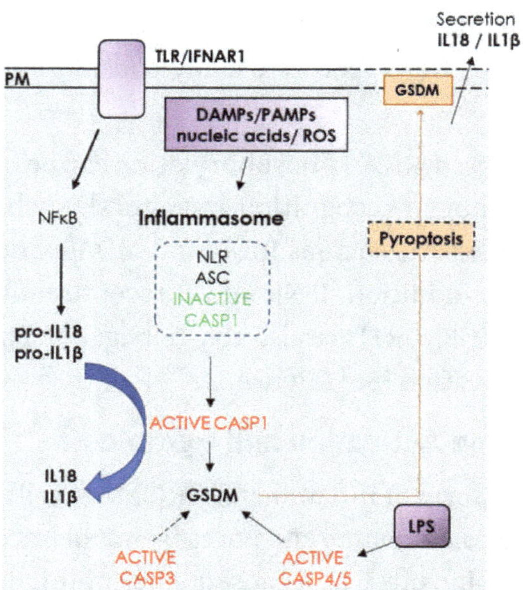

Figure 2. Inflammasome activation and pyroptosis. PM: plasma membrane, TLR: Toll-like receptor, IFNAR1: Interferon associated-receptor 1, DAMP: damage-associated molecular patterns, PAMPs: pathogen-associated molecular patterns, ROS: reactive oxygen species, NLR: NOD-like receptor, IL: interleukin, LPS: lipopolysaccharide.

The recent description of necroptosis and pyroptosis processes has amplified the complexity of RCD understanding. Contrary to general consensus so far, it is now clear that intrinsic as well as extrinsic regulated cell death can be immunogenic, thus participating in the establishment of the adaptive immune response [91,92]. This has been recently underlined by the description of the concomitant activation of apoptosis, necroptosis, and pyroptosis in a context of bacterial or viral infection in macrophages, leading to inflammatory cell death. The phenomenon has been named PANoptosis and would involve molecules of the three RCD pathways (i.e., CASP8, RIPK3, and CASP1) in a single complex called PANoptosome [93].

2.4. Autophagy-Dependent Cell Death

Autophagy-dependent cell death is a type of RCD that depends on components of the macroautophagy machinery [19]. Macroautophagy is a particular form of autophagy where double-membrane vesicles (autophagosomes) sequester a large part of organelles and cytoplasm, leading to their lysis and, in some cases, to cell death. Morphologically, dying cells present an accumulation of autophagosomes and autolysosomes in the cytoplasm, a feature extremely different from apoptotic or necrotic RCD. However, it seems increasingly clear that autophagy usually inhibits rather than induces cell death and has to be considered as a way for the cell to maintain homeostasis after stress signals (hypoxia, ROS, starvation, PRR activation, etc.) [94]. Even if in some cases inhibition of specific autophagy proteins can delay RCD [95], pharmacological or genetic inhibition of macroautophagy components usually accelerates the death of cells rather than protects them. Autophagy may degrade damaged mitochondria or pro-apoptotic complexes thus preventing cell death.

It is of interest to note that autophagy leads to the degradation by lysosomes of components of endogenous or exogenous origin, which are accessible in the cytoplasm. This has to be distinguished from vesicular trafficking, which starts in the plasma membrane and also leads to lysosomal degradation (i.e., phagocytosis or receptor-mediated endocytosis). Macroautophagy and vesicular trafficking pathways interact at numerous regulation points, especially in the late phases of the pathways. Autophagosomes or late endosomes fusion with lysosomes actually require the same machinery. Thus, numerous proteins involved in physiological or lytic autophagy are also essential for viral penetration, from receptor-mediated endocytosis to fusion of the viral envelope with the endosome membrane and subsequent liberation of the viral genome in the cytosol.

3. IMPACT OF ALPHAVIRAL INFECTION ON REGULATED CELL DEATH

During a viral infection, RCD is generally described as a defense mechanism, induced to limit virus replication and production, to prevent infection of neighboring cells and, to some extent, to participate in immune response induction. Cell death induced by alphavirus infection has been observed and studied in several cell types infected by different alphaviruses, mainly CHIKV, SFV, SINV, VEEV, and EEV.

3.1. Apoptotic Pathways in Alphavirus-Infected Cells

Infections of baby hamster kidney (BHK), rat prostatic adenocarcinoma (AT-3), and mouse neuroblastoma (N18) cells with SINV result in clear nuclear condensation and membrane blebbing, 24 h post-infection [96]. In the same context, SFV infection of AT-3 induces apoptotic features, correlated to strain virulence. Grandgirard et al. [97] have described, in rat embryonic fibroblasts, a potential caspase-dependent Bcl-2 cleavage in SFV- or SINV-infected cells leading to cell death and viral replication, even in a context of Bcl-2 overexpression. In 293T and BHK cells, the BH3-only protein Bad seems to participate in SINV-induced cell death, through its interaction with some specific anti-apoptotic Bcl-2 proteins, while the other non-binding members also regulate cell death [98]. The dynamics of mitochondria are also highly altered during apoptosis induced by VEEV infection of human astrocytoma cells U87MG [99]. Infection rapidly induces MOMP and ROS increase, followed by perinuclear localization and fission of mitochondria, and then mitophagy. Moreover, VEEV capsid co-localizes with mitochondria and could participate in mitochondria dysregulations.

Lin et al. showed a link between SINV induced cell death, oxidative stress, NF-κB, and Bcl-2 expression. In AT-3 and N18 infected cells, NF-κB activation and cell death were indeed inhibited either by antioxidant agents or Bcl-2 overexpression, with no effect on viral entry or replication however [100]. Likewise, MRC5 human fibroblasts present a SINV persistent infection, when manganese-superoxide dismutase (Mn-SOD) is over-expressed, confirming that oxidative pathways are implicated in the effects of SINV [101]. Chikungunya virus infection has been studied in the neuroblastoma cell line SH-SY5Y by Dhanwani et al. [102]. Intrinsic apoptosis features (CYC release, CASP3 activation, PARP cleavage) are observed 24 h and 36 h post-infection. Moreover, the infection is followed, 36 h and 48 h after by an elevation of ROS, a decrease of anti-oxidant enzymes expression and glutathione (GSH) depletion.

Another cellular response to stress induced by viral infection is the unfolded protein response (UPR) of the ER due to the accumulation of newly synthetized viral proteins in the ER, leading to translation blocking and intrinsic apoptosis [103]. Endoplasmic reticulum stress response has been described in flavivirus-infected cells, but little is known about alphaviruses. However, SFV envelope glycoproteins, but not capsid, seem able to induce and accelerate apoptotic cell death [104], while VEEV glycoproteins induce UPR and apoptosis in primary astrocytes [105,106].

Finally, intrinsic apoptosis may be triggered by alphavirus non-structural protein activity. Indeed, SINV, VEEV, and EEEV nsp2 and nsp3 have been shown to be responsible for viral cytopathic effects, enabling persistent infections when specifically mutated [107,108,109]. Frolov and colleagues have described the nuclear translocation of nsp2 and global transcriptional shutoff through RNA polymerase II degradation [110,111,112] and a subsequent nsp3-dependent translational shutoff for arthritogenic viruses [112], while capsid would play this role for encephalitic viruses [113]. However, CHIKV nsp2 seems to inhibit UPR as well through its transcriptional shutoff activity [114] and to interfere with the IFNβ signaling pathway [115,116].

However, Sarid et al. characterized a CASP8 and TNFα/TNFR1-dependent PC-12 RCD after infection by the SINV SVNI strain (neurovirulent and cytotoxic). Indeed, they described an upregulation of TNFα expression in infected cells and a cell death inhibition following cFlip overexpression [117]. Nava and colleagues [118] have shown, in SINV-infected BHK cells and in mice, an inhibition of death after treatment with the pan-caspase inhibitor zVAD-fmk and the CASP1 and CASP8 inhibitor CrmA (a serine proteinase inhibitor from Cowpox virus). In addition, by using SFV replicon vectors or a wild-type SFV strain, Kiiver et al. have neither shown any Bcl-2 protective action against virally induced cellular protein synthesis shutdown post-infection nor cell death. Moreover, AT-3 and BHK cells did not present any CYC release after infection [119]. In addition, unlike poxvirus, that blocks CASP1 and CASP8 [120,121], or herpesviruses [122], alphaviruses have never been described as inducing necroptosis.

Thus, on the one hand, alphaviruses appear to induce apoptosis through mitochondria, oxidative, and ER stress or the transcriptional and translational shutoff induced by nsp (intrinsic pathway). On the other hand, some studies suggest the implication of death receptors and CASP8 activation, without mitochondrial involvement (extrinsic pathway). This apparent discrepancy may be explained by the different apoptotic cells that have been observed (infected or neighbor cells) and the duration of infection (few hours to days post-infection). Indeed, Joubert et al. suggest that, in a first wave, CHIKV-infected cells die through intrinsic apoptosis (CASP9 positive cells) and that, in a second time, during antiviral response, infected, and neighbor cells die through the extrinsic pathway (death receptors and CASP8 activation). These secondary pathways seem to be independent of ER and oxidative stress [123,124].

Other explanations could be found also in two recent studies, which involve two newly described pathways in SFV-induced cell death. Using 3T9 MEFs, Urban et al. [125] characterized SFV-induced RCD. In their study, they first showed that RCD was triggered by SFV replication and not only by viral entry, as previously described [126]. Secondly,

SFV-induced cell death occurred in a Bak dependent MOMP, leading to apoptosome activation. Moreover, they excluded the involvement of TRAILR, Fas or TNFR1 in this process. Surprisingly, CASP8 and tBid seemed to be activated downstream of apoptosome, maybe through CASP6 [127], acting as an amplification loop. In fact, although CASP6 has long been considered as an executioner caspase based on its homology with CASP3 and CASP7, recent data suggest that CASP6 may actually be involved in RCD initiation [127,128]. Additional investigation is required to elucidate the function of CASP6 in mammalian cells. Secondly, in several SFV-infected cell types, El Maadidi et al. recently described a new mitochondrial platform, involving the innate immune factor MAV and the initiator CASP8, comparable to the death-inducing signaling complex (DISC) of the death receptors signaling. However, this complex does not involve FADD but another potential, not yet characterized adaptor. The platform is activated via the dsRNA sensors MDA5 or RIG-I and acts in parallel of the classical Bax/Bak dependent MOMP, also leading to CASP3 activation, independently of type I IFN signaling factors (IRF3, IRF7, IFNβ, PKR, etc.) and mitochondrial depolarization [129]. As MAVs platforms have been described in MAMs, it is tempting to hypothesize that the MAV/CASP8 should be localized in these subcellular structures thus interacting with other metabolic pathways (such as ROS, Ca2+, lipid, autophagy) during antiviral response [130].

Hence, it appears that several alphaviruses induce regulated cell death in numerous cell types, involving mitochondria depolarization, ER stress, and CASP8 activation, maybe in a time-dependent regulation or through alternative processes. However, after nearly 30 years of study, large parts of the molecular pathway leading an infected cell to death remain to be deciphered. It would be of interest in the future studies to focus on the cell types relevant for the viral tropism (e.g., skin cells, muscle cells, neural cells) and to favor innovative cell culture technics that mimic better the natural cell characteristics (e.g., 3D culture, explants, iPS (induced Pluripotent Stem cells)).

3.2. Inflammasome and Pyroptosis in Alphaviral Infection

Activation of the inflammasome pathway and pyroptosis has been intensively studied for flaviviruses infection, especially for their involvement in pathogenesis [131,132,133], but little is known for alphaviruses. Even if inflammasome pathways seem to participate in the pathology and the response against alphaviruses, the involvement of pyroptosis has never been described. In dermal fibroblasts, CHIKV and the flavivirus West Nile virus (WNV) both induce IL1β production and CASP1 activation through the AIM2 inflammasome sensor, but only CHIKV replication and propagation can be controlled by CASP1 [134]. In PBMC from CHIKV-infected patients, high levels of NLRP3, IL18, and CASP1 are found [135]. Moreover, in mice, NLRP3 activation is correlated to inflammatory symptoms such as bone damage and myositis. NLRP3 inhibition leads to a reduction of the inflammatory pathology induced by CHIKV but not by WNV [135]. The alphavirus Mayaro (MAYV) induces the expression of inflammasome proteins in macrophages, and inflammatory cytokines production through the NLRP3 sensor, activated by ROS and K+ efflux. In mice, NLRP3 is also involved in MAYV induced pathogenesis [136].

Thus, inflammasome activation has been mainly involved in global inflammatory response to alphavirus in vivo, but the molecular pathways activated in the cell remain to be described. Indeed, it is still unknown if pyroptosis may participate in inflammatory cytokines secretion during alphavirus infection, and, to our knowledge, there is no molecular study of this process.

4. INTERPLAY BETWEEN CELL DEATH AND ALPHAVIRAL REPLICATION AND SPREAD IN MAMMALS

Apoptosis appears to be a strong antiviral process. Indeed, Bcl-2 overexpression converts SINV infection from lytic to persistent in vitro [96] and in vivo [137]. Moreover, Bcl-2 seems to be able to restrict SFV replication by inhibiting early stages of infection and appears to prolong survival of productively infected cells [138].

As described above, autophagy usually blocks apoptosis, and viruses have developed strategies to take advantage of this property. The first connection between alphavirus infection and autophagy has been made by Liang and colleagues [139], when they identified Beclin-1 as a new Bcl-2-interacting protein through a yeast two-hybrids screening. Beclin-1 is a major factor of autophagy, involved in autophagosome initiation and maturation. In this study, Beclin-1 protected SINV-infected mice against fatal encephalitis, with a significantly lower viral replication rate in mice brains. The author correlated these observations with the previously observed protective role of Bcl-2 against in vivo SINV infection [96,137]. However, years after this study, Bcl-2 was described as a Beclin-1 inhibitor, thus participating in ER stress connected autophagy regulation [140]. Another crucial protein of the autophagy pathway, Atg5, protects mouse neurons from SINV-induced cell death [141], with no apparent impact on viral replication. Moreover, the adaptor protein p62 seems to be linked to viral capsid clearance by direct interaction and target of autophagosomes, thus promoting cell survival [141]. Finally, in SFV-infected cells, autophagosomes accumulate but autophagy modulation has no effect on viral replication, and this autophagosomes accumulation seems to be due to the inhibition of their degradation rather than an induction by SFV infection [142].

In HEK293 cells, CHIKV infection induces autophagy features (LC3 positive vesicles and electron microscopy observation). In this study, autophagy has a clear pro-viral role, increasing the number of infected cells and viral RNA in the cell culture supernatant [143]. Moreover, in vitro and in vivo, CHIKV infection has also been shown to induce an autophagy flux, through ER and oxidative stress [123,124]. In these models, autophagy limits (i) extrinsic and intrinsic RCD induced by CHIKV infection, (ii) mice lethality, and (iii) viral propagation. Autophagy, as a host response to infection, limits indeed the cytopathic effects of CHIKV and regulates the pathogenesis of acute chikungunya disease. However, during late phases of in vitro infection (48 h post-infection), a switch between autophagy and apoptosis is observed and cells die. Finally, in HeLa cells, autophagy promotes CHIKV infection and inhibits cell death. Indeed, in addition to p62-dependent capsid clearing, another autophagy receptor, NDP52, interacts with nsp2, localizes near the CHIKV replication complex and restricts cell shutoff thus promoting viral replication and cell survival [144]. Hence, alphaviruses may exploit autophagy to delay cell death through (i) direct inhibition of intrinsic and extrinsic apoptosis and (ii) a limitation of viral proteins production, allowing cell survival and a longer viral replication.

However, several Old World Alphaviruses, such as CHIKV, SFV, and RRV, seem able to activate the phosphatidylinositol-3-kinase (PI3K)–AKT–mTOR pathway, involved in cell survival and autophagy inhibition. Furthermore, inhibition of this pathway has a negative effect on viral replication [145]. This apparent discrepancy with the previous observations may indicate that, more than autophagy per se, cell survival is the key process which favors viral replication. Finally, it is of interest to note that, in some cases, apoptosis has been shown to enhance viral spread. Indeed, in their study, Krejbich-Trotot and colleagues [146] first confirmed the dual nature of the alphavirus-induced apoptosis (intrinsic and extrinsic) in HeLa and primary fibroblasts infected with CHIKV accompanied by CASP8 activation in neighbor cells. More interestingly still, inhibitors of blebbing or engulfment drastically reduced infection rates. Finally, they detected infective CHIKV in apoptotic corpses and in the macrophages which phagocyted them, leading to macrophages infection and viral production. As macrophages are refractory to CHIKV infection in vitro, this study highlights a possible role of apoptotic blebs in viral propagation. This phenomenon, called "apoptotic mimicry", is used by a large number of viruses to exploit the PS receptors present on numerous cells membranes, enhancing viral spread and limiting immune response [147].

5. IMPACT OF APOPTOSIS ON VIRUS PATHOGENESIS IN MAMMALS

5.1. Alphavirus Encephalitis

The first cell death analysis was documented in vivo, using SINV, VEEV, and SFV, three alphaviruses causing encephalitis. In SINV-infected mice, the apoptotic cells were detected principally in the brain and contained viral antigens, suggesting that apoptosis was correlated to neurovirulence [148]. The in vivo mouse infection of VEEV was also associated to cell death in brain, demonstrated by TUNEL assay (DNA fragmentation) and morphological changes [149].

Comparing SINV infections with SVNI (neurovirulent and cytotoxic) or SVA (avirulent and leading to persistent infection) strains in PC-12 cells and astrocytes, revealed that SVNI induces Bax overexpression while SVA induces Bcl-2 expression [150]. Intranasally SFV-infected rats develop encephalitis, where infiltrating leucocytes and neural precursor cells undergo apoptosis while productively infected neurons present necrotic features, apparently due to the local inflammation [151].

Hence, it appears that alphaviruses pathogenicity is linked to its cytopathic effects in infected cells, at least in the case of the encephalitic group.

5.2. Alphaviral Chronic Infection: What about Cell Death?

One characteristic of the Alphavirus genus is the ability of some of them (CHIKV, SINV, MAYV, RRV, etc.) to induce chronic pain, such as arthritis and myalgia, which may last for years, with detectable viral genome in the organism. This persistent infection implies that some cells may be chronically infected, and in some way able to delay or block cell death. However, despite an intense immune response observed in chronic patients, damaged synovial tissues present strong apoptosis features. Chikungunya virus has been found in synovial macrophages several months after infection but joints do not seem to be the viral reservoir [152]. In addition, RRV-infected human monocyte acute leukemia MM6 cell line presents very low replication rates, without innate immune control, and apoptosis features at late stages of infection. This indicates that monocytes could be persistently infected and participate in the chronic form of RRV or CHIKV [153]. Young et al. [154] propose that dermal and muscular fibroblasts, as well as myofibers, may survive the acute CHIKV infection and harbor persistent CHIKV RNA during chronic phase of the disease. Moreover, they observe that synovial cells are not infected in large numbers in vivo and suggest that synovial cells may be infected but do not survive.

How these cells survive remains to be understood. The mechanisms involved in alphaviral persistence are mostly unknown. They may depend on infected cell type and the highly complex interplay between virus, immune response, and different RCD pathways.

6. INTERPLAY BETWEEN CELL DEATH AND SUSCEPTIBILITY OF MOSQUITO SPECIES TO ARBOVIRUSES

The interplay between arbovirus and arthropods is still poorly understood and the primary point of study concerns flaviviruses (mainly Dengue virus (DENV)) in mosquito and/or in Drosophila, used as a genetic model for insect immunity. Hence, very little is known concerning the impact of alphavirus infection in mosquitoes and which factors may explain the tolerance versus resistance of different mosquito strains. It is classically admitted that alphaviruses do not induce any major pathology in their vectors. However, several lesions are observed in tissues which are critical for viral propagation and transmission. Indeed, after feeding on infected blood, cellular response in the midgut plays a decisive role in vector competence. EEEV infection of Culiseta melanura mosquito induces severe lesions in midgut epithelial cells and basal lamina, associated to viral spread [155]. Likewise, infection of more or less susceptible Culex tarsalis strains with WEEV revealed lesions and apparent necrotic cell death only in the sensitive mosquito's gut [156]. Transcriptomic analysis of Aedes Aegypti fed with CHIKV in blood or different buffers reveals the over-expression of matrix metallo proteinases (MMP) and other peptidases in the midgut, as well as the decrease of Collagen IV, a component of the basal lamina [157]. Intrathoracically SINV-injected Aedes Albopictus present colocalization of virus antigen with structural lesions and TUNEL positive cells in salivary glands [158] and midgut-associated visceral muscles [159].

Furthermore, organ-associated muscles respond differently to SINV [160]: 10 days post-infection, the virus has cleared from the midgut, is persistent in the hindgut, and unable to infect ovary associated muscle cells. High viral titers induce pathology limited to gut associated muscles and gut epithelium. Finally, in Aedes Aegypti mosquitoes, AeIAP1 (IAP ortholog) downregulation leads to a higher replication of SINV in the midgut, while AeDronc (CASP9 ortholog) inhibition is associated to a lower viral replication and dissemination towards salivary glands [161] (see Box 1 for RCD pathway description in insect).

Box 1. Comparative cell death pathways in mammal, Drosophila and Aedes. **PM:** plasma membrane; **CYC:** cytochrome C; **IAP:** Inhibitors of apoptosis; **Ae:** Aedes; **RHG:** Reaper, Hid and Grim; **IMD:** immune deficiency.

Recent knowledge concerning apoptosis in mosquitoes has been acquired through gene homology with Drosophila melanogaster. Apoptosis is under the control of initiator and effector caspases [162], expressed ubiquitously and synthesized as inactive procaspases. The Apaf-1-related killer (Ark) molecule [163] assembles itself into an apoptosome-like complex [164] to activate Dronc, but the role of CYC in the insect apoptosome is very controversial [165]. Two Bcl-2 orthologues have been identified in Drosophila: Buffy and Debcl [166], whose pro- or anti-apoptotic roles are not clear [166,167]. Finally, mitochondrial fission, through Drp1 activation, seems to be required for efficient cell death [168]. However, the pro-apoptotic activity of caspases is mainly regulated by members of insect IAP family [169,170]. IAP antagonists (dOmi/HtrA2 and RHG proteins in drosophila; Michelob_X in mosquito) are localized in the mitochondrial intermembrane space in living cells and released into the cytosol, but remain near the mitochondria, after an apoptotic stimulus [170,171], where they compete for caspase binding through their IAP-binding domain. Additionally, RHG proteins can induce DIAP1 ubiquitination and degradation [168,172].

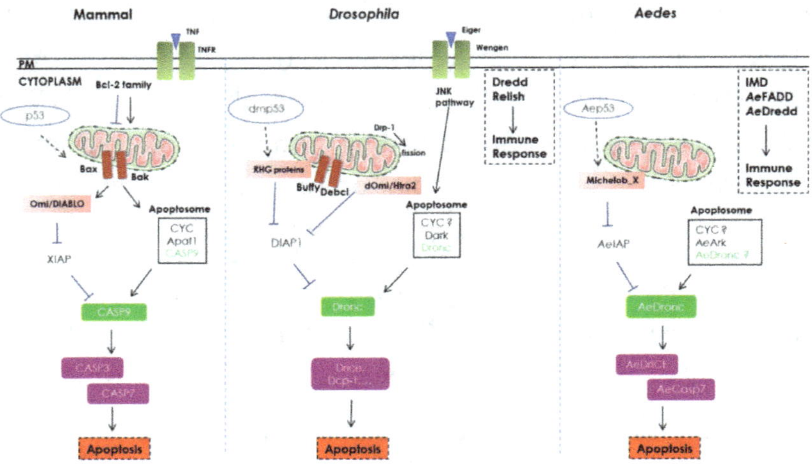

Interestingly, RHG and Mx promoters present different response elements regulated by transcription factors, such as dmp53, activated by developmental or environmental signals, leading to cell death [173]. In insects, no clear distinction can be made between intrinsic or extrinsic apoptotic pathways. Nevertheless orthologues of TNF (Eiger) and TNFR (Wengen) have been described in Drosophila [174] and induce cell death through a JNK-mediated pathway, requiring apoptosome components [175]. Immune response against pathogens is triggered by the NF-κB (Relish) pathway induction through the Immune Deficiency IMD/dFADD/Dredd (CASP8) pathway [176] in Drosophila and IMD/AeFADD/AeDredd in Aedes [177].

Finally, autophagy-dependent cell death is finely controlled in insects, and its role in development has been largely studied in Drosophila [178,179]. To our knowledge, pyroptosis and necroptosis have not yet been described in insects.

Thus, caspase activity may be required for dissemination of SINV from the midgut to the secondary organs by participating in the remodeling of the basal lamina, as suggested in baculovirus-infected lepidopteran, where caspase and MMP activity is necessary to cross the midgut barrier [180]. However, cell death modulation in vitro, in mosquito cells, does not seem to alter alphaviral replication. In fact, recombinant SINVs, expressing Reaper (Drosophila IAP inhibitor) or Michelob_X (Mx, Aedes IAP inhibitor), induce apoptosis in infected Aedes Albopictus C636 cells, with no inhibitory effect on viral production in the initial phase of infection. Moreover, in these conditions, inhibition of caspase activity has no effect on viral replication neither [181]. However, recombinant SINV expressing Reaper induces cell death in vivo in Aedes Aegypti's midgut, a delayed infection and propagation in the saliva [182]. More importantly, this last study also describes a rapid genetic selection of SINV variants in vivo against Reaper expression.

Hence, cell death, with tissue degradation features, seems important for alphavirus propagation in mosquito organism, with no clear effect on replication in vitro. However, as described above for mammal cells, RCD is also associated to efficient immune response against viruses and may be one of the key processes involved in mosquito resistance to virus.

How cell death is modulated in infected tolerant mosquito cells remains to be understood. Oxidative stress response may play an important role in the mosquito's response. Indeed, CHIKV infection induces upregulation of antioxidant pathways in mosquito midgut, which delays cell death [183]. During arbovirus infection, oxidative stress is actually detected in both mammal [101,123,184] and insect cells. Oxidative stress is defined by loss of homeostasis between accumulation of ROS and production of antioxidant enzymes such as superoxide dismutase (SOD), catalase (CAT) or glutathione transferase and reductase [185]. After blood feeding, the midgut is in contact with sugar, iron, heme and other components of vertebrate blood. Mosquitoes have developed protective adaptation against the damage caused by heme and iron uptake. Indeed, the heme can induce lipid peroxidation, protein degradation, and ultimately cell death. Once in the epithelial cells, these components are detoxified, and a strong antioxidant and protective response is engaged [186]. Concomitantly, pathogens present in the blood could take advantage of this antioxidant response, blocking cell death, to infect and replicate into midgut epithelial cells.

The pro-survival pathway PI3k–Akt–mTOR may also be involved in insect tolerance. Indeed, in drosophila, activation of the PI3k–Akt–mTOR pathway is associated to an increase of SINV infection, potentially through apoptosis and autophagy inhibition and a more efficient cap-dependent translation of viral genome [187]. Finally, the role of autophagy seems to be limited in CHIKV infection of mosquito cells. Even infection induces autophagy in Aag2 cells, every pharmacological modulation of autophagy (inducer or blocker) leads to a replication increase in mosquito cells [188].

Hence, the sensitivity may depend on the better resistance of midgut cells to oxidative stress induced by viral infection, leading to a delayed cell death but the involvement of autophagy in these regulations remains to be understood.

Few studies have been conducted to investigate the interplay between alphaviruses and their vector. However, recent findings in mosquito and drosophila underlie the role of p53 isoforms in cell response to oxidative stress and to DENV infection. The balance between a rapid apoptosis and a delayed, secondary necrosis may explain in part the differences between tolerant and resistant mosquitoes' strains. For more details see [177,189,190,191,192,193,194,195,196] and Figure 3 for suggested mechanisms which may be involved in alphavirus infection in sensitive versus resistant mosquitoes, extrapolated from other arboviruses.

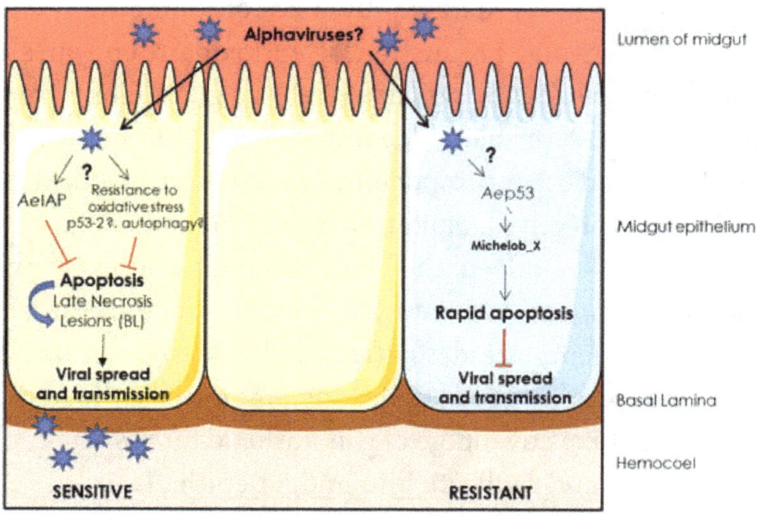

Figure 3. Possible interplay between cell death and alphavirus infection in Aedes mosquitoes: extrapolation from other arboviruses studies. BL: basal lamina, AeIAP: Aedes inhibitor of apoptosis protein. See Box 1 for RCD pathway description in insect.

In addition to the virus' ability to penetrate the cells of a particular mosquito (specific receptor, lipid membrane composition), the difference between resistant and tolerant strains may also lie in the rapidity of cell response to viral infection in the first targeted tissue, i.e., the mosquito gut. This response has to be apoptotic and not necrotic to ensure mosquito resistance. A p53/Mx- [189] and caspase- [177] dependent cell death has been linked to Aedes Aegypti resistance to DENV infection. Phagocytosis of dead cells and apoptotic bodies also seem to be important for a virus specific immune response, at least in drosophila [190].

Sensitive mosquitoes may tolerate viral infection through IAP-dependent apoptosis inhibition, as shown for the arbovirus bluetongue virus (BTV) [191,192]. Another tolerance mechanism to DENV seems to rely on resistance to oxidative stress through CAT protection in mosquito gut [193] or p53 isoforms regulation leading to cell survival in drosophila [194] and mosquito [195,196]. Delayed apoptosis may then lead to secondary necrosis, impairing a proper innate immune response, favoring basal lamina damage and viral spread [189].

7. CONCISE SUMMARY

The comparative analysis of cell response to alphavirus infection in mammals and mosquitoes highlights the intricate regulation of cell death across different animals confronting the same pathogen. The initial interaction between the virus and the first cells it encounters in both the host and the vector will have an impact on the development of the infection, affecting the immune response and causing cell death. Alphavirus-induced apoptosis in animals is also associated with the development of disease in organs that are infected subsequently. The differences between the intrinsic and extrinsic apoptotic pathways in mammals may be clarified by the presence of a secondary amplification loop involving the activation of CASP6, or by the recently discovered activation platform of CASP8, which directly connects viral RNA sensors and apoptosis. Nevertheless, additional research is required to investigate the exact pathways in many cell types and for diverse alphaviruses. Manipulating apoptosis in a laboratory setting does not appear to affect the rate at which a virus reproduces, but it can result in long-term presence of the virus. Autophagy may serve as a mechanism by which alphaviruses might prolong cell survival, hence facilitating reproduction for an extended duration. Alphavirus non-structural proteins 2 and 3 predominantly induce apoptosis in mammalian cells, whereas mosquito cells from susceptible strains do not exhibit the same response. Mammals are commonly regarded as unintentional hosts in arboviral infections, implying that nsp-induced cell death may likewise be unintentional and perhaps harmful for alphaviruses. Certain alphaviruses have undergone evolutionary changes that have resulted in their adaptation to mosquitos as their primary hosts, no longer depending on transmission to mammals [197,198,199]. The mechanisms by which mosquito cells with high tolerance are able to withstand arbovirus infection are still not understood. Furthermore, investigating the role of insect IAPs in combating oxidative stress and the potential involvement of p53 isoforms and autophagy in infection may provide insights into controlling alphaviruses. However, additional research is required to fully understand these mechanisms. Furthermore, it is necessary to have a better understanding of the mechanisms that result in chronic infection and the potential for infected cells to survive in the context of arthritogenic alphaviruses in mammals. The variations in cell types could potentially account for these inconsistencies, and additional research would be beneficial in understanding the mechanisms by which these cells evade cell death. Although it is widely acknowledged that alphavirus infection is typically asymptomatic in capable mosquitoes, some degree of tissue damage is detected and necessary in the intestinal epithelia and salivary glands to facilitate viral replication and transmission. Due to little research on alphaviruses about the connection between cell death and vector competence, we can only infer from recent findings on DENV. Furthermore, the mosquito's capacity to promptly destroy infected epithelial cells through apoptosis, rather than experiencing delayed cell death and subsequent necrosis, may elucidate the distinction between a mosquito that is resistant and one that is tolerant. Enhanced comprehension and subsequent modulation of vector tolerance could aid in the regulation of arboviral propagation, as recently proposed [15,18].

Alphaviruses, such as the chikungunya virus, are viruses that are now emerging or reemerging and pose a substantial worldwide public health threat. They are transmitted to humans and animals via blood-sucking arthropods, mostly mosquitoes. Although alphaviruses can induce serious illnesses in mammals, they do not appear to have any detrimental effects on the mosquito vector. Extensive research is being conducted to investigate the cellular and molecular interactions between Alphavirus and its host. While the significance of apoptosis in specific human diseases is clear, the exact influence of cell death on the result of mosquito infections has not been fully explained. This study aims to investigate the current knowledge on alphavirus-induced regulated cell death in both hosts and vectors. Additionally, we will analyze how this process may influence the tolerance or resistance of mosquitoes.

REFERENCES

1. Arthropod-borne and rodent-borne viral diseases. Report of a WHO Scientific Group. World Health Organ. Tech. Rep. Ser. 1985, 719, 1–116.
2. Weaver, S.C.; Barrett, A.D. Transmission cycles, host range, evolution and emergence of arboviral disease. Nat. Rev. MicroBiol. 2004, 2, 789–801.
3. Powers, A.M.; Roehrig, J.T. Alphaviruses. Methods Mol. Biol. 2011, 665, 17–38.
4. Suhrbier, A.; Jaffar-Bandjee, M.C.; Gasque, P. Arthritogenic alphaviruses--an overview. Nat. Rev. Rheumatol 2012, 8, 420–429.
5. Figueiredo, M.L.; Figueiredo, L.T. Emerging alphaviruses in the Americas: Chikungunya and Mayaro. Rev. Soc. Bras. Med. Trop 2014, 47, 677–683.
6. Baxter, V.K.; Heise, M.T. Immunopathogenesis of alphaviruses. Adv. Virus Res. 2020, 107, 315–382.

7. Schwartz, O.; Albert, M.L. Biology and pathogenesis of chikungunya virus. Nat. Rev. MicroBiol. 2010, 8, 491–500.

8. Konopka, J.L.; Penalva, L.O.; Thompson, J.M.; White, L.J.; Beard, C.W.; Keene, J.D.; Johnston, R.E. A two-phase innate host response to alphavirus infection identified by mRNP-tagging in vivo. PLoS Pathog. 2007, 3, e199.

9. MacDonald, G.H.; Johnston, R.E. Role of dendritic cell targeting in Venezuelan equine encephalitis virus pathogenesis. J. Virol. 2000, 74, 914–922.

10. Solignat, M.; Gay, B.; Higgs, S.; Briant, L.; Devaux, C. Replication cycle of chikungunya: A re-emerging arbovirus. Virology 2009, 393, 183–197.

11. da Silveira, I.D.; Petersen, M.T.; Sylvestre, G.; Garcia, G.A.; David, M.R.; Pavan, M.G.; Maciel-de-Freitas, R. Zika Virus Infection Produces a Reduction on Aedes aegypti Lifespan but No Effects on Mosquito Fecundity and Oviposition Success. Front. MicroBiol. 2018, 9, 3011.

12. Grimstad, P.R.; Ross, Q.E.; Craig, G.B., Jr. Aedes triseriatus (Diptera: Culicidae) and La Crosse virus. II. Modification of mosquito feeding behavior by virus infection. J. Med. Entomol. 1980, 17, 1–7.

13. Ciota, A.T.; Styer, L.M.; Meola, M.A.; Kramer, L.D. The costs of infection and resistance as determinants of West Nile virus susceptibility in Culex mosquitoes. BMC Ecol. 2011, 11, 23.

14. Oliveira, J.H.; Bahia, A.C.; Vale, P.F. How are arbovirus vectors able to tolerate infection? Dev. Comp. Immunol. 2020, 103, 103514.

15. Lambrechts, L.; Saleh, M.C. Manipulating Mosquito Tolerance for Arbovirus Control. Cell Host Microbe 2019, 26, 309–313.

16. Ruckert, C.; Ebel, G.D. How Do Virus-Mosquito Interactions Lead to Viral Emergence? Trends Parasitol 2018, 34, 310–321.

17. Powell, J.R. An Evolutionary Perspective on Vector-Borne Diseases. Front. Genet. 2019, 10.

18. Clem, R.J. Arboviruses and apoptosis: The role of cell death in determining vector competence. J. Gen. Virol. 2016, 97, 1033–1036.
19. Galluzzi, L.; Vitale, I.; Aaronson, S.A.; Abrams, J.M.; Adam, D.; Agostinis, P.; Alnemri, E.S.; Altucci, L.; Amelio, I.; Andrews, D.W.; et al. Molecular mechanisms of cell death: Recommendations of the Nomenclature Committee on Cell Death 2018. Cell Death Differ. 2018, 25, 486–541.
20. Conradt, B. Genetic control of programmed cell death during animal development. Annu Rev. Genet. 2009, 43, 493–523.
21. Vanden Berghe, T.; Vanlangenakker, N.; Parthoens, E.; Deckers, W.; Devos, M.; Festjens, N.; Guerin, C.J.; Brunk, U.T.; Declercq, W.; Vandenabeele, P. Necroptosis, necrosis and secondary necrosis converge on similar cellular disintegration features. Cell Death Differ. 2010, 17, 922–930.
22. Kepp, O.; Galluzzi, L.; Zitvogel, L.; Kroemer, G. Pyroptosis—A cell death modality of its kind? Eur. J. Immunol. 2010, 40, 627–630.
23. Xie, Y.; Hou, W.; Song, X.; Yu, Y.; Huang, J.; Sun, X.; Kang, R.; Tang, D. Ferroptosis: Process and function. Cell Death Differ. 2016, 23, 369–379.
24. Tang, D.; Kang, R.; Berghe, T.V.; Vandenabeele, P.; Kroemer, G. The molecular machinery of regulated cell death. Cell Res. 2019, 29, 347–364.
25. Rogers, C.; Fernandes-Alnemri, T.; Mayes, L.; Alnemri, D.; Cingolani, G.; Alnemri, E.S. Cleavage of DFNA5 by caspase-3 during apoptosis mediates progression to secondary necrotic/pyroptotic cell death. Nat. Commun 2017, 8, 14128.
26. Galluzzi, L.; Kepp, O.; Kroemer, G. Mitochondrial regulation of cell death: A phylogenetically conserved control. Microb Cell 2016, 3, 101–108.
27. Kalkavan, H.; Green, D.R. MOMP, cell suicide as a BCL-2 family business. Cell Death Differ. 2018, 25, 46–55.

28. Wei, M.C.; Zong, W.X.; Cheng, E.H.; Lindsten, T.; Panoutsakopoulou, V.; Ross, A.J.; Roth, K.A.; MacGregor, G.R.; Thompson, C.B.; Korsmeyer, S.J. Proapoptotic BAX and BAK: A requisite gateway to mitochondrial dysfunction and death. Science 2001, 292, 727–730.

29. Subburaj, Y.; Cosentino, K.; Axmann, M.; Pedrueza-Villalmanzo, E.; Hermann, E.; Bleicken, S.; Spatz, J.; Garcia-Saez, A.J. Bax monomers form dimer units in the membrane that further self-assemble into multiple oligomeric species. Nat. Commun 2015, 6, 8042.

30. Ren, D.; Tu, H.C.; Kim, H.; Wang, G.X.; Bean, G.R.; Takeuchi, O.; Jeffers, J.R.; Zambetti, G.P.; Hsieh, J.J.; Cheng, E.H. BID, BIM, and PUMA are essential for activation of the BAX- and BAK-dependent cell death program. Science 2010, 330, 1390–1393.

31. Luo, X.; Budihardjo, I.; Zou, H.; Slaughter, C.; Wang, X. Bid, a Bcl2 interacting protein, mediates cytochrome c release from mitochondria in response to activation of cell surface death receptors. Cell 1998, 94, 481–490.

32. Villunger, A.; Michalak, E.M.; Coultas, L.; Mullauer, F.; Bock, G.; Ausserlechner, M.J.; Adams, J.M.; Strasser, A. p53- and drug-induced apoptotic responses mediated by BH3-only proteins puma and noxa. Science 2003, 302, 1036–1038.

33. Yamada, K.; Yoshida, K. Mechanical insights into the regulation of programmed cell death by p53 via mitochondria. Biochim Biophys Acta Mol. Cell Res. 2019, 1866, 839–848.

34. Czabotar, P.E.; Lessene, G.; Strasser, A.; Adams, J.M. Control of apoptosis by the BCL-2 protein family: Implications for physiology and therapy. Nat. Rev. Mol. Cell Biol. 2014, 15, 49–63.

35. Oltvai, Z.N.; Milliman, C.L.; Korsmeyer, S.J. Bcl-2 heterodimerizes in vivo with a conserved homolog, Bax, that accelerates programmed cell death. Cell 1993, 74, 609–619.

36. Llambi, F.; Moldoveanu, T.; Tait, S.W.; Bouchier-Hayes, L.; Temirov, J.; McCormick, L.L.; Dillon, C.P.; Green, D.R. A unified model of mammalian BCL-2 protein family interactions at the mitochondria. Mol. Cell 2011, 44, 517–531.

37. Vervliet, T.; Parys, J.B.; Bultynck, G. Bcl-2 proteins and calcium signaling: Complexity beneath the surface. Oncogene 2016, 35, 5079–5092.

38. Scorrano, L.; Oakes, S.A.; Opferman, J.T.; Cheng, E.H.; Sorcinelli, M.D.; Pozzan, T.; Korsmeyer, S.J. BAX and BAK regulation of endoplasmic reticulum Ca2+: A control point for apoptosis. Science 2003, 300, 135–139.

39. Chen, Z.X.; Pervaiz, S. Involvement of cytochrome c oxidase subunits Va and Vb in the regulation of cancer cell metabolism by Bcl-2. Cell Death Differ. 2010, 17, 408–420.

40. Chen, Z.X.; Pervaiz, S. Bcl-2 induces pro-oxidant state by engaging mitochondrial respiration in tumor cells. Cell Death Differ. 2007, 14, 1617–1627.

41. Letai, A.; Bassik, M.C.; Walensky, L.D.; Sorcinelli, M.D.; Weiler, S.; Korsmeyer, S.J. Distinct BH3 domains either sensitize or activate mitochondrial apoptosis, serving as prototype cancer therapeutics. Cancer Cell 2002, 2, 183–192.

42. Vance, J.E. Phospholipid synthesis and transport in mammalian cells. Traffic 2015, 16, 1–18.

43. Phillips, M.J.; Voeltz, G.K. Structure and function of ER membrane contact sites with other organelles. Nat. Rev. Mol. Cell Biol. 2016, 17, 69–82.

44. Missiroli, S.; Patergnani, S.; Caroccia, N.; Pedriali, G.; Perrone, M.; Previati, M.; Wieckowski, M.R.; Giorgi, C. Mitochondria-associated membranes (MAMs) and inflammation. Cell Death Dis. 2018, 9, 329.

45. Li, P.; Nijhawan, D.; Budihardjo, I.; Srinivasula, S.M.; Ahmad, M.; Alnemri, E.S.; Wang, X. Cytochrome c and dATP-dependent formation of Apaf-1/caspase-9 complex initiates an apoptotic protease cascade. Cell 1997, 91, 479–489.

46. Verhagen, A.M.; Ekert, P.G.; Pakusch, M.; Silke, J.; Connolly, L.M.; Reid, G.E.; Moritz, R.L.; Simpson, R.J.; Vaux, D.L. Identification of DIABLO, a mammalian protein that promotes apoptosis by binding to and antagonizing IAP proteins. Cell 2000, 102, 43–53.

47. Martins, L.M. The serine protease Omi/HtrA2: A second mammalian protein with a Reaper-like function. Cell Death Differ. 2002, 9, 699–701.

48. Zamzami, N.; Marchetti, P.; Castedo, M.; Decaudin, D.; Macho, A.; Hirsch, T.; Susin, S.A.; Petit, P.X.; Mignotte, B.; Kroemer, G. Sequential reduction of mitochondrial transmembrane potential and generation of reactive oxygen species in early programmed cell death. J. Exp. Med. 1995, 182, 367–377.

49. Julien, O.; Wells, J.A. Caspases and their substrates. Cell Death Differ. 2017, 24, 1380–1389.

50. Nagata, S. DNA degradation in development and programmed cell death. Annu Rev. Immunol. 2005, 23, 853–875.

51. Martin, S.J.; Finucane, D.M.; Amarante-Mendes, G.P.; O'Brien, G.A.; Green, D.R. Phosphatidylserine externalization during CD95-induced apoptosis of cells and cytoplasts requires ICE/CED-3 protease activity. J. Biol. Chem. 1996, 271, 28753–28756.

52. Coleman, M.L.; Sahai, E.A.; Yeo, M.; Bosch, M.; Dewar, A.; Olson, M.F. Membrane blebbing during apoptosis results from caspase-mediated activation of ROCK I. Nat. Cell Biol. 2001, 3, 339–345.

53. Eckelman, B.P.; Salvesen, G.S.; Scott, F.L. Human inhibitor of apoptosis proteins: Why XIAP is the black sheep of the family. EMBO Rep. 2006, 7, 988–994.

54. Eckelman, B.P.; Salvesen, G.S. The human anti-apoptotic proteins cIAP1 and cIAP2 bind but do not inhibit caspases. J. Biol. Chem. 2006, 281, 3254–3260.

55. Kamber Kaya, H.E.; Ditzel, M.; Meier, P.; Bergmann, A. An inhibitory mono-ubiquitylation of the Drosophila initiator caspase Dronc functions in both apoptotic and non-apoptotic pathways. PLoS Genet. 2017, 13, e1006438.

56. Varfolomeev, E.; Goncharov, T.; Fedorova, A.V.; Dynek, J.N.; Zobel, K.; Deshayes, K.; Fairbrother, W.J.; Vucic, D. c-IAP1 and c-IAP2 are critical mediators of tumor necrosis factor alpha (TNFalpha)-

induced NF-kappaB activation. J. Biol. Chem. 2008, 283, 24295–24299.

57. Witt, A.; Vucic, D. Diverse ubiquitin linkages regulate RIP kinases-mediated inflammatory and cell death signaling. Cell Death Differ. 2017, 24, 1160–1171.

58. Martins, I.; Wang, Y.; Michaud, M.; Ma, Y.; Sukkurwala, A.Q.; Shen, S.; Kepp, O.; Metivier, D.; Galluzzi, L.; Perfettini, J.L.; et al. Molecular mechanisms of ATP secretion during immunogenic cell death. Cell Death Differ. 2014, 21, 79–91.

59. Huang, Q.; Li, F.; Liu, X.; Li, W.; Shi, W.; Liu, F.F.; O'Sullivan, B.; He, Z.; Peng, Y.; Tan, A.C.; et al. Caspase 3-mediated stimulation of tumor cell repopulation during cancer radiotherapy. Nat. Med. 2011, 17, 860–866.

60. Mehlen, P.; Rabizadeh, S.; Snipas, S.J.; Assa-Munt, N.; Salvesen, G.S.; Bredesen, D.E. The DCC gene product induces apoptosis by a mechanism requiring receptor proteolysis. Nature 1998, 395, 801–804.

61. Llambi, F.; Causeret, F.; Bloch-Gallego, E.; Mehlen, P. Netrin-1 acts as a survival factor via its receptors UNC5H and DCC. EMBO J. 2001, 20, 2715–2722.

62. Matsunaga, E.; Tauszig-Delamasure, S.; Monnier, P.P.; Mueller, B.K.; Strittmatter, S.M.; Mehlen, P.; Chedotal, A. RGM and its receptor neogenin regulate neuronal survival. Nat. Cell Biol. 2004, 6, 749–755.

63. Rabizadeh, S.; Oh, J.; Zhong, L.T.; Yang, J.; Bitler, C.M.; Butcher, L.L.; Bredesen, D.E. Induction of apoptosis by the low-affinity NGF receptor. Science 1993, 261, 345–348.

64. Bordeaux, M.C.; Forcet, C.; Granger, L.; Corset, V.; Bidaud, C.; Billaud, M.; Bredesen, D.E.; Edery, P.; Mehlen, P. The RET proto-oncogene induces apoptosis: A novel mechanism for Hirschsprung disease. EMBO J. 2000, 19, 4056–4063.

65. Tauszig-Delamasure, S.; Yu, L.Y.; Cabrera, J.R.; Bouzas-Rodriguez, J.; Mermet-Bouvier, C.; Guix, C.; Bordeaux, M.C.; Arumae, U.; Mehlen, P. The TrkC receptor induces apoptosis when the dependence receptor notion meets the neurotrophin paradigm. Proc. Natl. Acad. Sci. USA 2007, 104, 13361–13366.

66. Wang, H.; Boussouar, A.; Mazelin, L.; Tauszig-Delamasure, S.; Sun, Y.; Goldschneider, D.; Paradisi, A.; Mehlen, P. The Proto-oncogene c-Kit Inhibits Tumor Growth by Behaving as a Dependence Receptor. Mol. Cell 2018, 72, 413–425 e415.

67. Negulescu, A.M.; Mehlen, P. Dependence receptors—The dark side awakens. FEBS J. 2018, 285, 3909–3924.

68. Paradisi, A.; Maisse, C.; Bernet, A.; Coissieux, M.M.; Maccarrone, M.; Scoazec, J.Y.; Mehlen, P. NF-kappaB regulates netrin-1 expression and affects the conditional tumor suppressive activity of the netrin-1 receptors. Gastroenterology 2008, 135, 1248–1257.

69. Paradisi, A.; Maisse, C.; Coissieux, M.M.; Gadot, N.; Lepinasse, F.; Delloye-Bourgeois, C.; Delcros, J.G.; Svrcek, M.; Neufert, C.; Flejou, J.F.; et al. Netrin-1 up-regulation in inflammatory bowel diseases is required for colorectal cancer progression. Proc. Natl. Acad. Sci. USA 2009, 106, 17146–17151.

70. Aherne, C.M.; Collins, C.B.; Eltzschig, H.K. Netrin-1 guides inflammatory cell migration to control mucosal immune responses during intestinal inflammation. Tissue Barriers 2013, 1, e24957.

71. Chen, Z.; Chen, Y.; Zhou, J.; Li, Y.; Gong, C.; Wang, X. Netrin-1 reduces lung ischemia-reperfusion injury by increasing the proportion of regulatory T cells. J. Int. Med. Res. 2020, 48, 300060520926415.

72. von Karstedt, S.; Montinaro, A.; Walczak, H. Exploring the TRAILs less travelled: TRAIL in cancer biology and therapy. Nat. Rev. Cancer 2017, 17, 352–366.

73. Wajant, H.; Siegmund, D. TNFR1 and TNFR2 in the Control of the Life and Death Balance of Macrophages. Front. Cell Dev. Biol. 2019, 7, 91.

74. Chinnaiyan, A.M.; O'Rourke, K.; Tewari, M.; Dixit, V.M. FADD, a novel death domain-containing protein, interacts with the death domain of Fas and initiates apoptosis. Cell 1995, 81, 505–512.

75. Kischkel, F.C.; Lawrence, D.A.; Chuntharapai, A.; Schow, P.; Kim, K.J.; Ashkenazi, A. Apo2L/TRAIL-dependent recruitment of endogenous FADD and caspase-8 to death receptors 4 and 5. Immunity 2000, 12, 611–620.

76. Muzio, M.; Chinnaiyan, A.M.; Kischkel, F.C.; O'Rourke, K.; Shevchenko, A.; Ni, J.; Scaffidi, C.; Bretz, J.D.; Zhang, M.; Gentz, R.; et al. FLICE, a novel FADD-homologous ICE/CED-3-like protease, is recruited to the CD95 (Fas/APO-1) death--inducing signaling complex. Cell 1996, 85, 817–827.

77. Tummers, B.; Mari, L.; Guy, C.S.; Heckmann, B.L.; Rodriguez, D.A.; Ruhl, S.; Moretti, J.; Crawford, J.C.; Fitzgerald, P.; Kanneganti, T.D.; et al. Caspase-8-Dependent Inflammatory Responses Are Controlled by Its Adaptor, FADD, and Necroptosis. Immunity 2020, 52, 994–1006.e8.

78. Brenner, D.; Blaser, H.; Mak, T.W. Regulation of tumour necrosis factor signalling: Live or let die. Nat. Rev. Immunol. 2015, 15, 362–374.

79. Micheau, O.; Lens, S.; Gaide, O.; Alevizopoulos, K.; Tschopp, J. NF-kappaB signals induce the expression of c-FLIP. Mol. Cell Biol. 2001, 21, 5299–5305.

80. Grootjans, S.; Vanden Berghe, T.; Vandenabeele, P. Initiation and execution mechanisms of necroptosis: An overview. Cell Death Differ. 2017, 24, 1184–1195.

81. Li, J.; McQuade, T.; Siemer, A.B.; Napetschnig, J.; Moriwaki, K.; Hsiao, Y.S.; Damko, E.; Moquin, D.; Walz, T.; McDermott, A.; et al. The RIP1/RIP3 necrosome forms a functional amyloid signaling complex required for programmed necrosis. Cell 2012, 150, 339–350.

82. Dondelinger, Y.; Declercq, W.; Montessuit, S.; Roelandt, R.; Goncalves, A.; Bruggeman, I.; Hulpiau, P.; Weber, K.; Sehon, C.A.; Marquis, R.W.; et al. MLKL compromises plasma membrane integrity by binding to phosphatidylinositol phosphates. Cell Rep. 2014, 7, 971–981.

83. He, S.; Liang, Y.; Shao, F.; Wang, X. Toll-like receptors activate programmed necrosis in macrophages through a receptor-interacting kinase-3-mediated pathway. Proc. Natl. Acad. Sci. USA 2011, 108, 20054–20059.

84. Upton, J.W.; Kaiser, W.J.; Mocarski, E.S. DAI/ZBP1/DLM-1 complexes with RIP3 to mediate virus-induced programmed necrosis that is targeted by murine cytomegalovirus vIRA. Cell Host Microbe 2012, 11, 290–297.

85. Kaiser, W.J.; Sridharan, H.; Huang, C.; Mandal, P.; Upton, J.W.; Gough, P.J.; Sehon, C.A.; Marquis, R.W.; Bertin, J.; Mocarski, E.S. Toll-like receptor 3-mediated necrosis via TRIF, RIP3, and MLKL. J. Biol. Chem. 2013, 288, 31268–31279.

86. McComb, S.; Cessford, E.; Alturki, N.A.; Joseph, J.; Shutinoski, B.; Startek, J.B.; Gamero, A.M.; Mossman, K.L.; Sad, S. Type-I interferon signaling through ISGF3 complex is required for sustained Rip3 activation and necroptosis in macrophages. Proc. Natl. Acad. Sci. USA 2014, 111, E3206–E3213.

87. Man, S.M.; Karki, R.; Kanneganti, T.D. Molecular mechanisms and functions of pyroptosis, inflammatory caspases and inflammasomes in infectious diseases. Immunol. Rev. 2017, 277, 61–75.

88. Franchi, L.; Eigenbrod, T.; Munoz-Planillo, R.; Nunez, G. The inflammasome: A caspase-1-activation platform that regulates immune responses and disease pathogenesis. Nat. Immunol. 2009, 10, 241–247.

89. Shi, J.; Zhao, Y.; Wang, Y.; Gao, W.; Ding, J.; Li, P.; Hu, L.; Shao, F. Inflammatory caspases are innate immune receptors for intracellular LPS. Nature 2014, 514, 187–192.

90. Jiang, S.; Gu, H.; Zhao, Y.; Sun, L. Teleost Gasdermin E Is Cleaved by Caspase 1, 3, and 7 and Induces Pyroptosis. J. Immunol. 2019, 203, 1369–1382.

91. Yatim, N.; Cullen, S.; Albert, M.L. Dying cells actively regulate adaptive immune responses. Nat. Rev. Immunol. 2017, 17, 262–275.

92. Green, D.R.; Ferguson, T.; Zitvogel, L.; Kroemer, G. Immunogenic and tolerogenic cell death. Nat. Rev. Immunol. 2009, 9, 353–363.

93. Christgen, S.; Zheng, M.; Kesavardhana, S.; Karki, R.; Malireddi, R.K.S.; Banoth, B.; Place, D.E.; Briard, B.; Sharma, B.R.; Tuladhar, S.; et al. Identification of the PANoptosome: A Molecular Platform Triggering Pyroptosis, Apoptosis, and Necroptosis (PANoptosis). Front. Cell Infect. MicroBiol. 2020, 10, 237.

94. Boya, P.; Gonzalez-Polo, R.A.; Casares, N.; Perfettini, J.L.; Dessen, P.; Larochette, N.; Metivier, D.; Meley, D.; Souquere, S.; Yoshimori, T.; et al. Inhibition of macroautophagy triggers apoptosis. Mol. Cell Biol. 2005, 25, 1025–1040.

95. Galluzzi, L.; Baehrecke, E.H.; Ballabio, A.; Boya, P.; Bravo-San Pedro, J.M.; Cecconi, F.; Choi, A.M.; Chu, C.T.; Codogno, P.; Colombo, M.I.; et al. Molecular definitions of autophagy and related processes. EMBO J. 2017, 36, 1811–1836.

96. Levine, B.; Huang, Q.; Isaacs, J.T.; Reed, J.C.; Griffin, D.E.; Hardwick, J.M. Conversion of lytic to persistent alphavirus infection by the bcl-2 cellular oncogene. Nature 1993, 361, 739–742.

97. Grandgirard, D.; Studer, E.; Monney, L.; Belser, T.; Fellay, I.; Borner, C.; Michel, M.R. Alphaviruses induce apoptosis in Bcl-2-overexpressing cells: Evidence for a caspase-mediated, proteolytic inactivation of Bcl-2. EMBO J. 1998, 17, 1268–1278.

98. Moriishi, K.; Koura, M.; Matsuura, Y. Induction of Bad-mediated apoptosis by Sindbis virus infection: Involvement of pro-survival members of the Bcl-2 family. Virology 2002, 292, 258–271.

99. Keck, F.; Brooks-Faulconer, T.; Lark, T.; Ravishankar, P.; Bailey, C.; Salvador-Morales, C.; Narayanan, A. Altered mitochondrial

dynamics as a consequence of Venezuelan Equine encephalitis virus infection. Virulence 2017, 8, 1849–1866.

100. Lin, K.I.; Lee, S.H.; Narayanan, R.; Baraban, J.M.; Hardwick, J.M.; Ratan, R.R. Thiol agents and Bcl-2 identify an alphavirus-induced apoptotic pathway that requires activation of the transcription factor NF-kappa B. J. Cell Biol. 1995, 131, 1149–1161.

101. Yoshinaka, Y.; Takahashi, Y.; Nakamura, S.; Katoh, I.; Takio, K.; Ikawa, Y. Induction of manganese-superoxide dismutase in MRC-5 cells persistently infected with an alphavirus, sindbis. BioChem. Biophys Res. Commun. 1999, 261, 139–143.

102. Dhanwani, R.; Khan, M.; Bhaskar, A.S.; Singh, R.; Patro, I.K.; Rao, P.V.; Parida, M.M. Characterization of Chikungunya virus infection in human neuroblastoma SH-SY5Y cells: Role of apoptosis in neuronal cell death. Virus Res. 2012, 163, 563–572.

103. Iranpour, M.; Moghadam, A.R.; Yazdi, M.; Ande, S.R.; Alizadeh, J.; Wiechec, E.; Lindsay, R.; Drebot, M.; Coombs, K.M.; Ghavami, S. Apoptosis, autophagy and unfolded protein response pathways in Arbovirus replication and pathogenesis. Expert Rev. Mol. Med. 2016, 18, e1.

104. Barry, G.; Fragkoudis, R.; Ferguson, M.C.; Lulla, A.; Merits, A.; Kohl, A.; Fazakerley, J.K. Semliki forest virus-induced endoplasmic reticulum stress accelerates apoptotic death of mammalian cells. J. Virol. 2010, 84, 7369–7377.

105. Dahal, B.; Lin, S.C.; Carey, B.D.; Jacobs, J.L.; Dinman, J.D.; van Hoek, M.L.; Adams, A.A.; Kehn-Hall, K. EGR1 upregulation following Venezuelan equine encephalitis virus infection is regulated by ERK and PERK pathways contributing to cell death. Virology 2020, 539, 121–128.

106. Baer, A.; Lundberg, L.; Swales, D.; Waybright, N.; Pinkham, C.; Dinman, J.D.; Jacobs, J.L.; Kehn-Hall, K. Venezuelan Equine Encephalitis Virus Induces Apoptosis through the Unfolded Protein Response Activation of EGR1. J. Virol. 2016, 90, 3558–3572.

107. Frolov, I.; Agapov, E.; Hoffman, T.A., Jr.; Pragai, B.M.; Lippa, M.; Schlesinger, S.; Rice, C.M. Selection of RNA replicons capable of persistent noncytopathic replication in mammalian cells. J. Virol. 1999, 73, 3854–3865.

108. Petrakova, O.; Volkova, E.; Gorchakov, R.; Paessler, S.; Kinney, R.M.; Frolov, I. Noncytopathic replication of Venezuelan equine encephalitis virus and eastern equine encephalitis virus replicons in Mammalian cells. J. Virol. 2005, 79, 7597–7608.

109. Perri, S.; Driver, D.A.; Gardner, J.P.; Sherrill, S.; Belli, B.A.; Dubensky, T.W., Jr.; Polo, J.M. Replicon vectors derived from Sindbis virus and Semliki forest virus that establish persistent replication in host cells. J. Virol. 2000, 74, 9802–9807.

110. Garmashova, N.; Gorchakov, R.; Frolova, E.; Frolov, I. Sindbis virus nonstructural protein nsP2 is cytotoxic and inhibits cellular transcription. J. Virol. 2006, 80, 5686–5696.

111. Akhrymuk, I.; Kulemzin, S.V.; Frolova, E.I. Evasion of the innate immune response: The Old World alphavirus nsP2 protein induces rapid degradation of Rpb1, a catalytic subunit of RNA polymerase II. J. Virol. 2012, 86, 7180–7191.

112. Akhrymuk, I.; Frolov, I.; Frolova, E.I. Sindbis Virus Infection Causes Cell Death by nsP2-Induced Transcriptional Shutoff or by nsP3-Dependent Translational Shutoff. J. Virol. 2018, 92.

113. Garmashova, N.; Gorchakov, R.; Volkova, E.; Paessler, S.; Frolova, E.; Frolov, I. The Old World and New World alphaviruses use different virus-specific proteins for induction of transcriptional shutoff. J. Virol. 2007, 81, 2472–2484.

114. Fros, J.J.; Major, L.D.; Scholte, F.E.M.; Gardner, J.; van Hemert, M.J.; Suhrbier, A.; Pijlman, G.P. Chikungunya virus non-structural protein 2-mediated host shut-off disables the unfolded protein response. J. Gen. Virol. 2015, 96, 580–589.

115. Fros, J.J.; van der Maten, E.; Vlak, J.M.; Pijlman, G.P. The C-terminal domain of chikungunya virus nsP2 independently governs viral

RNA replication, cytopathicity, and inhibition of interferon signaling. J. Virol. 2013, 87, 10394–10400.

116. Goertz, G.P.; McNally, K.L.; Robertson, S.J.; Best, S.M.; Pijlman, G.P.; Fros, J.J. The Methyltransferase-Like Domain of Chikungunya Virus nsP2 Inhibits the Interferon Response by Promoting the Nuclear Export of STAT1. J. Virol. 2018, 92.

117. Sarid, R.; Ben-Moshe, T.; Kazimirsky, G.; Weisberg, S.; Appel, E.; Kobiler, D.; Lustig, S.; Brodie, C. vFLIP protects PC-12 cells from apoptosis induced by Sindbis virus: Implications for the role of TNF-alpha. Cell Death Differ. 2001, 8, 1224–1231.

118. Nava, V.E.; Rosen, A.; Veliuona, M.A.; Clem, R.J.; Levine, B.; Hardwick, J.M. Sindbis virus induces apoptosis through a caspase-dependent, CrmA-sensitive pathway. J. Virol. 1998, 72, 452–459.

119. Kiiver, K.; Merits, A.; Sarand, I. Novel vectors expressing anti-apoptotic protein Bcl-2 to study cell death in Semliki Forest virus-infected cells. Virus Res. 2008, 131, 54–64.

120. Zhou, Q.; Snipas, S.; Orth, K.; Muzio, M.; Dixit, V.M.; Salvesen, G.S. Target protease specificity of the viral serpin CrmA. Analysis of five caspases. J. Biol. Chem. 1997, 272, 7797–7800.

121. Cho, Y.S.; Challa, S.; Moquin, D.; Genga, R.; Ray, T.D.; Guildford, M.; Chan, F.K. Phosphorylation-driven assembly of the RIP1-RIP3 complex regulates programmed necrosis and virus-induced inflammation. Cell 2009, 137, 1112–1123.

122. Nailwal, H.; Chan, F.K. Necroptosis in anti-viral inflammation. Cell Death Differ. 2019, 26, 4–13.

123. Joubert, P.E.; Werneke, S.; de la Calle, C.; Guivel-Benhassine, F.; Giodini, A.; Peduto, L.; Levine, B.; Schwartz, O.; Lenschow, D.; Albert, M.L. Chikungunya-induced cell death is limited by ER and oxidative stress-induced autophagy. Autophagy 2012, 8, 1261–1263.

124. Joubert, P.E.; Werneke, S.W.; de la Calle, C.; Guivel-Benhassine, F.; Giodini, A.; Peduto, L.; Levine, B.; Schwartz, O.; Lenschow, D.J.;

Albert, M.L. Chikungunya virus-induced autophagy delays caspase-dependent cell death. J. Exp. Med. 2012, 209, 1029–1047.

125. Urban, C.; Rheme, C.; Maerz, S.; Berg, B.; Pick, R.; Nitschke, R.; Borner, C. Apoptosis induced by Semliki Forest virus is RNA replication dependent and mediated via Bak. Cell Death Differ. 2008, 15, 1396–1407.

126. Jan, J.T.; Griffin, D.E. Induction of apoptosis by Sindbis virus occurs at cell entry and does not require virus replication. J. Virol. 1999, 73, 10296–10302.

127. Cowling, V.; Downward, J. Caspase-6 is the direct activator of caspase-8 in the cytochrome c-induced apoptosis pathway: Absolute requirement for removal of caspase-6 prodomain. Cell Death Differ. 2002, 9, 1046–1056.

128. Zheng, M.; Karki, R.; Vogel, P.; Kanneganti, T.D. Caspase-6 Is a Key Regulator of Innate Immunity, Inflammasome Activation, and Host Defense. Cell 2020.

129. El Maadidi, S.; Faletti, L.; Berg, B.; Wenzl, C.; Wieland, K.; Chen, Z.J.; Maurer, U.; Borner, C. A novel mitochondrial MAVS/Caspase-8 platform links RNA virus-induced innate antiviral signaling to Bax/Bak-independent apoptosis. J. Immunol. 2014, 192, 1171–1183.

130. Vazquez, C.; Horner, S.M. MAVS Coordination of Antiviral Innate Immunity. J. Virol. 2015, 89, 6974–6977.

131. Pan, P.; Zhang, Q.; Liu, W.; Wang, W.; Lao, Z.; Zhang, W.; Shen, M.; Wan, P.; Xiao, F.; Liu, F.; et al. Dengue Virus M Protein Promotes NLRP3 Inflammasome Activation To Induce Vascular Leakage in Mice. J. Virol. 2019, 93.

132. Liu, T.; Tang, L.; Tang, H.; Pu, J.; Gong, S.; Fang, D.; Zhang, H.; Li, Y.P.; Zhu, X.; Wang, W.; et al. Zika Virus Infection Induces Acute Kidney Injury Through Activating NLRP3 Inflammasome Via Suppressing Bcl-2. Front. Immunol. 2019, 10, 1925.

133. Wang, Z.Y.; Zhen, Z.D.; Fan, D.Y.; Qin, C.F.; Han, D.S.; Zhou, H.N.; Wang, P.G.; An, J. Axl deficiency promotes the neuroinvasion of

Japanese encephalitis virus by enhancing IL-1alpha production from pyroptotic macrophages. J. Virol. 2020.

134. Ekchariyawat, P.; Hamel, R.; Bernard, E.; Wichit, S.; Surasombatpattana, P.; Talignani, L.; Thomas, F.; Choumet, V.; Yssel, H.; Despres, P.; et al. Inflammasome signaling pathways exert antiviral effect against Chikungunya virus in human dermal fibroblasts. Infect. Genet. Evol. 2015, 32, 401–408.

135. Chen, W.; Foo, S.S.; Zaid, A.; Teng, T.S.; Herrero, L.J.; Wolf, S.; Tharmarajah, K.; Vu, L.D.; van Vreden, C.; Taylor, A.; et al. Specific inhibition of NLRP3 in chikungunya disease reveals a role for inflammasomes in alphavirus-induced inflammation. Nat. MicroBiol. 2017, 2, 1435–1445.

136. de Castro-Jorge, L.A.; de Carvalho, R.V.H.; Klein, T.M.; Hiroki, C.H.; Lopes, A.H.; Guimaraes, R.M.; Fumagalli, M.J.; Floriano, V.G.; Agostinho, M.R.; Slhessarenko, R.D.; et al. The NLRP3 inflammasome is involved with the pathogenesis of Mayaro virus. PLoS Pathog. 2019, 15, e1007934.

137. Levine, B.; Goldman, J.E.; Jiang, H.H.; Griffin, D.E.; Hardwick, J.M. Bc1-2 protects mice against fatal alphavirus encephalitis. Proc. Natl. Acad. Sci. USA 1996, 93, 4810–4815.

138. Scallan, M.F.; Allsopp, T.E.; Fazakerley, J.K. bcl-2 acts early to restrict Semliki Forest virus replication and delays virus-induced programmed cell death. J. Virol. 1997, 71, 1583–1590.

139. Liang, X.H.; Kleeman, L.K.; Jiang, H.H.; Gordon, G.; Goldman, J.E.; Berry, G.; Herman, B.; Levine, B. Protection against fatal Sindbis virus encephalitis by beclin, a novel Bcl-2-interacting protein. J. Virol. 1998, 72, 8586–8596.

140. Pattingre, S.; Tassa, A.; Qu, X.; Garuti, R.; Liang, X.H.; Mizushima, N.; Packer, M.; Schneider, M.D.; Levine, B. Bcl-2 antiapoptotic proteins inhibit Beclin 1-dependent autophagy. Cell 2005, 122, 927–939.

141. Orvedahl, A.; MacPherson, S.; Sumpter, R., Jr.; Talloczy, Z.; Zou, Z.; Levine, B. Autophagy protects against Sindbis virus infection of the central nervous system. Cell Host Microbe 2010, 7, 115–127.

142. Eng, K.E.; Panas, M.D.; Murphy, D.; Karlsson Hedestam, G.B.; McInerney, G.M. Accumulation of autophagosomes in Semliki Forest virus-infected cells is dependent on expression of the viral glycoproteins. J. Virol. 2012, 86, 5674–5685.

143. Krejbich-Trotot, P.; Gay, B.; Li-Pat-Yuen, G.; Hoarau, J.J.; Jaffar-Bandjee, M.C.; Briant, L.; Gasque, P.; Denizot, M. Chikungunya triggers an autophagic process which promotes viral replication. Virol. J. 2011, 8, 432.

144. Judith, D.; Mostowy, S.; Bourai, M.; Gangneux, N.; Lelek, M.; Lucas-Hourani, M.; Cayet, N.; Jacob, Y.; Prevost, M.C.; Pierre, P.; et al. Species-specific impact of the autophagy machinery on Chikungunya virus infection. EMBO Rep. 2013, 14, 534–544.

145. Van Huizen, E.; McInerney, G.M. Activation of the PI3K-AKT Pathway by Old World Alphaviruses. Cells 2020, 9, 970.

146. Krejbich-Trotot, P.; Denizot, M.; Hoarau, J.J.; Jaffar-Bandjee, M.C.; Das, T.; Gasque, P. Chikungunya virus mobilizes the apoptotic machinery to invade host cell defenses. FASEB J. 2011, 25, 314–325.

147. Amara, A.; Mercer, J. Viral apoptotic mimicry. Nat. Rev. MicroBiol. 2015, 13, 461–469.

148. Lewis, J.; Wesselingh, S.L.; Griffin, D.E.; Hardwick, J.M. Alphavirus-induced apoptosis in mouse brains correlates with neurovirulence. J. Virol. 1996, 70, 1828–1835.

149. Jackson, A.C.; Rossiter, J.P. Apoptotic cell death is an important cause of neuronal injury in experimental Venezuelan equine encephalitis virus infection of mice. Acta Neuropathol. 1997, 93, 349–353.

150. Appel, E.; Katzoff, A.; Ben-Moshe, T.; Kazimirsky, G.; Kobiler, D.; Lustig, S.; Brodie, C. Differential regulation of Bcl-2 and Bax expression in cells infected with virulent and nonvirulent strains of sindbis virus. Virology 2000, 276, 238–242.

151. Sammin, D.J.; Butler, D.; Atkins, G.J.; Sheahan, B.J. Cell death mechanisms in the olfactory bulb of rats infected intranasally

with Semliki forest virus. Neuropathol. Appl. NeuroBiol. 1999, 25, 236–243.

152. Hoarau, J.J.; Jaffar Bandjee, M.C.; Krejbich Trotot, P.; Das, T.; Li-Pat-Yuen, G.; Dassa, B.; Denizot, M.; Guichard, E.; Ribera, A.; Henni, T.; et al. Persistent chronic inflammation and infection by Chikungunya arthritogenic alphavirus in spite of a robust host immune response. J. Immunol. 2010, 184, 5914–5927.

153. Krejbich-Trotot, P.; Belarbi, E.; Ralambondrainy, M.; El-Kalamouni, C.; Viranaicken, W.; Roques, P.; Despres, P.; Gadea, G. The growth of arthralgic Ross River virus is restricted in human monocytic cells. Virus Res. 2016, 225, 64–68.

154. Young, A.R.; Locke, M.C.; Cook, L.E.; Hiller, B.E.; Zhang, R.; Hedberg, M.L.; Monte, K.J.; Veis, D.J.; Diamond, M.S.; Lenschow, D.J. Dermal and muscle fibroblasts and skeletal myofibers survive chikungunya virus infection and harbor persistent RNA. PLoS Pathog. 2019, 15, e1007993.

155. Weaver, S.C.; Scott, T.W.; Lorenz, L.H.; Lerdthusnee, K.; Romoser, W.S. Togavirus-associated pathologic changes in the midgut of a natural mosquito vector. J. Virol. 1988, 62, 2083–2090.

156. Weaver, S.C.; Lorenz, L.H.; Scott, T.W. Pathologic changes in the midgut of Culex tarsalis following infection with Western equine encephalomyelitis virus. Am. J. Trop. Med. Hyg. 1992, 47, 691–701.

157. Dong, S.; Behura, S.K.; Franz, A.W.E. The midgut transcriptome of Aedes aegypti fed with saline or protein meals containing chikungunya virus reveals genes potentially involved in viral midgut escape. BMC Genom. 2017, 18, 382.

158. Kelly, E.M.; Moon, D.C.; Bowers, D.F. Apoptosis in mosquito salivary glands: Sindbis virus-associated and tissue homeostasis. J. Gen. Virol. 2012, 93, 2419–2424.

159. Bowers, D.F.; Coleman, C.G.; Brown, D.T. Sindbis virus-associated pathology in Aedes albopictus (Diptera: Culicidae). J. Med. Entomol. 2003, 40, 698–705.

160. Vo, M.; Linser, P.J.; Bowers, D.F. Organ-associated muscles in Aedes albopictus (Diptera: Culicidae) respond differentially to Sindbis virus. J. Med. Entomol. 2010, 47, 215–225.

161. Wang, H.; Gort, T.; Boyle, D.L.; Clem, R.J. Effects of manipulating apoptosis on Sindbis virus infection of Aedes aegypti mosquitoes. J. Virol. 2012, 86, 6546–6554.

162. Denton, D.; Aung-Htut, M.T.; Kumar, S. Developmentally programmed cell death in Drosophila. Biochim. Biophys. Acta 2013, 1833, 3499–3506.

163. Zhou, L.; Song, Z.; Tittel, J.; Steller, H. HAC-1, a Drosophila homolog of APAF-1 and CED-4 functions in developmental and radiation-induced apoptosis. Mol. Cell 1999, 4, 745–755.

164. Yu, X.; Wang, L.; Acehan, D.; Wang, X.; Akey, C.W. Three-dimensional structure of a double apoptosome formed by the Drosophila Apaf-1 related killer. J. Mol. Biol. 2006, 355, 577–589.

165. Dorstyn, L.; Kumar, S. A cytochrome c-free fly apoptosome. Cell Death Differ. 2006, 13, 1049–1051.

166. Doumanis, J.; Dorstyn, L.; Kumar, S. Molecular determinants of the subcellular localization of the Drosophila Bcl-2 homologues DEBCL and BUFFY. Cell Death Differ. 2007, 14, 907–915.

167. Igaki, T.; Miura, M. Role of Bcl-2 family members in invertebrates. Biochim. Biophys. Acta 2004, 1644, 73–81.

168. Clavier, A.; Rincheval-Arnold, A.; Colin, J.; Mignotte, B.; Guenal, I. Apoptosis in Drosophila: Which role for mitochondria? Apoptosis 2016, 21, 239–251.

169. Cooper, D.M.; Granville, D.J.; Lowenberger, C. The insect caspases. Apoptosis 2009, 14, 247–256.

170. Liu, Q.; Clem, R.J. Defining the core apoptosis pathway in the mosquito disease vector Aedes aegypti: The roles of iap1, ark, dronc, and effector caspases. Apoptosis 2011, 16, 105–113.

171. Challa, M.; Malladi, S.; Pellock, B.J.; Dresnek, D.; Varadarajan, S.; Yin, Y.W.; White, K.; Bratton, S.B. Drosophila Omi, a mitochondrial-localized IAP antagonist and proapoptotic serine protease. EMBO J. 2007, 26, 3144–3156.

172. Chai, J.; Yan, N.; Huh, J.R.; Wu, J.W.; Li, W.; Hay, B.A.; Shi, Y. Molecular mechanism of Reaper-Grim-Hid-mediated suppression of DIAP1-dependent Dronc ubiquitination. Nat. Struct. Biol. 2003, 10, 892–898.

173. Brodsky, M.H.; Nordstrom, W.; Tsang, G.; Kwan, E.; Rubin, G.M.; Abrams, J.M. Drosophila p53 binds a damage response element at the reaper locus. Cell 2000, 101, 103–113.

174. Igaki, T.; Kanda, H.; Yamamoto-Goto, Y.; Kanuka, H.; Kuranaga, E.; Aigaki, T.; Miura, M. Eiger, a TNF superfamily ligand that triggers the Drosophila JNK pathway. EMBO J. 2002, 21, 3009–3018.

175. Tafesh-Edwards, G.; Eleftherianos, I. JNK signaling in Drosophila immunity and homeostasis. Immunol. Lett. 2020, 226, 7–11.

176. Kleino, A.; Silverman, N. The Drosophila IMD pathway in the activation of the humoral immune response. Dev. Comp. Immunol. 2014, 42, 25–35.

177. Ocampo, C.B.; Caicedo, P.A.; Jaramillo, G.; Ursic Bedoya, R.; Baron, O.; Serrato, I.M.; Cooper, D.M.; Lowenberger, C. Differential expression of apoptosis related genes in selected strains of Aedes aegypti with different susceptibilities to dengue virus. PLoS ONE 2013, 8, e61187.

178. Tettamanti, G.; Casartelli, M. Cell death during complete metamorphosis. Philos. Trans. R. Soc. Lond. B Biol. Sci. 2019, 374, 20190065.

179. Gohel, R.; Kournoutis, A.; Petridi, S.; Nezis, I.P. Molecular mechanisms of selective autophagy in Drosophila. Int. Rev. Cell Mol. Biol. 2020, 354, 63–105.

180. Means, J.C.; Passarelli, A.L. Viral fibroblast growth factor, matrix metalloproteases, and caspases are associated with enhancing

systemic infection by baculoviruses. Proc. Natl. Acad. Sci. USA 2010, 107, 9825–9830.

181. Wang, H.; Blair, C.D.; Olson, K.E.; Clem, R.J. Effects of inducing or inhibiting apoptosis on Sindbis virus replication in mosquito cells. J. Gen. Virol. 2008, 89, 2651–2661.

182. O'Neill, K.; Olson, B.J.; Huang, N.; Unis, D.; Clem, R.J. Rapid selection against arbovirus-induced apoptosis during infection of a mosquito vector. Proc. Natl. Acad. Sci. USA 2015, 112, E1152–E1161.

183. Tchankouo-Nguetcheu, S.; Khun, H.; Pincet, L.; Roux, P.; Bahut, M.; Huerre, M.; Guette, C.; Choumet, V. Differential protein modulation in midguts of Aedes aegypti infected with chikungunya and dengue 2 viruses. PLoS ONE 2010, 5.

184. Camini, F.C.; da Silva Caetano, C.C.; Almeida, L.T.; da Costa Guerra, J.F.; de Mello Silva, B.; de Queiroz Silva, S.; de Magalhaes, J.C.; de Brito Magalhaes, C.L. Oxidative stress in Mayaro virus infection. Virus Res. 2017, 236, 1–8.

185. Felton, G.W.; Summers, C.B. Antioxidant systems in insects. Arch. Insect BioChem. Physiol. 1995, 29, 187–197.

186. Whiten, S.R.; Eggleston, H.; Adelman, Z.N. Ironing out the Details: Exploring the Role of Iron and Heme in Blood-Sucking Arthropods. Front. Physiol. 2017, 8, 1134.

187. Patel, R.K.; Hardy, R.W. Role for the phosphatidylinositol 3-kinase-Akt-TOR pathway during sindbis virus replication in arthropods. J. Virol. 2012, 86, 3595–3604.

188. Brackney, D.E.; Correa, M.A.; Cozens, D.W. The impact of autophagy on arbovirus infection of mosquito cells. PLoS Negl. Trop. Dis. 2020, 14, e0007754.

189. Liu, B.; Behura, S.K.; Clem, R.J.; Schneemann, A.; Becnel, J.; Severson, D.W.; Zhou, L. P53-mediated rapid induction of apoptosis conveys resistance to viral infection in Drosophila melanogaster. PLoS Pathog. 2013, 9, e1003137.

190. Lamiable, O.; Arnold, J.; de Faria, I.; Olmo, R.P.; Bergami, F.; Meignin, C.; Hoffmann, J.A.; Marques, J.T.; Imler, J.L. Analysis of the Contribution of Hemocytes and Autophagy to Drosophila Antiviral Immunity. J. Virol. 2016, 90, 5415–5426.

191. Vermaak, E.; Maree, F.F.; Theron, J. The Culicoides sonorensis inhibitor of apoptosis 1 protein protects mammalian cells from apoptosis induced by infection with African horse sickness virus and bluetongue virus. Virus Res. 2017, 232, 152–161.

192. Li, Q.; Li, H.; Blitvich, B.J.; Zhang, J. The Aedes albopictus inhibitor of apoptosis 1 gene protects vertebrate cells from bluetongue virus-induced apoptosis. Insect Mol. Biol. 2007, 16, 93–105.

193. Oliveira, J.H.M.; Talyuli, O.A.C.; Goncalves, R.L.S.; Paiva-Silva, G.O.; Sorgine, M.H.F.; Alvarenga, P.H.; Oliveira, P.L. Catalase protects Aedes aegypti from oxidative stress and increases midgut infection prevalence of Dengue but not Zika. PLoS Negl. Trop. Dis. 2017, 11, e0005525.

194. Robin, M.; Issa, A.R.; Santos, C.C.; Napoletano, F.; Petitgas, C.; Chatelain, G.; Ruby, M.; Walter, L.; Birman, S.; Domingos, P.M.; et al. Drosophila p53 integrates the antagonism between autophagy and apoptosis in response to stress. Autophagy 2019, 15, 771–784.

195. Chen, T.H.; Wu, Y.J.; Hou, J.N.; Chiang, Y.H.; Cheng, C.C.; Sifiyatun, E.; Chiu, C.H.; Wang, L.C.; Chen, W.J. A novel p53 paralogue mediates antioxidant defense of mosquito cells to survive dengue virus replication. Virology 2018, 519, 156–169.

196. Chen, T.H.; Wu, Y.J.; Hou, J.N.; Chiu, C.H.; Chen, W.J. The p53 gene with emphasis on its paralogues in mosquitoes. J. MicroBiol. Immunol. Infect. 2017, 50, 747–754.

197. Nasar, F.; Palacios, G.; Gorchakov, R.V.; Guzman, H.; Da Rosa, A.P.; Savji, N.; Popov, V.L.; Sherman, M.B.; Lipkin, W.I.; Tesh, R.B.; et al. Eilat virus, a unique alphavirus with host range restricted to insects by RNA replication. Proc. Natl. Acad. Sci. USA 2012, 109, 14622–14627.

198. Hermanns, K.; Zirkel, F.; Kopp, A.; Marklewitz, M.; Rwego, I.B.; Estrada, A.; Gillespie, T.R.; Drosten, C.; Junglen, S. Discovery of a novel alphavirus related to Eilat virus. J. Gen. Virol. 2017, 98, 43–49.

199. Elrefaey, A.M.; Abdelnabi, R.; Rosales Rosas, A.L.; Wang, L.; Basu, S.; Delang, L. Understanding the Mechanisms Underlying Host Restriction of Insect-Specific Viruses. Viruses 2020, 12, 964.

14 SIGNIFICANCE OF ENDOCYTOSIS IN THE BIOLOGICAL FUNCTION OF CEDAR VIRUS FUSION PROTEIN

1. INTRODUCTION: AN OVERVIEW

Cedar virus (CedV) is a member of the Henipavirus genus in the Paramyxoviridae family. It was initially discovered in 2012 from urine samples of bats in an Australian Pteropus colony [1]. Although CedV is genetically similar to the highly pathogenic Hendra (HeV) and Nipah viruses (NiV), it has only generated asymptomatic infections in small animal models thus far [1,2]. Hence, scientific investigation has concentrated on elucidating the molecular mechanisms that cause variations in the pathogenicity of these closely related viruses. An interesting characteristic of CedV is its diminished capacity of the immunomodulatory phosphoprotein P to neutralize the interferon response in cell culture [1,3]. Additional distinctions lie in the receptor use of the attachment proteins. The widespread presence of ephrins as cell entry receptors in many organs and their great similarity across different species lead to a wide range of vulnerable hosts and a broad tropism for many cell types. This is crucial for the zoonotic nature and the development of disease in henipaviruses. Highly pathogenic HeV and NiV viruses use ephrin-B2, which is found in endothelial cells and lung tissue, as well as ephrin-B3, which is mainly present in the central nervous system, to enter cells [4,5,6,7]. However, CedV is unable to use ephrin-B3 and instead binds to ephrin-B1, which is expressed in various tissues including salivary glands, esophagus, and lung [8]. Hence, current research has examined how the specific receptor use of the CedV attachment protein contributes to its diminished ability to cause disease [8,9]. In addition to attaching to receptors, the interaction between the attachment protein G and the viral fusion protein F is necessary for the virus to enter the host cell and disseminate throughout the body. A crucial process for the biological function of fusion proteins and consequently, viral infectivity, is the enzymatic splitting of the precursor protein F0 into the two subunits F1 and F2 [10]. Notably, the process of activating the HeV and NiV F protein by proteolysis is significantly different from that of other paramyxoviruses in terms of where it occurs within the cell and the specific protease that is involved. Following transportation through the secretory system, the recently produced HeV and NiV F protein precursors necessitate endocytosis from the surface of the cell in order to interact with the activating host cell protease and subsequently become biologically active. After cleavage occurs in the endosomal compartment, the components are recycled back to the cell surface. Then, the mature fusogenic F1+F2 heterodimers are incorporated into newly emerging virions. This process has been supported by studies [11,12,13,14,15,16,17]. In summary, both viral envelope proteins play a crucial role in determining the ability of a virus to cause disease. They must work together to facilitate the fusion of the virus with the cell membrane, allowing the virus to enter the cell. Additionally, they are involved in the fusion of infected cells with neighboring cells, leading to the formation of syncytia and the spread of the virus. The process of trafficking through early and late endosomes before merging with cellular membranes is crucial for the entry of other viruses, including influenza virus [18,19], ebolaviruses [20,21], and flaviviruses [22,23]. However, this process is not necessary for NiV entry [16]. Furthermore, other viruses and their glycoproteins exploit endosomal pathways to facilitate their reproduction within infected cells [24,25]. The viral envelope glycoprotein of human immunodeficiency virus 1 (HIV-1) undergoes endocytosis as part of its replication cycle. This process is believed to be a way for the virus to avoid detection by the host's immune system by decreasing its presence on the surface of infected cells (as discussed in [26,27]). Furthermore, the transportation of the HIV-1 envelope glycoprotein through the endocytic recycling compartment has recently been identified as a crucial process for its inclusion in virus particles [28]. Endocytosis of herpesvirus glycoproteins has been proposed to have a functional role in cell-cell fusion and the formation of infectious particles. This is achieved by transporting the glycoproteins to the intracellular region where virus assembly occurs [25, 29, 30]. A new discovery reveals that the endocytic trafficking of HeV F protein, rather than its proteolytic cleavage, is a key step for effective assembly of HeV virus-like particles (VLPs) [31]. In addition to its significance in the viral replication cycle, endocytosis plays a crucial role in several cellular activities. Endocytosis is a cellular process that involves the internalization of the plasma membrane and extracellular molecules from the cell surface into internal membrane compartments. It plays a crucial role in various biological events, including maintaining the composition of the plasma membrane and transporting specific cargo molecules from the cell surface to the interior. This process is essential for the proper functioning of cells [32]. Out of the various forms, clathrin-mediated endocytosis (CME) is the most extensively studied and understood [33,34]. The CME (Clathrin-Mediated Endocytosis) of a transmembrane protein is a tightly regulated process that mainly relies on the interaction between the cytoplasmic domain of the protein and intracellular adaptor proteins (AP). These adaptor proteins selectively choose transmembrane cargo and facilitate its entry into clathrin-coated pits by associating with the clathrin lattice. The adaptor complexes identify particular sorting motifs in the cytoplasmic tail of the cargo protein, which are often based on tyrosine or leucine [35,36,37,38]. A common pattern observed in the cytoplasmic tails of transmembrane proteins that interact with certain cytosolic adaptor complex proteins is the presence of a sorting motif called the tyrosine-based YXXΦ motif. In this motif, the letter X can represent any amino acid, while the letter Φ represents an amino acid with a large hydrophobic residue. Several studies have reported that YXXΦ motifs are recognized by AP2, which serves to select cargo from the plasma membrane. Additionally, AP2 enhances binding to clathrin, ultimately leading to clathrin-mediated endocytosis (CME) of the selected transmembrane cargo [35,36,37,38,39,40,41]. Adaptor complex proteins have additional functions beyond facilitating endocytic absorption from the cell surface. Additionally, there have been reports of their involvement in other transportation pathways, such as from early endosomes to recycling or late endosomes, and also to lysosomes by adapting to certain motifs [38,42]. The clathrin-mediated endocytic absorption of HeV and NiV F proteins is highly reliant on a distinct YXXΦ motif found in the cytoplasmic tail's membrane-proximal region, as shown in Table 1 [11,12,43]. As a result, when this pattern was disturbed, it had noticeable impacts on the biological function of the proteins [11,12]. However, another di-tyrosine pattern located in the C-terminal section of the tail had minimal affects on NiV F endocytosis [11].

Table 1. Boldface and underlined letters highlight (potential) endocytosis signals. Numbers above the sequence indicate the amino acid position. Cytoplasmic tails for NiV and HeV F proteins range from amino acid positions 519 to 546, cytoplasmic tail for CedV F is predicted for amino acid positions 517 to 557. Basic aa's that have been shown to be of importance for fusion protein functionality are highlighted in orange.

	519	546
NiV F	...EKKRNTYSRLEDRRVRPTSSGDLYYIGT	
HeV F	...EKKRGNYSRLDDRQVRPVSNGDLYYIGT	
	517	557
CedV F	...KSKHSYKYNKFIDDPDYYNDYKRERINGKASKSNNIYYVGD	

The cytoplasmic tail of the CedV F protein contains several tyrosine residues that could potentially function as endocytosis motifs. Furthermore, there is another di-tyrosine motif present, in addition to the C-terminal di-tyrosine motif that is also seen in HeV and NiV F protein. Two degenerate motifs of the YxxΦ are clearly evident: a YxxN motif, which has been proven to be a functional endocytosis motif for the spike protein of the Porcine Epidemic Diarrhea Virus [44], and a YxxN motif, which has been identified as an endocytosis signal in the Measles virus hemagglutinin [45]. Therefore, the objective of this study was to examine the influence of the tyrosine-based motifs on the protein's cell surface expression, endocytosis, and biological activity. The findings of our study indicate that the uptake of CedV F protein is facilitated by a signal-mediated process. Additionally, our results reveal that a YXXΦ motif near the cell membrane and a di-tyrosine motif at the protein's C-terminus are crucial for its biological activity.

2. MATERIALS AND METHODS

2.1. Cell Lines, Transfection

Vero76, MDCK-2 and HeLa cells (Collection of Cell Lines in Veterinary Medicine, Friedrich-Loeffler-Institut, FLI; CCLV-RIE 0228, 1061 and 0082, respectively) were maintained in Dulbecco's modified Eagle's medium (DMEM) supplemented with fetal calf serum (10%; DMEM10) and incubated at 37 °C. Vero76 and MDCK-2 cells were reverse transfected by using the Lipofectamine 3000 reagent (Invitrogen, Schwerte, Germany) following the manufacturer's protocol. HeLa cells were reverse transfected by using the METAFECTENE® transfection reagent (Biontex, Munich, Germany) according to the manufacturer's instructions.

2.2. Plasmids and Site-Directed Mutagenesis

The open reading frame (ORF) of CedV F and G (GenBank accession no. NC_025351.1) were synthesized by GeneArt (Thermo Fisher Scientific Inc., Regensburg, Germany) and subcloned into the pCAGGS expression vector. The ORF of the CedV F gene was codon-optimized for expression in human cells and HA-tagged at the C-terminus and will be designated as wild-type (wt) in the manuscript. Selected tyrosine residues within the cytoplasmic domain of CedV F protein were mutated to alanine by site-directed mutagenesis using the QuikChange Lightning Site-Directed Mutagenesis Kit (Agilent, Waldbronn, Germany) and primers designed according to manufacturer's instructions.

2.3. Generation of Polyclonal Antibodies

To obtain a polyclonal antiserum raised against the CedV F protein, a rabbit was immunized subcutaneously three times at three-week intervals with CedV F protein. The antigen was produced in 293T cells as described earlier [46]. Briefly, plasmid DNA encoding for CedV F protein was used to transfect 293T cells. At 48 h post-transfection (p.t.), the supernatant was purified through a 20% sucrose cushion at 96,000× g for 2 h at 4 °C. After washing for 30 min at 155,000× g, pellets were resuspended in Tris-sodium chloride buffer. Immunization experiments were assessed and approved by the competent authority for animal welfare issues of the Federal State of Mecklenburg-Western Pomerania, Germany (license number: LALLF 7221.3-2-042/17).

2.4. Antibody Uptake Assay

MDCK-2 cells were reverse transfected with plasmids encoding for parental and mutant CedV F proteins (1 µg/well) in a 24 well plate as mentioned above. At 24 h p.t., confluent monolayers were washed and incubated without prior fixation with the polyclonal anti-CedV F rabbit serum (1:200 in 0.35% bovine serum albumin (BSA) in PBS++ (PBS with 0.5 mM MgCl, 1 mM CaCl2) (0.35% BSA/PBS++)). After incubation at 4 °C for one hour, cells were intensely washed and then either kept on ice or incubated with pre-warmed medium for 30 min at 37 °C to allow endocytic uptake. After fixing the cells with 4% paraformaldehyde (PFA), surface-bound primary antibodies were detected by an Alexa-Fluor (AF) 488-labeled secondary antibody (1:50 in 0.35% BSA/PBS++; LifeTechnologies, Darmstadt, Germany) for 90 min at 4 °C. After permeabilization with methanol-acetone (1:1) for 10 min, AF 568-labelled secondary antibodies (1:500 in 0.35% BSA/PBS++; LifeTechnologies, Darmstadt, Germany) were added to stain internalized primary antibodies. Images were acquired with an Eclipse Ti-S inverted microscope system and were processed with the ImageJ software version 1.45 s [47].

2.5. Metabolic Labeling

For pulse-chase analysis, MDCK-2 cells transiently expressing CedV F proteins were incubated at 24 h p.t. for 15 min with medium lacking cysteine and methionine, followed by incubation with medium containing [35S]cysteine and -methionine (Hartmann Analytic, Braunschweig, Germany) at a final concentration of 100 µCi/mL for 10 min (pulse). Then, labeling medium was replaced by a serum-free nonradioactive medium, and cells were incubated at 37 °C for 2 h (chase). Afterwards, cells were washed with PBS, followed by cell lysis in radioimmunoprecipitation assay (RIPA) buffer (1% Triton X-100, 1% sodium deoxycholate, 0.1% SDS, 0.15 M NaCl, 10 mM EDTA, 50 units/mL aprotinin, and 20 mM Tris-HCl, pH 7.5). Cell lysates were centrifuged for 45 min at 20,000× g, and CedV F proteins were immunoprecipitated using a polyclonal anti-HA antibody (H6908; 1:500; Sigma, Darmstadt, Germany). Protein A-Sepharose CL-4B (Sigma, Darmstadt, Germany) suspension was added for another 45 min followed by several washes of the immune complexes with RIPA buffer. After suspension in sample buffer, proteins were separated on a 12% polyacrylamide gel under reducing conditions. Dried gels were subjected to autoradiography and analyzed with a CR35 Dark Box Image analyzer (Duerr Medical, Bietigheim-Bissingen, Germany).

2.6. Colocalization Studies

For colocalization studies of CedV F and mutants with early endosomes, antibody uptake assays were performed as described above with modifications. Briefly, at 24 h p.t., CedV F-expressing HeLa cells were incubated with the polyclonal rabbit anti-CedV F serum (1:200) for 1 h at 4 °C. After washing, cells were shifted to 37 °C for 5 or 30 min to allow endocytosis to occur. Then, surface-bound primary antibodies were blocked with a horseradish peroxidase (HRP)-conjugated secondary antibody (1:50; LifeTechnologies, Darmstadt, Germany). After fixation with 2% PFA for 20 min and permeabilization with 0.2% Triton X-100 in PBS, a mouse anti-human early endosomal antigen 1 (EEA-1) antibody (1:50; BD Biosciences, Heidelberg, Germany) was added for 1 h at 4 °C for staining of early endosomes. Internalized primary rabbit antibodies were stained with AF 488-conjugated secondary antibodies (1:500; LifeTechnologies, Darmstadt, Germany). Primary mouse antibodies bound to EEA-1 were detected with AF 568-conjugated secondary antibodies (1:500; LifeTechnologies, Darmstadt, Germany). Cell nuclei were counterstained with 4′,6-Diamidin-2-phenylindol (DAPI). Representative images were recorded with a confocal laser scanning microscope (Leica SP5) and processed with the ImageJ software version 1.45 s [47].

2.7. Surface Biotinylation and Western Blot Analysis

MDCK-2 cells were transfected with plasmid DNA encoding either CedV F or mutant CedV F protein genes. At 24 h p.t., cells were washed and incubated twice with 2 mg/mL EZ-Link® Sulfo-NHS-LC-Biotin (ThermoFisher Scientific, Waltham, MA, USA) for 20 min at 4 °C. Following cell lysis, F proteins were immunoprecipitated overnight using NeutrAvidin Agarose Resin (ThermoFisher Scientific, Waltham, MA, USA) according to the manufacturer's protocol. Proteins were separated on a 10% SDS-gel under reducing or non-reducing conditions and then transferred onto nitrocellulose. Biotinylated CedV F proteins were detected by incubation with the HA-tag specific rabbit antibody H6908 (dilution 1:2,000 in PBS-Tween (0.05%)) followed by labeling with anti-rabbit HRP-conjugated secondary antibodies (1:5.000). Under non-reducing conditions, HRP-labeled Concanavalin A (ConA; Sigma-Aldrich, Darmstadt, Germany, 1:1,000 in PBS containing 0.05% (v/v) TWEEN 20, 1 mM CaCl2, 1 mM MnCl2, and 1 mM MgCl2 for 16 h at 20 °C) was used to stain NeutrAvidin-precipitated surface glycoproteins, which served as a loading control and as a reference to compare CedV F0 protein intensities. Proteins were visualized using a chemiluminescent substrate (Clarity Western ECL substrate, Bio-Rad, Feldkirchen, Deutschland) and the Bio-Rad Molecular Imager Chemi DocTM XRS+ in combination with the Image Lab software (Bio-Rad, Feldkirchen, Deutschland, Version 6.0.1).

2.8. Quantification of Endocytosis Rate by MESNA Reduction

MDCK-2 cells transiently expressing either CedV F or mutant F proteins were surface labeled with cleavable EZ-Link sulfo-NHS-SS-biotin (ThermoFisher Scientific, Waltham, MA, USA) at 4 °C. Then, wells were flooded with pre-warmed DMEM (37 °C) and shifted to 37 °C for either 5, 15, 30, or 90 min to allow endocytosis to occur. After rapid cooling to 4 °C, biotin still bound to the cell surface was cleaved by a membrane-impermeable reducing agent, 2-mercaptoethane-sulfonic acid sodium salt (MESNA; 50 mM in MESNA buffer (50 mM Tris, pH 8.5, 100 mM NaCl, 2.5 mM CaCl2)) a total of three times for 20 min. One sample was kept on ice and was neither incubated at 37 °C nor cleaved with MESNA and thus served as the surface biotinylation control to determine the total amount of biotinylated protein (control). After cell lysis in RIPA buffer, CedV F and CedV mutant F proteins were immunoprecipitated using the H6908 antibody as described above and separated by SDS-PAGE under non-reducing conditions. Transfer to nitrocellulose was followed by the detection of biotinylated proteins with HRP- labeled streptavidin and a chemiluminescent substrate. Protein bands were quantified using ImageLab software (version 6.0.1). Mean internalization rates per min were calculated as the ratio of band intensities from internalized, intracellular biotinylated protein after 5, 15, 30, and 90 min and total surface-labeled F protein (control, 50%) divided by 5, 15, 30, and 90 min, respectively.

2.9. Fusion Assay

A total of 4 × 105 Vero76 cells or 3 × 105 MDCK-2 cells were co-transfected with expression plasmids encoding for CedV G (0.5 µg/well) and either F or mutant F proteins (0.5 µg/well). At 48 h p.t., cells were fixed with ethanol and stained with 1:10-diluted Giemsa staining solution. Representative images (20× magnification) were recorded with an inverted microscope.

2.10. Luciferase Reporter Gene-Based Fusion Assay

Vero76 cells were reverse transfected with expression plasmids encoding for CedV G and either F or mutant F proteins as described above and seeded in 24-well plates. In addition, pCITE Renilla (200 ng/well), a vector containing the renilla luciferase gene under the control of the T7 promoter, was co-transfected. In parallel, Vero76 cells were reverse transfected with the pCAGGS T7 polymerase vector. At 27 h p.t., Vero76 cells expressing the T7 polymerase were washed twice with EDTA-containing PBS (5 mM), detached and added to the cells pre-transfected with pCAGGS CedV F and G and pCITE Renilla to allow fusion to proceed. After 3 h, cells were lysed using Lysis Buffer (pjk, GmbH, Kleinblittersdorf, Germany) and luciferase activity was measured by adding Renilla Glow substrate (pjk GmbH, Kleinblittersdorf, Germany) according to the manufacturer's instructions. Samples were tested in duplicate in three independent experiments. Reporter activity detected upon co-transfection of the parental CedV F with CedV G protein was set to one, which served as a reference point for fusion activity. Reporter activities measured for the fusion of mutant CedV F proteins were used to calculate their fusion activity in relation to the parental protein. Background activity of the luciferase reporter was assessed with cells transfected with pCAGGS CedV G and pCITE Renilla only overlaid with T7 polymerase expressing cells.

3. RESULTS

Endocytosis of HeV and NiV F protein precursors represents a critical step to gain biological activity, and thus, viral infectivity. In order to investigate whether CedV F protein similarly undergoes endocytosis from the cell surface, we performed a qualitative antibody uptake assay as described previously [11]. Therefore, MDCK-2 cells were transfected with plasmid DNA encoding for the CedV F protein. At 24 h p.t. without prior fixation, cell-surface expressed CedV F protein was labeled with CedV F protein-specific antibodies. Next, cells were either kept on ice or shifted to 37 °C for 30 min to allow endocytosis to occur. Surface-bound antibodies were then detected with an AF 488-conjugated secondary antibody that was added in excess to saturate surface-bound primary antibodies. Subsequent permeabilization allowed the staining of intracellular F protein-antibody complexes with an AF 568-conjugated antibody. After incubation at 37 °C, cells expressing the CedV F protein showed both green surface staining and multiple red fluorescent intracellular vesicles indicating endocytosis of the labeled CedV F protein (Figure 1). In contrast, we observed only green fluorescent signals of F proteins in cells that were kept on ice suffering a temperature-related inhibition of endocytosis.

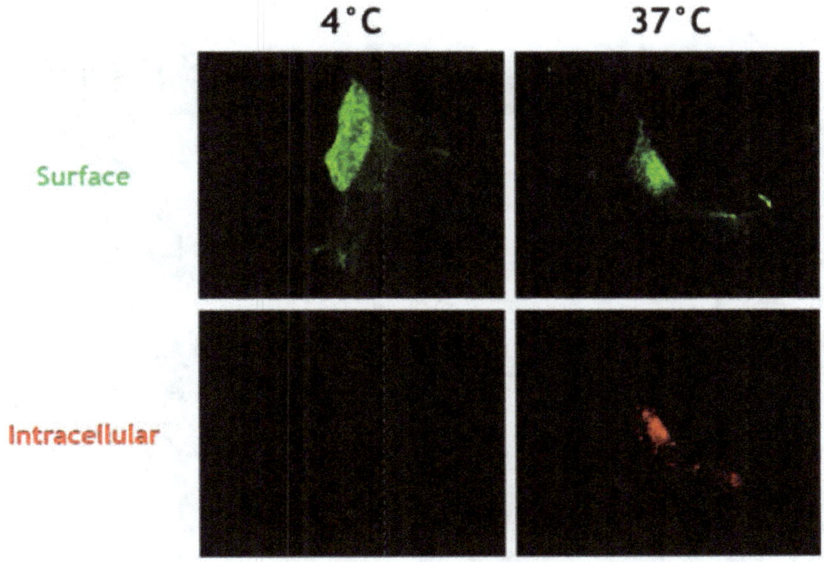

Figure 1. Endocytosis of CedV F protein in MDCK-2 cells. At 24 h p.t., CedV F-expressing MDCK-2 cells were incubated with a CedV F-specific rabbit antiserum to label surface-expressed F proteins. Then, the cells were shifted to 37 °C for 30 min to allow endocytosis to occur. AlexaFluor (AF) 488-conjugated secondary antibodies visualized surface-bound primary antibodies. After fixation and permeabilization, AF 568-conjugated secondary antibodies were used to stain internalized primary antibody-CedV F protein complexes. Magnification, ×60. n = 3.

To evaluate the functional importance of the tyrosine-based motifs within the CedV F protein cytoplasmic tail for endocytosis, we generated seven F protein mutants by tyrosine to alanine substitution (Figure 2). We constructed mutants encoding either single (e1–e5) or multiple (e3, e6) mutations as well as one mutant devoid of all putative endocytosis signals (e7) to investigate potentially additive effects of these mutated motifs (Figure 2). Apart from a classical YXXΦ motif in the membrane-proximal region (Y524XXF, Figure 2), we chose to also mutate the tyrosine residue at amino acid position 522 (Y522XXN, Figure 2) since a similar motif present in the cytoplasmic domain of Measles virus hemagglutinin has been described earlier to affect basolateral sorting and internalization of the protein [45].

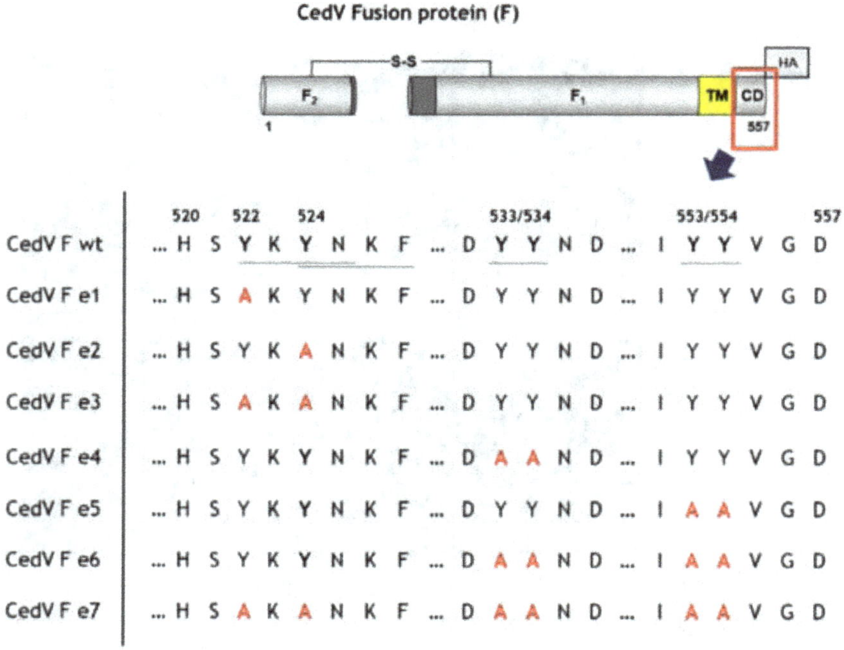

Figure 2. Schematic overview of the cytoplasmic tail mutants of CedV F protein generated in this study. Tyrosine-based putative endocytosis motifs are underlined. F1: CedV F protein subunit F1; F2: CedV F protein subunit F2; TM: transmembrane domain; CD: cytoplasmic domain; HA: HA-tag at the C-terminus of the F1 subunit; S-S: di-sulfide bond. wt: wild-type.

To assess whether these mutations have an impact on F protein synthesis, we compared expression efficiencies using a pulse-chase analysis. At 24 h p.t., MDCK-2 cells were labeled metabolically with [35] cysteine and -methionine for 15 min. After pulse-labeling, cells were incubated for 2 h followed by immunoprecipitation and SDS-PAGE under reducing conditions. We observed similar amounts of expression for parental and mutant F proteins within 2 h demonstrating that mutations within the cytoplasmic tail do not impair protein synthesis (Figure 3). However, although mutant e7 was expressed at similar levels, proteolytic processing was much less efficient. We generated additional F protein constructs combining selected mutations of particular motifs (e8 = e3 + e4; e9 = e3 + e5). Since these mutants displayed no differences in proteolytic processing to F wt, they were not included in subsequent analyses. Together, these findings suggest that proteolytic activation of the CedV F protein is markedly impaired upon the simultaneous disruption of all tyrosine-based motifs.

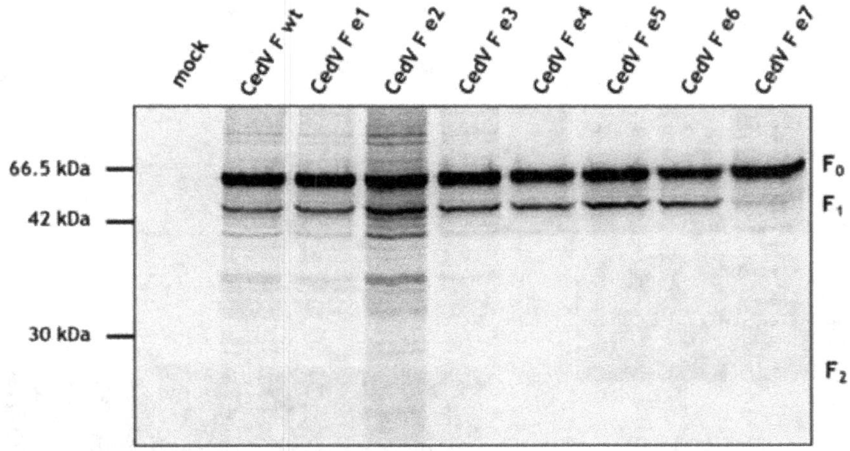

Figure 3. Effect of cytoplasmic tail mutations on CedV F protein expression and cleavage. At 24 h p.t., MDCK-2 cells expressing different CedV F proteins were metabolically labeled for 15 min (pulse) and then incubated for 2 h in serum-free nonradioactive medium (chase). After immunoprecipitation of F proteins from cell lysates and separation on a 12% SDS-gel under reducing conditions, samples were analyzed by autoradiography. n = 2; wt: wild-type.

HeV and NiV F proteins are known to be cleaved within the endosomal compartment. In analogy, blocking clathrin-mediated endocytosis by addition of chlorpromazine also inhibited proteolytic processing of CedV F protein demonstrating that endocytosis is a prerequisite for CedV F cleavage. Thus, inefficient cleavage of CedV mutant e7 might result from an inability to undergo endocytosis. In order to study the effects of single and multiple cytoplasmic tail mutations on CedV F protein endocytosis, we performed another antibody uptake experiment as described above using all generated mutants. After a period of 30 min at 37 °C, we observed intracellular red vesicles for all mutants except mutant e7 with all tyrosine-based motifs disrupted (Figure 4a). Disruption of a single motif such as YXXN (mutant e1) or YXXF (mutant e2), the di-tyrosine motifs (mutants e4, e5) as well as the combination of several mutated tyrosine residues (mutants e3, e6) displayed no difference in comparison to the fluorescent signals of the parental CedV F protein (Figure 1). The qualitative finding of red fluorescent vesicles indicates the internalization of mutants e1 to e6 from the cell surface while endocytosis of mutant e7 was strongly impaired after 30 min of endocytosis.

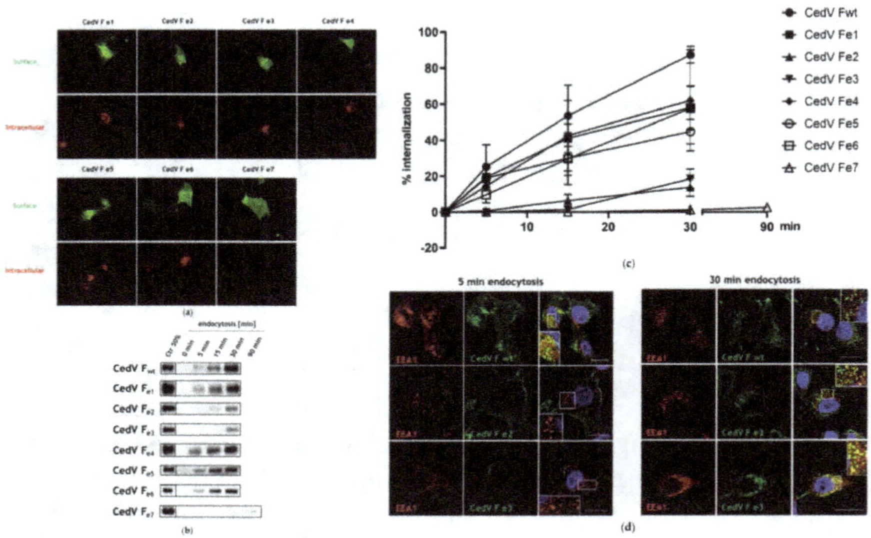

Figure 4. Endocytosis of CedV F proteins in MDCK-2 cells. (a) Antibody uptake assay of mutant CedV F proteins. MDCK-2 cells were transfected with plasmids encoding the indicated CedV F protein mutants. At 24 h p.t., a CedV F-specific rabbit antiserum was used to label surface-expressed F proteins at 4 °C. Then, the cells were shifted to 37 °C for 30 min to allow endocytosis to occur. AF 488-conjugated secondary antibodies visualized surface-bound primary antibodies. After fixation and permeabilization, AF 568-conjugated secondary antibodies were used to stain internalized primary antibody-CedV F protein complexes. Magnification, ×60. n = 2. (b) At 24 h p.t., CedV F-expressing cells were surface-labeled with cleavable NHS-SS biotin at 4 °C followed by a shift to 37 °C for the indicated times allowing endocytosis to occur. Residual biotin was then cleaved from the cell surface using MESNA. To quantify the amount of surface-biotinylated proteins that underwent endocytosis in a certain time, samples were compared to the total amount of surface-biotinylated control cells (Ctr) that were neither incubated at 37 °C nor treated with MESNA. Following cell lysis, F proteins were immunoprecipitated and samples were separated under non-reducing conditions. After transfer to nitrocellulose, biotinylated proteins were detected with peroxidase-conjugated streptavidin and chemiluminescence. The control lane represents 50% of

the total amount of biotinylated F proteins (Ctr 50%). One representative blot is shown for each CedV F protein variant. (n = 3; n = 2 for mutants e1, e3, and e4). wt: wild-type (c) CedV F protein internalization in percentage (%) per minute. To quantify the internalization rate, the percentages of internalized F protein amounts measured in the experiment shown in Figure 4b are displayed as a function of the time of incubation at 37 °C. Error bars represent the standard error of the mean. (d) Intracellular localization of wt and mutant CedV F proteins in MDCK-2 cells after 5 and 30 min at 37 °C. An antibody uptake assay of CedV F proteins was performed as described above with slight modifications. After the endocytosis step, surface-bound primary antibodies were blocked using a peroxidase-labeled secondary antibody while internalized primary antibodies were detected with AF 488-conjugated rabbit-specific secondary antibodies after fixation and permeabilization. Likewise, early endosomes were visualized with a primary antibody against the early endosomal antigen-1 (EEA1) and a mouse-specific AF568-conjugated secondary antibody. Scale bars indicate 20 μm. Representative images from two independent experiments are displayed (n = 2). Inserts show magnifications of indicated areas. Magnification, ×63.

To further understand the effects of single and multiple mutations on endocytosis, we next aimed to quantify the internalization of the different CedV F proteins and compare their endocytosis rate with the wt CedV F protein in a semi-quantitative biotin internalization assay. Moreover, this assay allowed us to demonstrate that the internalization of F proteins is not induced by antibody cross-linking. Briefly, we performed a surface biotinylation assay using a cleavable NHS-SS biotin derivative. Using the membrane-impermeable reducing agent MESNA, residual biotin that was not internalized from the cell surface after incubation at 37 °C for different periods can be cleaved. Cell lysis, immunoprecipitation of F proteins, and Western blot using Streptavidin-HRP were performed with subsequent detection and quantification of intracellular biotinylated F proteins. The amount of internalized protein was quantified by comparison to F-expressing cells that were neither incubated at 37 °C nor treated with MESNA, therefore, representing the total amount of biotinylated F proteins (Ctr lane, Figure 4b). Since internalization rates did not appear to be linear over time, endocytosis of biotinylated proteins is displayed additionally as a function of the time in Figure 4c. More than 80% of the parental CedV F protein was internalized after 30 min (Figure 4b,c). Similarly, mutant e1 carrying a substitution of a single tyrosine residue at aa position 522 as well as the di-tyrosine motif mutants e4, e5, and e6 displayed only a marginal decrease in internalization compared to the parental F protein. Interestingly, mutation of the YXXΦ motif (mutant e2 and e3) clearly decreased internalization with weak signals detectable after a period of 15 to 30 min at 37 °C (Figure 4b). However, the strongest effects were observed for mutant e7. Here, biotinylated mutant e7 was only detected after 90 min of incubation at 37 °C (Figure 4b) leading to a drastically reduced internalization rate over time in contrast to the parental F protein and the other mutants (Figure 4c). An antibody uptake assay of CedV F proteins including a co-staining of endosomal marker protein early endosomal antigen-1 (EEA1) in HeLa cells confirmed the delayed uptake of mutants e2 and e3. After 5 min at 37 °C, we observed no internalization of CedV F mutants e2 and e3 and thus, no co-localization with EEA1 (Figure 4d). However, both mutants co-localized with EEA1 after 30 min. In contrast, the parental F protein showed intracellular staining and co-localization with EEA1 as early as 5 min after the shift to 37 °C (Figure 4d).

Since differences in the internalization rate might directly affect cell surface expression, we next performed a surface biotinylation assay of F-expressing MDCK-2 cells followed by immunoprecipitation of biotinylated proteins and detection of the HA-tagged proteins in Western blot. In accordance with the antibody uptake assay and the MESNA reduction, we observed that all F proteins reached the cell surface (Figure 5). However, comparing the cell surface expression of all F proteins under non-reducing conditions, the amount of F0 differed distinctly between the parental CedV F protein and some mutant F proteins (Figure 5a). In all biotinylation assays performed mutant e1 displayed a cell surface expression similar to the parental CedV F protein, while surface expression of mutants e2–e5 was slightly increased. In contrast, the average level of cell surface expression of mutants, e6 and e7 was substantially higher. For mutant e7, we detected up to 2.5-

fold more F0 on the cell surface than for the parental F protein. Under reducing conditions (Figure 5b), both the F0 precursor and the subunit F1 were detected for all F proteins analyzed, indicating that all of them were proteolytically cleaved despite mutations in their cytoplasmic tails. For the parental F protein as well as for mutant e1–e6, quantification of band intensities revealed that more cleavage product F1 than inactive precursor F0 is found at the cell surface. However, mutant e7 rather displayed a reversed cleavage ratio, with less F1 present on the cell surface indicating a reduced amount of cleaved F protein. Although internalization of mutants e2 and e3 was shown to be clearly delayed in the biotin internalization assay, proteolytic activation and surface expression of these mutants still seemed to reach levels comparable to the parental F protein after 24 h p.t.

Finally, to assess whether any of the observed effects influence the biological activity of the F proteins, syncytium formation was analyzed in a fusion assay. At 48 h p.t., co-expression of the parental CedV F and CedV G proteins resulted in the formation of multinucleated syncytia in MDCK2 cells (Figure 6a) and Vero76 cells (Figure 6b). Mutant CedV F e1 and e4 induced syncytium formation comparable to the parental F protein. Surprisingly, despite their reduced endocytosis rate but parental F-like cell surface expression, co-expression of mutant e2 or e3 with CedV G led to a hyperfusogenic phenotype forming syncytia that were markedly increased in number and size (Figure 6a,b). Interestingly, fusogenicity of mutant e5 and e6 displaying an increased surface expression compared to the parental F protein was slightly enhanced while fusion activity of endocytosis-deficient mutant e7 was strongly impaired in both cell lines tested (Figure 6a,b). In addition, we quantified these differences using a luciferase reporter gene-based fusion assay in Vero76 cells after 24 h p.t. The hyperfusogenic phenotype of mutant e2 and e3 displayed a 4- and 6.5-fold increase in measurable luciferase reporter activity compared to the parental F protein (Figure 6c).

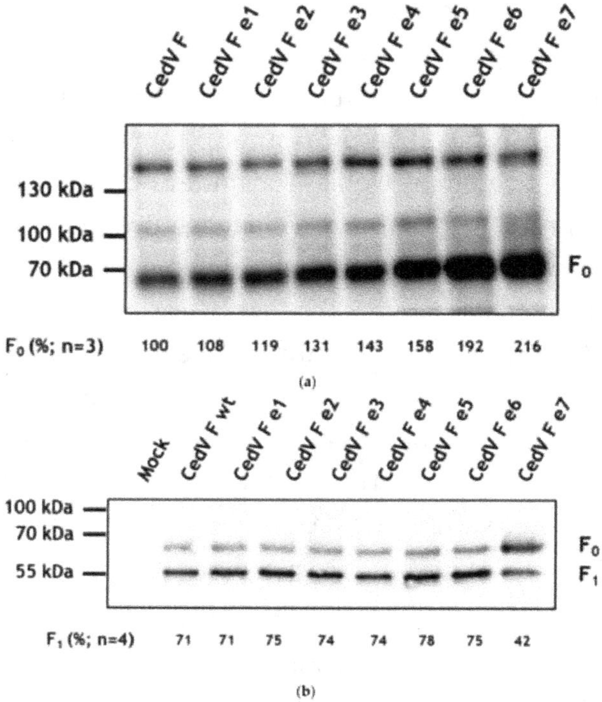

Figure 5. Cell surface expression of CedV F proteins. At 24 h p.t., MDCK-2 cells expressing F proteins were surface-labeled with biotin on ice. After cell lysis, biotinylated proteins were immunoprecipitated using NeutrAvidin beads and subjected to SDS-PAGE under non-reducing (a) and reducing (b) conditions. Precipitated F proteins were visualized using an antibody against the HA-tag (H6908), HRP-labeled secondary antibodies, and chemiluminescence. In (a), ConA staining is used as a loading control. Representative blots are shown from (a) three or (b) four independent experiments. The amount of F0 and F1 protein (in (a) % of CedV F0 protein with the parental F protein set to 100% or (b) % of F1 protein) is calculated as the mean of three or four independent experiments, respectively. wt: wild-type.

Co-expression of CedV G protein with mutant e5 still resulted in a 2.5-fold increase in reporter activity in Vero76 cells (Figure 6c). A marked decrease in reporter activity was confirmed for mutant e7. Since fusion activity is sensitive to the cell surface expression of both glycoproteins and to exclude that observed effects in fusogenicity were due to an altered cell-surface expression of CedV G protein, we performed a surface biotinylation assay in MDCK-2 cells co-transfected with CedV

G and wt or mutant CedV F proteins. Both F and G can be detected at the cell surface. Irrespective of the (mutant) F protein combination, the G expression at the cell surface seems to be similar suggesting that observed effects in fusion activity are not related to differences in CedV G cell surface expression. Taken together, these findings demonstrate that a membrane-proximal YXXΦ, as well as a C-terminal di-tyrosine motif in the CedV F protein cytoplasmic tail, are of functional relevance for endocytosis and biological activity of the protein.

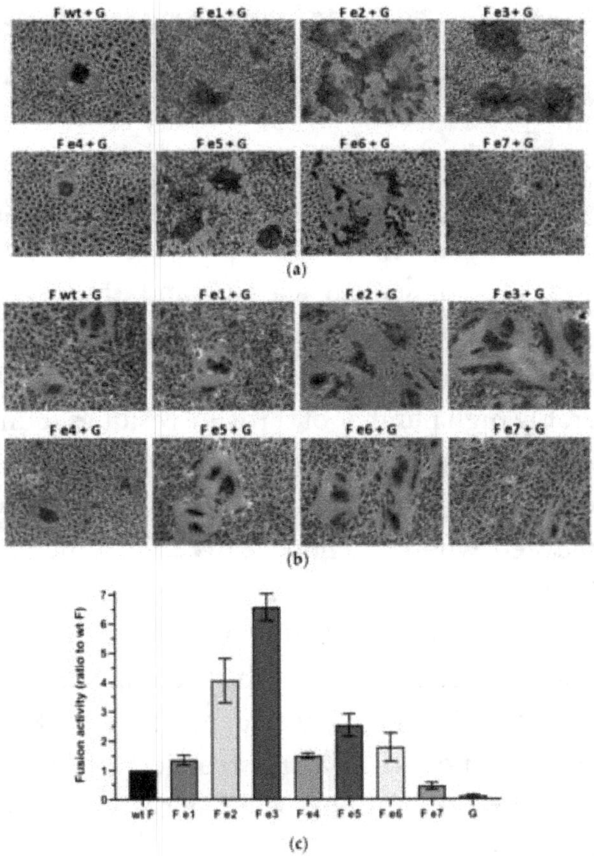

Figure 6. Effect of cytoplasmic tail mutations on CedV glycoprotein-mediated fusion activity. Syncytium formation in (a) MDCK-2 and (b) Vero76 cells co-expressing CedV F and G proteins were visualized by Giemsa staining at 48 h p.t. Magnification, ×20. n = 3; wt: wild-type (c) Quantitative reporter gene assay. Vero76 cells were co-transfected with plasmids encoding for the CedV glycoproteins F and G as well as a plasmid containing the luciferase gene under the control of a T7 promoter (pCITE

Renilla). At 24 h p.t., Vero76 cells expressing the T7 polymerase were layered on the glycoprotein-expressing cells and incubated for 3 h at 37 °C. Then, cells were lysed, and luciferase activity measured using a luminometer. Samples were tested in duplicates in three independent experiments. Reporter activity measured for the parental CedV F protein (wt F) co-transfected with CedV G protein was set to 1 serving as a reference point for fusion activity. Bars represent the fusion activities of the different (mutant) CedV F proteins in relation to the fusion activity of the parental protein and include the standard error of the mean (SEM). Background activity of the luciferase reporter was assessed with cells transfected with pCAGGS CedV G and pCITE Renilla only, layered with T7 polymerase expressing cells. n = 3.

4. DISCUSSION

Endocytic uptake from the cell surface plays a particular role in the replication cycle of highly pathogenic HeV and NiV [11,12,13]. Previous work has shown the importance of endocytosis for the maturation of F proteins to gain biological activity and thus, viral infectivity [11,12,13,16]. Given the fact that endocytic uptake of HeV and NiV F proteins depends on tyrosine-based motifs present in their cytoplasmic tail, the comparably high number of tyrosine residues found in the CedV F protein was of interest for this study. We first aimed to investigate whether these residues indeed mediate endocytosis of the protein. Second, we analyzed whether endocytosis is of functional relevance to the CedV F protein maturation. Taken together, our data clearly indicate that endocytic uptake of CedV F protein is signal-mediated and important for the biological activity of the protein. Furthermore, among the putative endocytosis signals, we identified a YXXΦ motif and a C-terminal di-tyrosine motif to have the strongest effects on endocytosis, cell surface expression, and fusion activity of CedV F protein.

The YXXΦ motif known to facilitate clathrin-mediated endocytosis of many transmembrane proteins is present in the membrane-proximal region of the cytoplasmic tail of both high and low pathogenic henipavirus F proteins. Mutation of the YXXΦ motif in the F protein of low pathogenic CedV resulted in a strongly decreased internalization rate for mutant e2 and e3, suggesting endocytic uptake of CedV F protein to be signal-mediated and largely dependent on this particular motif and most likely on Y524. For highly pathogenic HeV and NiV, several studies have similarly shown that an intact YXXΦ motif greatly contributes to the internalization of F proteins with a marked reduction in endocytosis rate upon motif disruption [11,12,13,43]. However, while the delayed endocytic uptake of NiV F protein led to a reduced fusogenicity of the mutant, the respective mutation in the HeV F protein rather resulted in an enhanced cell-to-cell fusion upon interaction with the viral attachment protein G [11,12]. Due to strong reduction but not the absence of endocytosis, the authors suggested an accumulation of recycled, fusogenic HeV F protein at the cell surface over time to account for the observed effects. For the respective CedV F mutants, we similarly found a clear delay in internalization (Figure 4b), but only moderate differences in total surface expression compared to the parental CedV F protein (Figure 5). However, we did indeed measure a strong 4- to 6.5-fold increase in the fusion activity of these two mutants in contrast to the parental CedV F protein, which cannot exclusively be explained by the only slightly increased cell surface expression. Although triggering of fusion is usually thought to primarily involve the ectodomain of the fusion protein, hyper- and hypofusogenicity were also observed in NiV F cytoplasmic tail mutants due to specific mutations in a membrane-proximal polybasic KKR motif [48]. The authors explained their findings by a mutation-dependent inside-out signaling mechanism resulting in conformational changes in the NiV F ectodomain followed by measurable effects on the fusion activity of the protein [48]. Overall, such conformational changes of the F-ectodomain due to mutations in the cytoplasmic tail can also affect the avidity of F and G interaction and thus, the coordinated processes required for cell–cell fusion [48,49,50]. Further, it is believed that mechanisms resulting in virus entry (viral—cellular membrane fusion) and cell–cell-fusion (fusion of neighboring cell membranes) are closely related [51]. Thus, cell–cell fusion levels often correlate to viral entry levels as shown for several Hendra and Nipah glycoprotein mutants [48,52,53,54]. In contrast, some hyperfusogenic G mutants with a modified O-glycosylation were described to display reduced entry levels but similar cell–cell fusion levels [55]. Also, a headless NiV G mutant readily triggered cell–cell fusion but pseudotyped NiV virions did not enter cells [56]. Hence, the underlying mechanism for the observed hyperfusogenicity of the CedV F mutants e2 and e3 and here the potential role of Y524 will have to be addressed in future studies.

Apart from a functional YXXΦ motif, the cytoplasmic tail of CedV F protein contains two additional di-tyrosine based motifs. Mutation of the di-tyrosine motif at amino acid position Y533/534A (mutant e4) showed no effect in any of the assays performed. However, mutation of the C-terminal di-tyrosine motif (Y553/554A; mutant e5 and e6) led to a detectable increase of cell surface expression with slightly enhanced fusion activity. On the one hand, this phenotype could result from a marginally reduced internalization rate. Alternatively, this di-tyrosine motif may affect the dynamics of endosomal trafficking and recycling of CedV F proteins, which could subsequently alter the availability of fusion-active F protein on the cell surface, thus having an impact on fusion activity. In a previous study, the lack of the di-tyrosine motif has been considered to delay the recycling of fusogenic NiV F proteins to the cell surface [43]. A tail-truncated NiV F variant lacking the C-terminal di-tyrosine motif was markedly downregulated in constitutive surface expression while F protein endocytosis and endosomal cleavage remained unaffected. Consequently, the intact di-tyrosine motif was hypothesized to act as a potential cytoplasmic recycling motif affecting intracellular trafficking [43]. Interestingly, both the YXXΦ and the C-terminal di-tyrosine motif have been described to be of importance for NiV F protein trafficking and sorting in polarized microvascular endothelial cells [57], in polarized epithelial cells [58], and in polarized neuronal cells [59].

The presence of intact tyrosine-based endocytosis and/or sorting motif in the cytoplasmic tail of transmembrane proteins has been described to be important for the replication cycle, infectivity, and virulence of many viruses such as herpes-, corona- and retroviruses [24,25,29,30,44,60,61]. Apart from signal-mediated intracellular trafficking of viral proteins to the sites of viral assembly, these motifs were shown to affect cell surface expression of different viral transmembrane proteins [24,25,29,30,44,60,61]. For instance, mutating the endocytosis signal of the envelope protein of simian immunodeficiency virus led to more efficient incorporation of envelope proteins into budding virions with enhanced infectivity due to increased levels of envelope protein expressed on the cell surface [61]. In the case of the CedV F protein, mutation of the di-tyrosine motif had stronger effects on the cell surface expression than mutation of the classical YXXΦ motif. However, at this point, it is not clear whether the di-tyrosine mutants will be incorporated more efficiently into virions due to an increase in cell surface expression. Mutation of the di-tyrosine motif in the NiV F cytoplasmic tail almost completely abrogated NiV VLP budding despite the wt-like cell surface expression of the protein [62]. Additionally, the presence of an intact and functional YXXΦ endocytosis motif in the cytoplasmic tail of NiV and HeV F protein has been considered to play a critical role in endocytic trafficking and recycling, and thus, in efficient viral assembly and particle release [31,62]. Similar findings have been reported for the envelope protein of HIV-1 where recent evidence suggests that trafficking through the recycling endosome is required for efficient incorporation of viral envelope proteins into virus particles [28]. Considering the lack of an intact YXXΦ motif, the reduced endocytosis rate, and the level of cell surface expression similar to the parental F protein, it will be interesting to see whether assembly and budding of mutant CedV F e2 and e3 is impaired or rather enhanced.

The most significant effects on cell surface expression, internalization rate, and fusion activity were noted for CedV mutant e7, in which all putative tyrosine-based endocytosis motifs were mutated. Importantly, the combination of all motif mutations led to an almost abrogated fusion activity of the protein with the strongest delay in internalization, which is in accordance with previous findings for a NiV F protein mutant disrupted of all tyrosine-based motifs [11]. Though CedV F mutant e7 displayed the highest cell surface expression, the combination of all mutations led to a rather reversed cleavage ratio in contrast to the other mutants. These findings point towards a reduced proteolytic F activation, which might result from the marked decrease in the endocytosis rate. Considering what is known for the F proteins of highly pathogenic HeV and NiV that are cleaved within the endosomal compartment [12,13,14,15,17], it seems likely that the proteolytic activation of CedV F protein is quite similar in terms of subcellular localization. However, future studies are needed to unravel shared commonalities and potential differences in this process.

5. CONCISE SUMMARY

To summarize, our data clearly demonstrate that the CedV F protein is internalized from the cell surface, following the same process as the HeV and NiV F proteins. Moreover, the process of internalizing the CedV F protein by endocytosis is regulated by signals and is crucial for acquiring biological functionality. Further research should thoroughly examine proteolytic activation to identify other similarities and/or differences in the subcellular distribution and protease use between highly and minimally pathogenic henipaviruses. Additionally, it is necessary to examine the impact of tyrosine-based motifs on the sorting and endocytic trafficking of the CedV F protein. This analysis should include investigating their capacity to influence the assembly of viral particles and their possible influence on virus-cell fusion and infectivity.

Endocytosis has a specific function in the proteolytic activation of the fusion (F) protein precursors of highly pathogenic henipaviruses, specifically Hendra (HeV) and Nipah virus (NiV). These proteins need to be taken up by endocytosis from the cell surface in order to be cut by cellular proteases in the endosomal compartment. After that, they are recycled back to the plasma membrane to be included in budding virions or to facilitate cell-cell fusion. The internalization process primarily relies on a tyrosine-based consensus motif for endocytosis found in the cytoplasmic tail of HeV and NiV F. To determine if the F protein of Cedar virus (CedV), a henipavirus with low pathogenicity that has many tyrosine residues in its cytoplasmic domain, undergoes signal-mediated endocytosis from the cell surface through tyrosine-based motifs in its cytoplasmic tail, and if this endocytosis is important for its biological function, we conducted an investigation. Consequently, tyrosine-based signals were altered, and the alterations were evaluated to determine their impact on F cell surface expression, endocytosis, and biological activity. The presence of a YXXΦ motif at the membrane and a di-tyrosine motif at the C-terminal are crucial for the expression of the cell surface and the rate at which endocytosis occurs. Moreover, our observations provide compelling evidence for the crucial significance of endocytosis in the biological function of the CedV F protein.

REFERENCES

1. Marsh, G.A.; de Jong, C.; Barr, J.A.; Tachedjian, M.; Smith, C.; Middleton, D.; Yu, M.; Todd, S.; Foord, A.J.; Haring, V.; et al. Cedar Virus: A Novel Henipavirus Isolated from Australian Bats. PLoS Pathog. 2012, 8, e102836.
2. Schountz, T.; Campbell, C.; Wagner, K.; Rovnak, J.; Martellaro, C.; DeBuysscher, B.L.; Feldmann, H.; Prescott, J. Differential Innate Immune Responses Elicited by Nipah Virus and Cedar Virus Correlate with Disparate In Vivo Pathogenesis in Hamsters. Viruses 2019, 11, 291.
3. Lieu, K.G.; Marsh, G.A.; Wang, L.F.; Netter, H.J. The non-pathogenic Henipavirus Cedar paramyxovirus phosphoprotein has a compromised ability to target STAT1 and STAT2. Antivir. Res. 2015, 124, 69–76.
4. Negrete, O.A.; Levroney, E.L.; Aguilar, H.C.; Bertolotti-Ciarlet, A.; Nazarian, R.; Tajyar, S.; Lee, B. EphrinB2 is the entry receptor for Nipah virus, an emergent deadly paramyxovirus. Nature 2005, 436, 401–405.
5. Bonaparte, M.I.; Dimitrov, A.S.; Bossart, K.N.; Crameri, G.; Mungall, B.A.; Bishop, K.A.; Choudhry, V.; Dimitrov, D.S.; Wang, L.F.; Eaton, B.T.; et al. Ephrin-B2 ligand is a functional receptor for Hendra virus and Nipah virus. Proc. Natl. Acad. Sci. USA 2005, 102, 10652–10657.
6. Negrete, O.A.; Wolf, M.C.; Aguilar, H.C.; Enterlein, S.; Wang, W.;

Muhlberger, E.; Su, S.V.; Bertolotti-Ciarlet, A.; Flick, R.; Lee, B. Two key residues in ephrinB3 are critical for its use as an alternative receptor for Nipah virus. PLoS Pathog. 2006, 2, e7.

7. Hafner, C.; Schmitz, G.; Meyer, S.; Bataille, F.; Hau, P.; Langmann, T.; Dietmaier, W.; Landthaler, M.; Vogt, T. Differential gene expression of Eph receptors and ephrins in benign human tissues and cancers. Clin. Chem. 2004, 50, 490–499.

8. Pryce, R.; Azarm, K.; Rissanen, I.; Harlos, K.; Bowden, T.A.; Lee, B. A key region of molecular specificity orchestrates unique ephrin-B1 utilization by Cedar virus. Life Sci. Alliance 2020, 3.

9. Laing, E.D.; Navaratnarajah, C.K.; Cheliout Da Silva, S.; Petzing, S.R.; Xu, Y.; Sterling, S.L.; Marsh, G.A.; Wang, L.F.; Amaya, M.; Nikolov, D.B.; et al. Structural and functional analyses reveal promiscuous and species specific use of ephrin receptors by Cedar virus. Proc. Natl. Acad. Sci. USA 2019, 116, 20707–20715.

10. Azarm, K.D.; Lee, B. Differential Features of Fusion Activation within the Paramyxoviridae. Viruses 2020, 12, 161.

11. Vogt, C.; Eickmann, M.; Diederich, S.; Moll, M.; Maisner, A. Endocytosis of the Nipah virus glycoproteins. J. Virol. 2005, 79, 3865–3872.

12. Meulendyke, K.A.; Wurth, M.A.; McCann, R.O.; Dutch, R.E. Endocytosis plays a critical role in proteolytic processing of the Hendra virus fusion protein. J. Virol. 2005, 79, 12643–12649.

13. Diederich, S.; Moll, M.; Klenk, H.D.; Maisner, A. The nipah virus fusion protein is cleaved within the endosomal compartment. J. Biol. Chem. 2005, 280, 29899–29903.

14. Pager, C.T.; Dutch, R.E. Cathepsin L is involved in proteolytic processing of the Hendra virus fusion protein. J. Virol. 2005, 79, 12714–12720.

15. Pager, C.T.; Craft, W.W., Jr.; Patch, J.; Dutch, R.E. A mature and fusogenic form of the Nipah virus fusion protein requires proteolytic processing by Cathepsin L. Virology 2006, 346, 251–257.

16. Diederich, S.; Thiel, L.; Maisner, A. Role of endocytosis and cathepsin-mediated activation in Nipah virus entry. Virology 2008, 375, 391–400.

17. Diederich, S.; Sauerhering, L.; Weis, M.; Altmeppen, H.; Schaschke, N.; Reinheckel, T.; Erbar, S.; Maisner, A. Activation of the Nipah virus fusion protein in MDCK cells is mediated by cathepsin B within the endosome-recycling compartment. J. Virol. 2012, 86, 3736–3745.

18. Matlin, K.S.; Reggio, H.; Helenius, A.; Simons, K. Infectious entry pathway of influenza virus in a canine kidney cell line. J. Cell Biol. 1981, 91, 601–613.

19. Yoshimura, A.; Ohnishi, S. Uncoating of Influenza-Virus in Endosomes. J. Virol. 1984, 51, 497–504.

20. Nanbo, A.; Imai, M.; Watanabe, S.; Noda, T.; Takahashi, K.; Neumann, G.; Halfmann, P.; Kawaoka, Y. Ebolavirus Is Internalized into Host Cells via Macropinocytosis in a Viral Glycoprotein-Dependent Manner. PLoS Pathog. 2010, 6, e1001121.

21. Aleksandrowicz, P.; Marzi, A.; Biedenkopf, N.; Beimforde, N.; Becker, S.; Hoenen, T.; Feldmann, H.; Schnittler, H.J. Ebola Virus Enters Host Cells by Macropinocytosis and Clathrin-Mediated Endocytosis. J. Infect. Dis. 2011, 204, S957–S967.

22. Chu, J.J.H.; Ng, M.L. Infectious entry of West Nile virus occurs through a clathrin-mediated endocytic pathway. J. Virol. 2004, 78, 10543–10555.

23. Blanchard, E.; Belouzard, S.; Goueslain, L.; Wakita, T.; Dubuisson, J.; Wychowski, C.; Rouille, Y. Hepatitis C virus entry depends on clathrin-mediated endocytosis. J. Virol. 2006, 80, 6964–6972.

24. Marsh, M.; Pelchen-Matthews, A. Endocytosis in viral replication. Traffic 2000, 1, 525–532.

25. Favoreel, H.W. The why's of Y-based motifs in alphaherpesvirus envelope proteins. Virus Res. 2006, 117, 202–208.

26. Checkley, M.A.; Luttge, B.G.; Freed, E.O. HIV-1 envelope glycoprotein biosynthesis, trafficking, and incorporation. J. Mol. Biol. 2011, 410, 582–608.

27. Postler, T.S.; Desrosiers, R.C. The tale of the long tail: The cytoplasmic domain of HIV-1 gp41. J. Virol. 2013, 87, 2–15.

28. Kirschman, J.; Qi, M.; Ding, L.; Hammonds, J.; Dienger-Stambaugh, K.; Wang, J.J.; Lapierre, L.A.; Goldenring, J.R.; Spearman, P. HIV-1 Envelope Glycoprotein Trafficking through the Endosomal Recycling Compartment Is Required for Particle Incorporation. J. Virol. 2018, 92.

29. Albecka, A.; Laine, R.F.; Janssen, A.F.J.; Kaminski, C.F.; Crump, C.M. HSV-1 Glycoproteins Are Delivered to Virus Assembly Sites Through Dynamin-Dependent Endocytosis. Traffic 2016, 17, 21–39.

30. Beitia Ortiz de Zarate, I.; Cantero-Aguilar, L.; Longo, M.; Berlioz-Torrent, C.; Rozenberg, F. Contribution of endocytic motifs in the cytoplasmic tail of herpes simplex virus type 1 glycoprotein B to virus replication and cell-cell fusion. J. Virol. 2007, 81, 13889–13903.

31. Cifuentes-Munoz, N.; Sun, W.; Ray, G.; Schmitt, P.T.; Webb, S.; Gibson, K.; Dutch, R.E.; Schmitt, A.P. Mutations in the Transmembrane Domain and Cytoplasmic Tail of Hendra Virus Fusion Protein Disrupt Virus-Like-Particle Assembly. J. Virol. 2017, 91.

32. Conner, S.D.; Schmid, S.L. Regulated portals of entry into the cell. Nature 2003, 422, 37–44.

33. Kaksonen, M.; Roux, A. Mechanisms of clathrin-mediated endocytosis. Nat. Rev. Mol. Cell Biol. 2018, 19, 313–326.

34. Robinson, M.S. Forty Years of Clathrin-coated Vesicles. Traffic 2015, 16, 1210–1238.

35. Bonifacino, J.S.; Traub, L.M. Signals for sorting of transmembrane proteins to endosomes and lysosomes. Annu. Rev. Biochem. 2003, 72, 395–447.

36. Ohno, H.; Stewart, J.; Fournier, M.C.; Bosshart, H.; Rhee, I.; Miyatake, S.; Saito, T.; Gallusser, A.; Kirchhausen, T.; Bonifacino,

J.S. Interaction of tyrosine-based sorting signals with clathrin-associated proteins. Science 1995, 269, 1872–1875.

37. Reider, A.; Wendland, B. Endocytic adaptors—Social networking at the plasma membrane. J. Cell. Sci. 2011, 124, 1613–1622.
38. Park, S.Y.; Guo, X.L. Adaptor protein complexes and intracellular transport. Biosci. Rep. 2014, 34, 381–390.
39. Traub, L.M. Sorting it out: AP-2 and alternate clathrin adaptors in endocytic cargo selection. J. Cell Biol. 2003, 163, 203–208.
40. Le Roy, C.; Wrana, J.L. Clathrin- and non-clathrin-mediated endocytic regulation of cell signalling. Nat. Rev. Mol. Cell Biol. 2005, 6, 112–126.
41. Traub, L.M. Tickets to ride: Selecting cargo for clathrin-regulated internalization. Nat. Rev. Mol. Cell Biol. 2009, 10, 583–596.
42. Ohno, H. Clathrin-associated adaptor protein complexes. J. Cell Sci. 2006, 119, 3719–3721.
43. Weis, M.; Maisner, A. Nipah virus fusion protein: Importance of the cytoplasmic tail for endosomal trafficking and bioactivity. Eur. J. Cell Biol. 2015, 94, 316–322.
44. Hou, Y.; Meulia, T.; Gao, X.; Saif, L.J.; Wang, Q. Deletion of both the Tyrosine-Based Endocytosis Signal and the Endoplasmic Reticulum Retrieval Signal in the Cytoplasmic Tail of Spike Protein Attenuates Porcine Epidemic Diarrhea Virus in Pigs. J. Virol. 2019, 93.
45. Moll, M.; Klenk, H.D.; Herrler, G.; Maisner, A. A single amino acid change in the cytoplasmic domains of measles virus glycoproteins H and F alters targeting, endocytosis, and cell fusion in polarized Madin-Darby canine kidney cells. J. Biol. Chem. 2001, 276, 17887–17894.
46. Stroh, E.; Fischer, K.; Schwaiger, T.; Sauerhering, L.; Franzke, K.; Maisner, A.; Groschup, M.H.; Blohm, U.; Diederich, S. Henipavirus-like particles induce a CD8 T cell response in C57BL/6 mice. Vet. Microbiol. 2019, 237, 108405.

47. Schneider, C.A.; Rasband, W.S.; Eliceiri, K.W. NIH Image to ImageJ: 25 years of image analysis. Nat. Methods 2012, 9, 671–675.

48. Aguilar, H.C.; Matreyek, K.A.; Choi, D.Y.; Filone, C.M.; Young, S.; Lee, B. Polybasic KKR motif in the cytoplasmic tail of Nipah virus fusion protein modulates membrane fusion by inside-out signaling. J. Virol. 2007, 81, 4520–4532.

49. Aguilar, H.C.; Matreyek, K.A.; Filone, C.M.; Hashimi, S.T.; Levroney, E.L.; Negrete, O.A.; Bertolotti-Ciarlet, A.; Choi, D.Y.; McHardy, I.; Fulcher, J.A.; et al. N-glycans on Nipah virus fusion protein protect against neutralization but reduce membrane fusion and viral entry. J. Virol. 2006, 80, 4878–4889.

50. Carter, J.R.; Pager, C.T.; Fowler, S.D.; Dutch, R.E. Role of N-linked glycosylation of the Hendra virus fusion protein. J. Virol. 2005, 79, 7922–7925.

51. White, J.M.; Delos, S.E.; Brecher, M.; Schornberg, K. Structures and mechanisms of viral membrane fusion proteins: Multiple variations on a common theme. Crit. Rev. Biochem. Mol. Biol. 2008, 43, 189–219.

52. Biering, S.B.; Huang, A.; Vu, A.T.; Robinson, L.R.; Bradel-Tretheway, B.; Choi, E.; Lee, B.; Aguilar, H.C. N-Glycans on the Nipah virus attachment glycoprotein modulate fusion and viral entry as they protect against antibody neutralization. J. Virol. 2012, 86, 11991–12002.

53. Bradel-Tretheway, B.G.; Liu, Q.; Stone, J.A.; McInally, S.; Aguilar, H.C. Novel Functions of Hendra Virus G N-Glycans and Comparisons to Nipah Virus. J. Virol. 2015, 89, 7235–7247.

54. Liu, Q.; Bradel-Tretheway, B.; Monreal, A.I.; Saludes, J.P.; Lu, X.; Nicola, A.V.; Aguilar, H.C. Nipah virus attachment glycoprotein stalk C-terminal region links receptor binding to fusion triggering. J. Virol. 2015, 89, 1838–1850.

55. Stone, J.A.; Nicola, A.V.; Baum, L.G.; Aguilar, H.C. Multiple Novel Functions of Henipavirus O-glycans: The First O-glycan Functions

Identified in the Paramyxovirus Family. PLoS Pathog. 2016, 12, e1005445.

56. Liu, Q.; Stone, J.A.; Bradel-Tretheway, B.; Dabundo, J.; Benavides Montano, J.A.; Santos-Montanez, J.; Biering, S.B.; Nicola, A.V.; Iorio, R.M.; Lu, X.; et al. Unraveling a three-step spatiotemporal mechanism of triggering of receptor-induced Nipah virus fusion and cell entry. PLoS Pathog. 2013, 9, e1003770.

57. Erbar, S.; Maisner, A. Nipah virus infection and glycoprotein targeting in endothelial cells. Virol. J. 2010, 7, 305.

58. Weise, C.; Erbar, S.; Lamp, B.; Vogt, C.; Diederich, S.; Maisner, A. Tyrosine residues in the cytoplasmic domains affect sorting and fusion activity of the Nipah virus glycoproteins in polarized epithelial cells. J. Virol. 2010, 84, 7634–7641.

59. Mattera, R.; Farias, G.G.; Mardones, G.A.; Bonifacino, J.S. Co-assembly of viral envelope glycoproteins regulates their polarized sorting in neurons. PLoS Pathog. 2014, 10, e1004107.

60. Day, J.R.; Munk, C.; Guatelli, J.C. The membrane-proximal tyrosine-based sorting signal of human immunodeficiency virus type 1 gp41 is required for optimal viral infectivity. J. Virol. 2004, 78, 1069–1079.

61. Yuste, E.; Reeves, J.D.; Doms, R.W.; Desrosiers, R.C. Modulation of Env content in virions of simian immunodeficiency virus: Correlation with cell surface expression and virion infectivity. J. Virol. 2004, 78, 6775–6785.

62. Johnston, G.P.; Contreras, E.M.; Dabundo, J.; Henderson, B.A.; Matz, K.M.; Ortega, V.; Ramirez, A.; Park, A.; Aguilar, H.C. Cytoplasmic Motifs in the Nipah Virus Fusion Protein Modulate Virus Particle Assembly and Egress. J. Virol. 2017, 91.

15 THE PRESENCE OF THE HOST FACTOR ERLIN-1 IS NECESSARY FOR THE OPTIMAL INFECTION OF HEPATITIS C VIRUS

1. INTRODUCTION: AN OVERVIEW

Hepatitis C virus (HCV) is a virus with a protective outer layer that is classified as a member of the Flaviviridae family [1]. Approximately 71 million individuals are believed to be chronically infected, with a significant portion of them likely to acquire cirrhosis, end-stage liver disease, and hepatocellular carcinoma [2]. The HCV genome is a positive-sense single-stranded RNA of approximately 9.6 kilobases (kb) in length. It consists of a single open reading frame (ORF) that encodes a polyprotein of around 3000 amino acids [3]. The nontranslated regions (NTR) located at the 5′ and 3′ ends of the open reading frame (ORF) are comprised of crucial sequences that contribute to the stability, translation, and replication of RNA [4,5,6]. The translation of the polyprotein is facilitated by a well-organized internal ribosomal entry site (IRES) found in the 5' non-translated region (NTR). This polyprotein undergoes co- and post-translational processing by both host and viral proteases, resulting in the production of three structural proteins (core, E1, and E2) and seven non-structural proteins (p7, NS2, NS3, NS4A, NS4B, NS5A, and NS5B) [7]. The process of HCV entering hepatocytes begins with the interaction between proteins found in the viral membrane and receptors in the plasma membrane of hepatocytes. These factors are responsible for the binding of the virus to the cells. This process has been extensively described in reference [8]. The contacts between HCV and the host cells cause the internalization of the virus through a process called receptor-mediated endocytosis. This is followed by the fusion of the viral and cellular membranes, allowing the viral genome to be released into the cytosol. The HCV RNA is subsequently conveyed to intracellular membranes originating from the endoplasmic reticulum (ER), where primary translation and polyprotein processing take place. The RNA replication process is carried out by the viral replicase complex in conjunction with cellular components. This process begins with the creation of negative-sense RNA intermediates, which serve as templates for the synthesis of new positive-sense RNA molecules. Subsequently, recently produced HCV genomes serve as templates for translation, additional cycles of RNA replication, or inclusion into offspring virus. Lipid droplets (LD) play a crucial role in the assembly of the HCV virus [10]. This assembly process is highly organized and involves several intermediate steps, such as the recruitment of NS5A and core proteins into LDs, the encapsidation of HCV RNA into core particles, and the envelopment of HCV RNA-containing core particles by the cell membrane [11,12,13,14,15]. Hepatitis C virus (HCV) exits the cell by entering the endoplasmic reticulum (ER) lumen and then utilizing the cellular secretory pathway. Due to the close connection between the endoplasmic reticulum (ER) and various stages of its life cycle, it is not unexpected that multiple proteins associated with the ER and ER-related proteins have been found to be host factors that regulate HCV infection. Examples of such proteins include signal peptide peptidase [16] and Sigma-1 receptor [17]. Erlin-1 and erlin-2 are proteins found in the endoplasmic reticulum membrane that are linked with lipid rafts. They are part of a broader family of proteins that have a conserved stomatin, prohibitin, flotillin, HflK/C (SPFH) domain. This domain is believed to have a role in organizing membrane microdomains [18,19,20]. Both erlins have a close relationship since they have around 80% similarity in their amino acid composition [20,21]. Additionally, they are evolutionarily conserved and have similar proteins in Caenorhabditis elegans and Arabidopsis thaliana [21]. Erlin proteins are found in detergent resistant membranes (DRM), where they create large complexes consisting of erlin homo- and hetero-oligomers, along with other cellular proteins [22]. The initial findings indicated that erlin-1 and erlin-2 proteins play a role in the degradation of inositol 1,4,5-triphosphate (IP3) receptors (IP3Rs) in the endoplasmic reticulum (ER) through a process known as endoplasmic reticulum associated degradation (ERAD) [23,24]. Subsequent investigations indicated that the erlin-2 protein is necessary for the breakdown of the cholesterol biosynthesis enzyme HMG-CoA reductase, which is stimulated by sterols [25]. Additionally, erlin-2 is involved in the conversion of amyloid β-peptide (Aβ) precursor (APP) into Aβ by γ-secretase in the brain [26]. In addition to their role in the ERAD pathway, erlin proteins have demonstrated the ability to regulate cholesterol homeostasis. These proteins bind to cholesterol and interact with the sterol regulatory element binding protein (SREBP)-Scap-Insig complex. This interaction limits the activation of SREBP and results in the buildup of lipids and cholesterol inside cells [27]. In a recent study, Inoue and Tsai established the initial connection between erlin proteins and viral infections [28]. The study shown that the presence of both erlin 1 and erlin 2 proteins is necessary for the infection of polyomavirus SV40. These proteins facilitate the movement of B12 transmembrane J-proteins to particular locations in the endoplasmic reticulum (ER), which is essential for the transfer of SV40 from the ER to the cytosol. This process enables the virus to establish an infection [28]. Given the physiological roles and ER localization of erlin proteins, as well as the reliance of HCV on lipid metabolism and the ER for its life cycle, we choose to examine the possible involvement of erlin proteins in HCV infection. This study presents the finding that the erlin-1 protein controls the start of HCV RNA replication, the buildup of viral proteins, and consequently, the generation of infectious virus. This discovery includes erlin-1 as one of the necessary host components for effective HCV infection.

2. MATERIALS AND METHODS

2.1. Cells, Plasmids, Antibodies and Reagents

The origin of Huh-7 [29], Huh-7.5.1 [29], Huh-7.5.1 subclone 2 (Huh-7.5.1c2) [15] and HEK-293T [30] cells have been described previously. All cells were maintained in Dulbecco's modified Eagle's medium (DMEM)

(Cellgro; Mediatech, Herndon, VA, USA) supplemented with 10% fetal calf serum (FCS) (Cellgro), 10 mM HEPES (Invitrogen, Carlsbad, CA, USA), 100 units/mL penicillin, 100 mg/mL streptomycin, and 2 mM L-glutamine (Invitrogen) in 5% CO2 at 37 °C. The sub-genomic and full-length JFH-1 stable replicon Huh-7 cell lines were cultured in medium supplemented with 400 or 200 μg/mL of G418, respectively, as described previously [15]. The JFH-1 genome-containing plasmid has been previously described [31]. The JFH-1 Rluc/SGR wt or JFH-1 Rluc/SGR GND plasmids carry bicistronic constructs containing the luciferase reporter gene in the first cistron and wild-type (wt) or replication-deficient (encoding a GDD-to-GND mutation in NS5B) JFH-1 sub-genomic replicon in the second cistron, respectively [32]. The rabbit polyclonal antibody for the detection of cellular erlin-2 protein was in-house generated and affinity purified against the full-length immunogen [27]. The rabbit polyclonal antibody HPA011252 against erlin-1 protein was obtained from Sigma-Aldrich (St. Louis, MO, USA). The recombinant human IgG anti-E2, the mouse monoclonal 9E10 anti-NS5A and the rabbit polyclonal MS5 anti-NS5A antibodies were kindly provided by M. Law (Scripps Research, La Jolla, CA, USA), C. M. Rice (Rockefeller University, New York, NY, USA) and M. Houghton (University of Alberta, Edmonton, AB, Canada), respectively. The monoclonal mouse antibodies against EEA1, HCV core (clone C7-50) and NS3 (clone 2E3) proteins were obtained from BD Transduction Laboratories (Franklin Lakes, NJ), Santa Cruz Biotechnology (Santa Cruz, CA, USA) and BioFront Technologies (Tallahassee, FL, USA), respectively. The HCV RNA replication inhibitor 2′-C-methyladenosine (2 mAd) was used at 10 μM final concentration and was a gift from W. Zhong (Gilead Sciences, Foster City, CA, USA). The 3-[4,5-dimethylthiazol-2-yl]-2,5-diphenyltetrazolium bromide (MTT) was purchased from Sigma Aldrich. Protease and phosphatase inhibitors were purchased from Roche (Indianapolis, IN, USA).

2.2. Silencing of Erlin Proteins by siRNA Transfection

siRNAstargetinghumanERLIN1(siErlin1.5:CCACAAATAGGAGCAGCAT [27]) or ERLIN2 (siErlin 2.3: GCCTCTCCGGTACTAACAT [27]) individually, or ERLIN1 and ERLIN2 simultaneously (siErlin 1&2:

AGAAGCAATGGCCTGGTAC [27]), and the non-targeting control siRNA (siCtrol: ACTGTCACAAGTACCTACA [24]), as well as the siRNA targeting HCV genome (siHCV: ACCTCAAAGAAAAACCAAA [17]) were all previously described. All the siRNAs were purchased from Integrated DNA Technologies (IDT, San Diego, CA, USA). Typically, Huh-7 cells were plated at a density of 1 × 105 cells per well on a 6-well plate. 24 h later cells were transfected with 20 pmol of the corresponding siRNA per well using Dharmafect 4 transfection reagent following manufacturer's instructions (GE Healthcare Dharmacon Inc, Pittsburgh, PA, USA). In brief, 20 pmol of siRNA was added into 200 μL of OptiMEM containing tube and 1 μL of Dharmafect 4 reagent was added into a second tube containing another 200 μL of OptiMEM. These mixtures were incubated for 5 min at room temperature before they were mixed together and incubated for another 20 min at room temperature. During the incubation time the culture medium of the Huh-7 cells was discarded and 1.6 mL of complete DMEM medium was added to each well. After 20 min of incubation the siRNA-Dharmafect complexes-containing medium (400 μL) was added to the corresponding wells. Fresh medium was applied after overnight incubation and the cells were assayed at the indicated times after transfection as indicated in each experiment. Cell viability was determined by evaluating cell biomass at different times after siRNA transfection by crystal violet staining and colorimetry at 570 nm, as previously described [33]. As a complementary approach, the cytotoxic effect of siRNAs was evaluated by quantifying the mitochondrial activity of siRNA-transfected cells (in MTT assays) and by determining the cellular respiration capacity. The cellular respiration capacity was calculated as the mitochondrial activity relative to the total protein content of the cells, measured by BCA in wells that were transfected in parallel.

2.3. Preparation of Viral Stocks and Infections

The original JFH-1 virus was generated by transfection of an in vitro transcribed full-length JFH-1 HCV RNA into Huh-7 cells and viral stocks were produced by inoculation of fresh Huh-7 cells at a multiplicity of infection (moi) of 0.01 as described [29]. Production of high titer cell culture adapted JFH-1 day 183 virus (D183) [34] was achieved by inoculation of highly susceptible Huh-7.5.1c2 at low moi. These D183 virus stocks were used in all low and high moi infection experiments (moi = 0.2 or 3) as described previously [15]. Intracellular infectious HCV particles were obtained by 5 freeze-thaw cycles of infected cells as described [35]. The infectivity titers present in culture supernatants and cell extracts were determined by end-point dilution and fluorescence focus forming unit (FFU) assay in Huh-7.5.1 cells, as described previously [36].

2.4. Analysis of HCV Cell Entry Using HCVpp

Retroviral particles pseudotyped with the JFH-1 E1 and E2 envelope proteins or the VSV G glycoprotein, as control, were produced in HEK293T cells [37]. For HCV entry experiments, Huh-7 cells (2 × 104 cells) were plated on 48-well plates and twenty hours later they were transfected with the indicated siRNAs as described above. Thirty-six hours later, cells were inoculated with pseudotyped retroviral particles (HCVpp or VSVpp) diluted to produce similar luciferase activity levels. Forty-eight hours later, cells were lysed and luciferase activity was detected and quantitated in a luminometer using a commercial kit (Luciferase Assay Kit; Promega, Madison, WI, USA). Luciferase results were normalized to cell density and relative infection values were calculated as a percentage of cells transfected with control siRNA. In parallel, cultures transfected with siRNAs were used: (i) to analyze knockdown efficiency by Western-blotting (WB) and (ii) to confirm the reduced susceptibility of erlin-1-deficient cells to HCV infection.

2.5. In Vitro Transcription and HCV RNA Transfection

The subgenomic JFH-1 replicon plasmid bearing a luciferase reporter gene (JFH-1 Rluc/SGR wt) and the corresponding replication-deficient construct (JFH-1 Rluc/SGR GND) have been described previously [32]. After digestion with XbaI restriction enzyme the linearized plasmids were in vitro transcribed using the T7 MEGAscript (Ambion, Austin, TX, USA) kit following the manufacturer's instructions. siRNAs targeting the indicated genes were transfected into Huh-7 cells as described above.

Twenty hours later the cells were trypsinized and cells were seeded in wells of 12-well plate at a density of 1 × 105 cells per well. The following day in vitro transcribed subgenomic HCV RNAs were introduced into those cells by transfection using TransIT mRNA Transfection Kit (Mirus Bio LLC, Madison, WI, USA) as recommended by the manufacturer. At the indicated time points the cells were lysed. An aliquot of each lysate was used to quantitate the luciferase activity in a luminometer using a commercial kit (Promega). Another aliquot from each lysate obtained at six hours post-transfection was used to quantitate the HCV RNA levels by reverse transcription real-time quantitative PCR (RT-qPCR) as described below. The HCV RNA values were used to control for differences in the transfection efficiency between samples and were used to normalize the luciferase values. Luciferase results were normalized to cell density at each time point and the relative values were calculated as a percentage of cells transfected with control siRNA. In parallel, cultures transfected with siRNAs were used: (i) to analyze knockdown efficiency by WB and (ii) to confirm the reduced susceptibility of erlin-1-deficient cells to HCV infection.

2.6. siRNA Transfection Experiments in Acutely HCV Infected Cells

Huh-7 cells were transfected with the indicated siRNAs as described above. Thirty-six hours later, transfected cells were infected with JFH-1 D183 virus at low or high moi (0.2 or 3, respectively) as described before [15]. Five hours after HCV inoculation, cells were washed twice with PBS and medium was replaced. At the indicated times after infection supernatants were collected and extracellular infectivity was determined by end-point dilution in Huh-7.5.1 cell line. Cell extracts were harvested as indicated above and the intracellular infectivity, HCV RNA, and viral proteins were analyzed by titration, RT-qPCR and WB, respectively, as described [38]. Erlin protein down-regulation efficiency was confirmed by WB and densitometry analysis at the time post-infection indicated in each figure legend.

2.7. siRNA Transfection Experiments in Persistently Infected Cells

Huh-7 cells were infected with JFH-1 virus at a low moi = 0.01 as described [29]. Infected cells were passaged for 3 weeks before siRNAs were transfected. Three days after siRNA transfection supernatants were discarded, cells were washed twice with PBS and medium was replenished. Twenty-four hours later supernatants were collected and progeny virus released during those last twenty-four hours was determined by end-point dilution in Huh-7.5.1 cells as described above. Cell extracts were harvested and the intracellular infectivity, HCV RNA and proteins and erlin protein down-regulation were analyzed by titration, RT-qPCR and WB, respectively.

2.8. Protein Analysis by WB

Cell extracts were prepared in RIPA buffer supplemented with protease and phosphatase inhibitors and, after protein quantitation by bicinchorinic acid (BCA) assay (Thermo Scientific, Waltham, MA, USA), equivalent amounts of total protein (typically 20–25 μg) were separated in polyacrylamide-SDS gels by electrophoresis and they were transferred to Immobilon membranes (Millipore, Billerica, MA, USA) for blotting. The membranes were blocked for 1 h at room temperature with PBS-5% milk and incubated overnight at 4 °C with the primary antibodies diluted in 1% milk-0.1% Tween 20 in PBS. Primary antibodies were used at the following dilutions: EEA1 at 1:2000; erlin-1, erlin-2 and NS3 at 1:1000; NS5A at 1:500; and Core at 1:200 dilution. After washing at least 4 times for 15 min with 0.1% Tween 20 in PBS, membranes were incubated with a 1:10,000 dilution of goat-anti-rabbit or goat-anti-mouse IgG conjugated to horseradish peroxidase prepared in 1% milk-0.1% Tween 20 in PBS. After washing at least 4 more times with 0.1% Tween 20 in PBS, membranes were developed using the SuperSignal-West Pico or Femto chemiluminescence reagents (Thermo Scientific) and autoradiography. Densitometry of non-saturated films was performed using Fiji/Image J analysis software [39] and the results, shown beneath each panel, are expressed relative to control cells using EEA1 protein expression for normalization and as loading control. Two-fold serial dilutions of siCtrol samples were used to produce standards for relative quantitation.

2.9. RNA Analysis

Total RNA was extracted from cells using the guanidinium isothiocyanate extraction method [29] after adding 20 μg of glycogen (Roche) per sample as a carrier. The content of HCV RNA, ERLIN1 and ERLIN2 mRNAs and GAPDH mRNA (for normalization) in each sample was quantitated in a two-step RT-qPCR assays using the High-Capacity cDNA Reverse Transcription Kit and the Maxima SYBR Green/ROX qPCR Master Mix (2X) (Thermo Scientific). 10-fold serial dilution of plasmids containing each target sequence was used to prepare the standard curves that were used with each corresponding pair of primers: HCV (5′-TCTGCGGAACCGGTGAGTA-3′ and 5′-TCAGGCAGTACCACAAGG-3′); ERLIN1 (5′-GGGGTTGGTGGC TGTCCTGC-3′ and 5′-TAGCCTGGT CCACTGGGGCT-3′); ERLIN2 (5′-TGTGCACACGCTTCAAGAGGTCTA-3′ and 5′-AATGACCAGCC CAGGGGCCA-3′) and GAPDH (5′-GAAGGTGAA GGTCGGAGTC-3′ and 5′-GAAGATGGTGATGGGATTTC-3′). Results were normalized using GAPDH mRNA levels and were displayed in the figures as relative values compared to control cells.

2.10. Confocal Analysis

Huh-7 cells were transfected with siRNAs and infected at high moi (moi = 3) with JFH-1 D183 virus as described above. The day before fixation cells were seeded in glass bottom 96-well plates (Thermo Scientific) for confocal analysis. At the indicated times after virus inoculation, cells were incubated in 4% paraformaldehyde for 20 min at room temperature, washed 3 times with PBS and processed for immunofluorescence analysis and LD staining as previously described [38]. In brief, fixed cells were first incubated for 1 h at room temperature in blocking buffer (1xPBS, 10% FBS, 3% BSA, 0.3% Triton X-100), washed 3 times with PBS and then incubated for 1 h with a mixture of primary antibodies at the appropriate concentrations that were prepared in binding buffer (1xPBS, 3% BSA, 0.3% Triton X-100).

The primary antibodies used for immunofluorescence were: a purified mouse monoclonal anti-core (clone C7-50) at 1:200 dilution and a rabbit polyclonal anti-NS5A (MS5) at 1:1000 dilution. After 3 washes with PBS, cells were incubated for one more hour in binding buffer containing a mixture of fluorophore-conjugated secondary antibodies at 1:1000 dilution (Invitrogen) and Hoechst dye (Invitrogen) for nuclei staining. LD staining was performed by incubating the cells for 30 min at room temperature with the LipidTox reagent (Invitrogen) at 1:500 dilution in PBS. Images of 1024 × 1024 pixels at 8-bit gray scale were acquired with a 40× objective (pixel size 0.345 μm) on a LSM 710 confocal laser scanning microscope (Zeiss, Dublin, CA, USA). All images were taken with exactly the same settings. Fiji/ImageJ software analysis package [39] was used to quantitate the fluorescence intensity signals in each image.

2.11. Statistical Analysis

GraphPad Prism v.5.0a software (GraphPad Software, San Diego, CA, USA) was used to analyze the data, prepare the graphs and perform all statistical analysis. All experimental results (quantitative RT-qPCR analysis, MTT assays, titration assays, luciferase activity assays, and quantitative fluorescence intensity analysis) were displayed in the graphs as the mean ± standard deviation (S.D.). The normal distribution of data was confirmed using Kolmogorov-Smirnov test in experiments with sample sizes $n \geq 6$. Experiments where normality test could not be performed because of sample size limitations ($n = 3$) normal distribution of the data was assumed. The differences among means between multiple (more than two) groups were analyzed by One-way ANOVA or Two-way ANOVA followed by Dunnett's or Bonferroni's Multiple Comparison Test as indicated in each figure legend. One-way ANOVA was used for single variable analysis in Figure 1D,E, Figure 2, Figure 3D and Figure 4, Figure 5, Figure 6, Figure 7, Figure 8 and Figure 9, while Two-way ANOVA was used in kinetic experiments for two variable analysis in Figure 1A,C and Figure 3B. The statistical significance was set as: * $p < 0.05$; ** $p < 0.01$; *** $p < 0.001$.

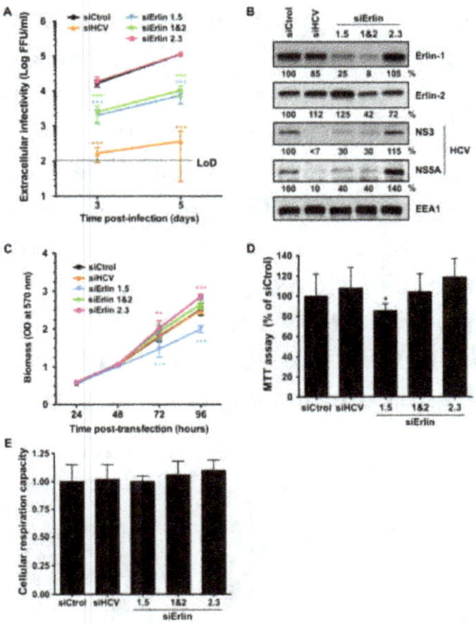

Figure 1. Erlin-1 protein down-regulation impairs efficient HCV infection. Huh-7 cells were transfected with individual siRNAs targeting erlin-1 (siErlin 1.5), erlin-2 (siErlin 2.3), both erlins simultaneously (siErlin 1&2), HCV (siHCV) or a non-targeting siRNA control (siCtrol) as indicated in Material and Methods section. Transfected cells were infected 36 h later with JFH-1 D183 virus at low multiplicity of infection (moi = 0.2). (A) The extracellular infectivity present in supernatants of infected cells was determined on days three and five post-inoculation by titration assay. Infectivity titers are represented as the FFUs per ml of supernatant in logarithmic scale and are displayed as the average and standard deviation (Mean; SD; n = 3). The horizontal black dotty line represents the limit of detection (LoD) of the assay. (B) Three days after virus inoculation cellular erlin-1, erlin-2 and EEA1 (as loading control) and viral NS3 and NS5A protein expression was determined in HCV-infected cell extracts by WB. Relative protein levels were determined by densitometry and are shown below each panel. The ratio of intensities of each protein and EEA1 in siCtrol-transfected cells was set as 100 and it was used to calculate the relative amount of each protein in each sample. Results in panels A and B are representative of two independent experiments, each one performed in triplicate. (C) Cells plated in 96-well plate format were transfected with siRNAs and the cell biomass was quantitated by crystal violet staining at different times post-transfection. Data are displayed as the average and standard deviation of two independent experiments with three replica wells per condition at each time point in each experiment (Mean; SD; n = 6). (D) Cells plated in 96-well plate format were transfected and infected

as described above and were used to perform MTT assays three days after HCV inoculation. Data are displayed as the average and standard deviation (Mean; SD; n = 7) relative to the levels in siCtrol condition, set as 100. Results are representative of two independent experiments with seven replica wells per condition in each experiment. (E) The cellular respiration capacity was calculated as described in Material and Methods section. Data are displayed as the average and standard deviation (Mean; SD; n = 3) relative to that of siCtrol, set as 1. Statistical significance was determined using: Two-way ANOVA followed by a Bonferroni posttest for panels A and C (** $p < 0.01$; *** $p < 0.001$), and One-way ANOVA followed by the Dunnett's Multiple Comparison Test for panels D and E (* $p < 0.05$).

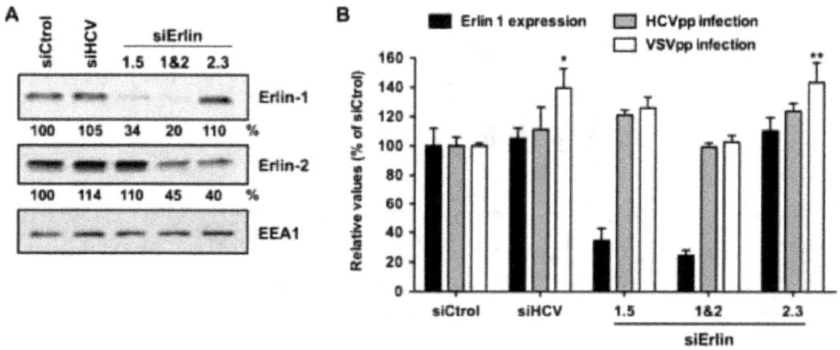

Figure 2. Erlin-1 protein down-regulation does not inhibit HCV cell-entry. Huh-7 cells were transfected with individual siRNAs as described in Material and Methods. (A) 36 h later one set of wells was harvested for erlin protein analysis by WB. Relative protein expression levels were determined by densitometry as described in the legend of Figure 1 and they are shown below each panel. (B) Another set of wells was inoculated with HCVpp (grey bars) or particles harboring the VSV-G glycoprotein (VSVpp; white bars) as control. 48 h after inoculation the intracellular luciferase activity was measured. Results are displayed as percentage of the luciferase activity present in siCtrol-transfected cell extracts for each virus particle. Data are displayed as the average and standard deviation (Mean; SD; n = 3). The luciferase values (units per well) in siCtrol-transfected cell extracts were 9 × 103 ± 5.1 × 102 and 1.2 × 104 ± 2 × 102 for HCVpp and VSVpp, respectively. These results are representative of two independent experiments, each one performed in triplicate. The average erlin-1 protein expression level of the two experiments is shown in black bars. One-way ANOVA followed by Dunnett's Multiple Comparison Test analysis of the HCVpp data did not show any statistically significant effect on the luciferase levels among different siRNAs. Instead, the entry of VSVpp was significantly higher in siHCV- and siErlin 2.3-transfected cells (* $p < 0.05$; ** $p < 0.01$).

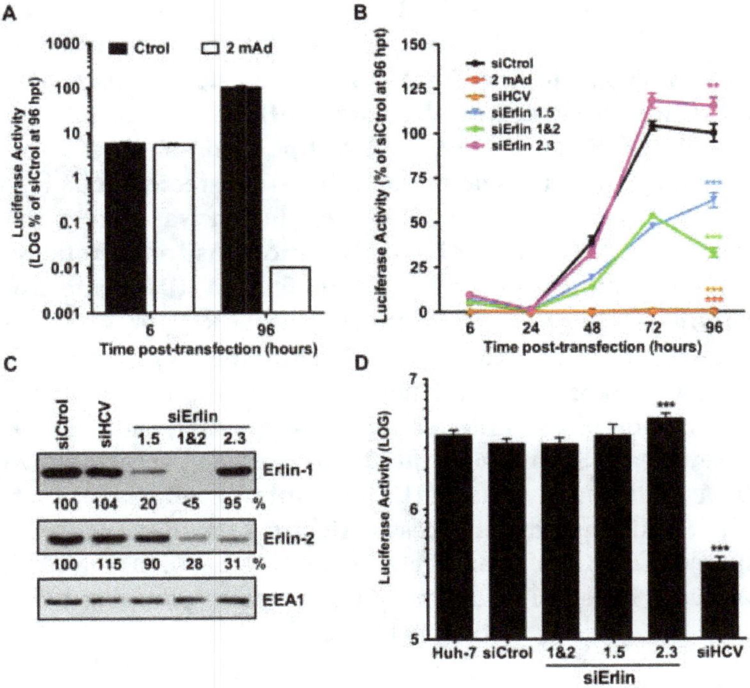

Figure 3. Erlin-1 protein down-regulation impairs the initiation of HCV RNA replication without affecting HCV IRES-dependent translation. Huh-7 cells were transfected with individual siRNAs as described in Material and Methods. 48 h later cells were transfected with in vitro transcribed bicistronic renilla luciferase (Rluc)-containing wt or replication-deficient (GND mutant) sub-genomic (SGR) HCV RNAs. As control for the inhibition of HCV RNA replication process, transfected cells were treated with the HCV polymerase inhibitor 2′-C-methyladenosine (2 mAd). (A) Quantitation of renilla luciferase activity derived from Rluc/SGR wt RNA in the presence (white bars) or absence (black bars) of 2 mAd at early (6 h) and late (96 h) time points. Data are displayed as the average and standard deviation (Mean; SD; n = 3) of luciferase values relative to that of control cells at 96 h post-transfection (hpt), that was set as 100. (B) Time course of renilla luciferase activity derived from RLuc/SGR wt RNA in siRNA-transfected cells. For each independent experiment, Rluc activity was normalized to cell density and was displayed as a percentage of that determined in siCtrol-transfected cell extracts at 96 hpt. Data shown are averages and standard deviation of two independent experiments, each one performed in triplicate (Mean; SD; n = 6). The luciferase activity in siCtrol-transfected cell extracts at 96 h post-transfection was 1.3 × 107 ± 9.1 × 105 light units per well. To assess the specificity of Rluc activity,

cells were treated with 2 mAd (red line) or were transfected with siRNAs targeting directly HCV RNA (orange line). Statistical analysis performed using Two-way ANOVA followed by Bonferroni posttest showed statistically significant differences in: 2 mAd-treated, siHCV-, siErlin 1.5- and siErlin 1&2-transfected cells at 48-, 72- and 96-hpt, and siErlin 2.3-transfected cells at 72- and 96-hpt compared to siCtrol-transfected cells (** $p < 0.01$; *** $p < 0.001$). (C) Erlin protein down-regulation was determined at the time of Rluc/SGR RNA transfection by WB and densitometry (shown below each panel) as described in the legend of Figure 1. (D) Luciferase activity of replication-deficient (GND mutant) Rluc/SGR RNA was determined at six hours post-transfection. For each independent experiment Rluc activity was normalized to cell density. Data shown are averages and standard deviation of two independent experiments, each one performed in triplicate (Mean; SD; n = 6). Statistical analysis performed using One-way ANOVA followed by Dunnett's Multiple Comparison Test showed statistically significant differences in siErlin 2.3- and siHCV-transfected cells compared to siCtrol-transfected cells (*** $p < 0.001$). No differences were observed in erlin-1 downregulated cells compared to siCtrol-transfected cells.

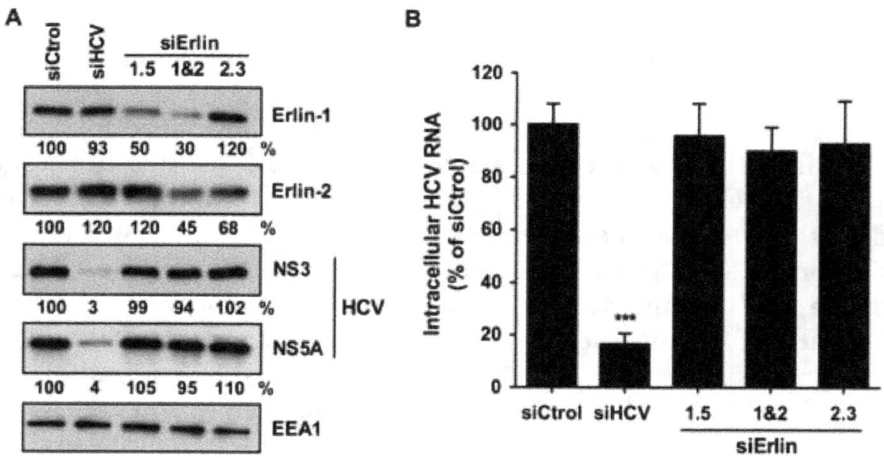

Figure 4. Erlin-1 protein down-regulation does not interfere with ongoing HCV RNA replication. JFH-1 sub-genomic replicon (SGR) bearing Huh-7 cells were transfected with siRNAs as described in Material and Methods. Three days after transfection cells were harvested for analysis. (A) Cellular erlin-1, erlin-2 and EEA1, and HCV NS3 and NS5A protein accumulation were determined by WB and densitometry (shown below each panel) as described in the legend of Figure 1. (B) The intracellular HCV RNA content was quantitated in infected cell extracts three days after siRNA transfection by RT-qPCR. Data were normalized relative to GAPDH mRNA

levels in the same samples and are displayed as percentage of the HCV RNA present in siCtrol-transfected cells. The HCV RNA content in siCtrol-transfected cell extracts was 1.4 × 106 ± 1.2 × 105 copies per μg of total RNA. Data shown are averages of three independent experiments, each one performed in triplicate (Mean; SD; n = 9). Only the HCV RNA level in siHCV-transfected cells was significantly lower than that in siCtrol-transfected cells as determined by One-way ANOVA followed by Dunnett's Multiple Comparison Test (* p < 0.001).**

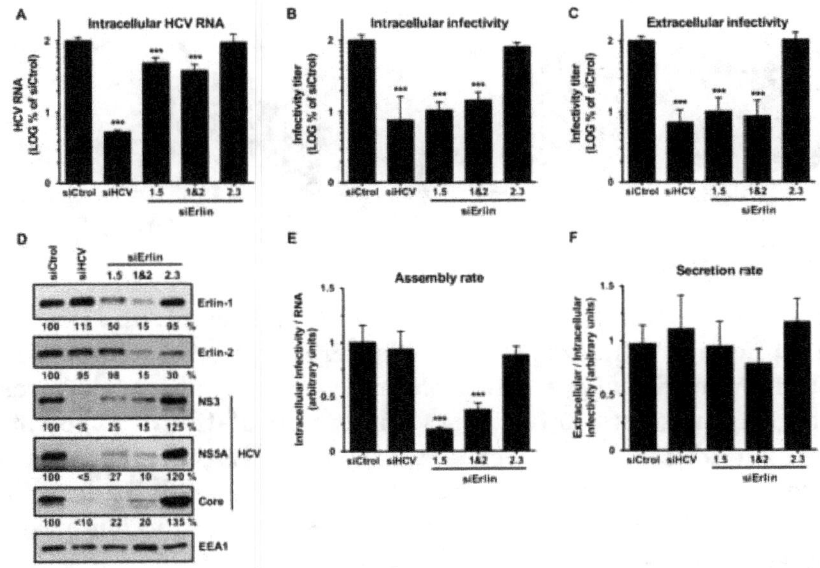

Figure 5. Erlin-1 protein down-regulation interferes with HCV in single-cycle infection experiments. Huh-7 cells were transfected with siRNAs as described in Material and Methods. 36 h later transfected cells were inoculated with JFH-1 D183 virus at high multiplicity of infection (moi = 3). 48 h after virus inoculation cellular extracts were prepared for intracellular RNA (A), infectivity (B) and protein (D) analysis and supernatants were collected for extracellular infectivity (C) analysis. Assembly (E) and secretion (F) rates were calculated using data from panels A, B and C. The RNA and infectivity results are displayed as percentage of the levels in siCtrol-transfected cells. The HCV RNA and the intracellular and extracellular infectivity levels in siCtrol-transfected cells were: 1.2 × 107 ± 1.3 × 106 HCV RNA copies per μg of total RNA, 3.5 × 104 ± 6.1 × 103 ffus per well, and 1.3 × 105 ± 2 × 104 ffus per ml of supernatant, respectively. Data shown are averages of three independent experiments, each one performed in triplicate (Mean; SD; n = 9). One-way ANOVA followed by Dunnett's Multiple Comparison Test was used to determine the statistical significance (* p < 0.001).**

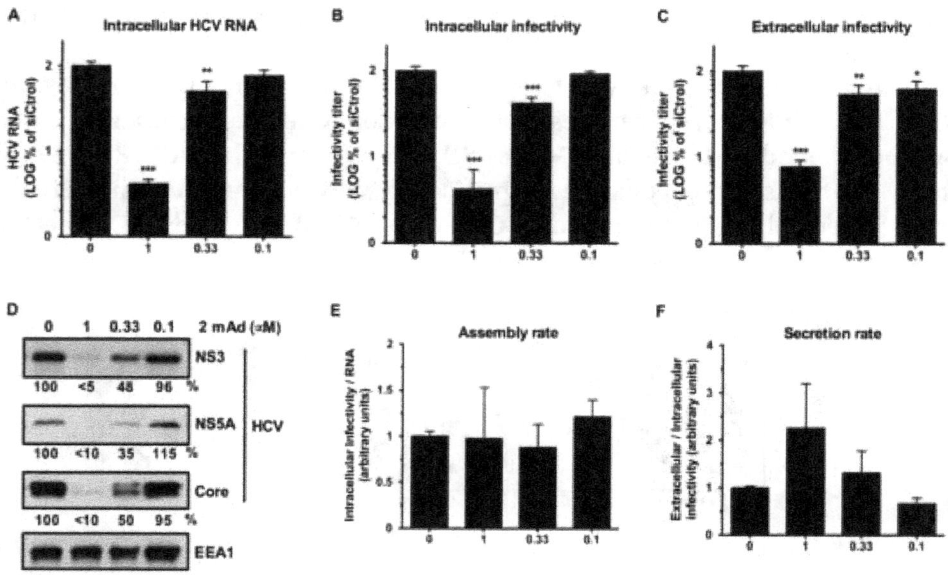

Figure 6. Dose-dependent reduction on HCV parameters upon treatment with a replication inhibitor. Huh-7 cells were inoculated with JFH-1 D183 virus at high multiplicity of infection (moi = 3) in the absence or presence of increasing amounts (0, 0.1, 0.33 and 1 μM) of the HCV polymerase inhibitor 2 mAd. 48 h later supernatants were collected, cell extracts were harvested and samples were subjected to the same analysis described in Figure 5. Note the proportional and dose response reduction of all viral parameters (RNA (A), infectivities (B,C) and protein (D)) in 2 mAd-treated cells and the absence of any significant effect in the assembly (E) and secretion (F) rates. The HCV RNA and the intracellular and extracellular infectivity levels in untreated cells were: 2 × 107 ± 2 × 106 HCV RNA copies per μg of total RNA, 2 × 104 ± 2.5 × 103 ffus per well, and 1.6 × 105 ± 2.5 × 104 ffus per ml of supernatant, respectively. Data shown are averages of two independent experiments, each one performed in duplicate (Mean; SD; n = 4). One-way ANOVA followed by Dunnett´s Multiple Comparison Test analysis was used for statistical significance analysis (* p < 0.05; ** p < 0.01; *** p < 0.001).

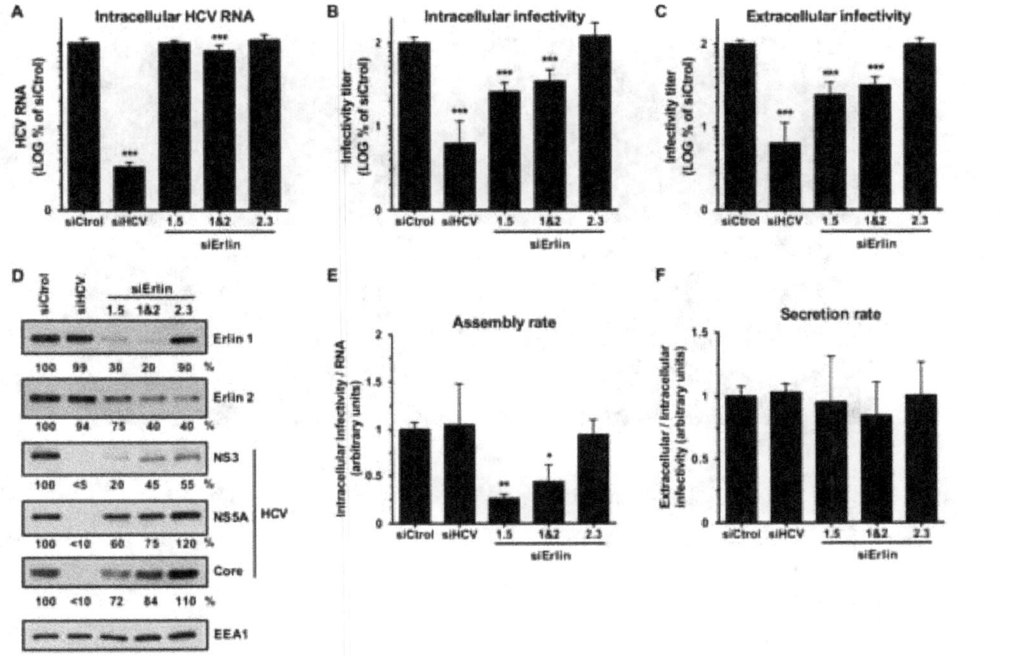

Figure 7. Erlin-1 protein down-regulation impairs infectious virus production in an ongoing HCV infection cell culture system. Persistently infected Huh-7 cells were transfected with siRNAs as described in Material and Methods. Four days after siRNA transfection cellular extracts were prepared for intracellular RNA (A), infectivity (B) and protein (D) analysis and supernatants were collected for extracellular infectivity (C) analysis. Assembly (E) and secretion (F) rates were calculated using data from panels A, B and C. The RNA and infectivity results are displayed as percentage of the levels in siCtrol-transfected cells. The HCV RNA and the intracellular and extracellular infectivity levels in siCtrol-transfected cells were: 5.8 × 105 ± 7 × 104 HCV RNA copies per µg of total RNA, 2.3 × 103 ± 2.8 × 102 ffus per well, and 1 × 104 ± 1.1 × 103 ffus per mL of supernatant, respectively. Data shown are averages of three independent experiments, each one performed in triplicate (Mean; SD; n = 9). One-way ANOVA followed by Dunnett's Multiple Comparison Test was used to determine the statistical significance (* p < 0.05; ** p < 0.01; *** p < 0.001).

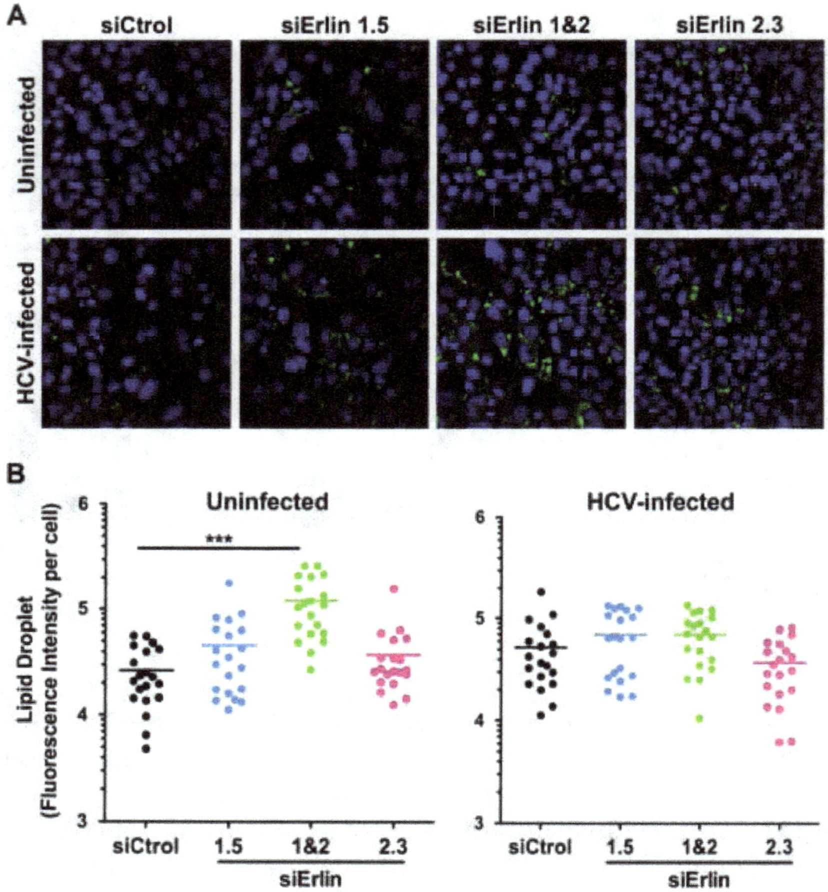

Figure 8. Erlin-1 protein down-regulation increases LD content in Huh-7 cells. siRNA-transfected Huh-7 cells were inoculated with JFH-1 D183 virus at high multiplicity of infection (moi = 3) and the intracellular LD content was analyzed 48 h after infection as described in Material and Methods. (A) Representative confocal image sections of intracellular LD accumulation (in green) in siRNA-transfected HCV-infected and uninfected cells. Nuclei (in blue) were counterstained with Hoechst dye. (B) Quantitation of the total LD fluorescence intensity signal per cell from confocal image sections. ImageJ software was used to quantitate the total LD fluorescence intensity signal in twenty individual cells of each condition. Each dot in the graph represents the LD fluorescence intensity of an individual cell and the horizontal lines show the average LD fluorescence intensity in each group of data. One-way ANOVA followed by Dunnett's Multiple Comparison Test was used to determine the statistical significance (* $p < 0.001$).**

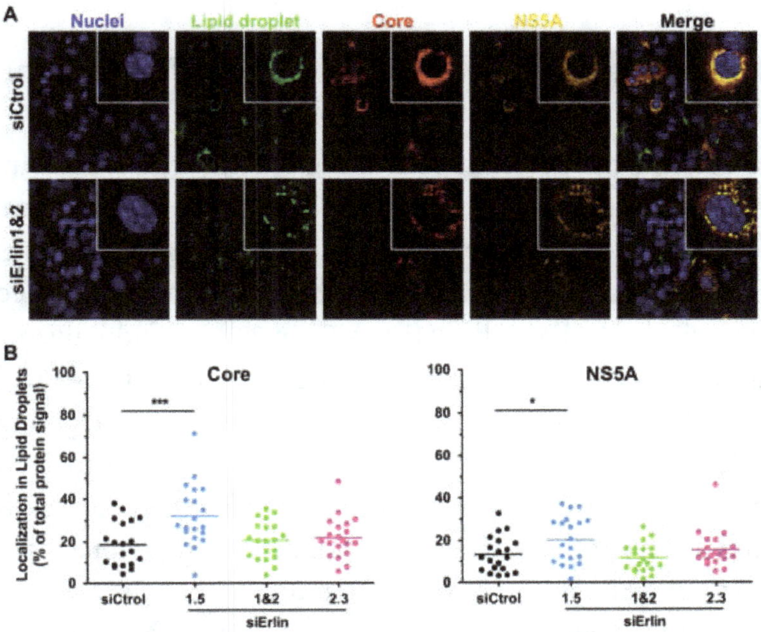

Figure 9. Erlin-1 protein down-regulation does not impair the localization of HCV core and NS5A proteins surrounding LDs. siRNA-transfected Huh-7 cells were inoculated with JFH-1 D183 virus at high multiplicity of infection (moi = 3) and the intracellular HCV protein and LD content was analyzed as described in Material and Methods. (A) Representative confocal images of the intracellular localization of HCV core (in red) and NS5A (in orange) proteins surrounding LDs (in green) in ERLIN down-regulated (siErlin 1&2) and control (siCtrol) HCV-infected cells. Nuclei (in blue) were counterstained with Hoechst dye. The panels on the right side show merged images of the four channels of each image. White boxes in the upper right side of each panel show zoom-in images of single cells for a more detailed observation. (B) The fluorescence intensity signal of core and NS5A proteins associated to LDs was quantitated in twenty randomly selected HCV-infected cells of each condition. Each dot in the graph represents the percentage of the fluorescence intensity signal of core or NS5A protein associated to LDs in a given cell, while the horizontal lines show the average of all data points in each group. One-way ANOVA followed by Dunnett's Multiple Comparison Test was used to determine the statistical significance (* $p < 0.05$; *** $p < 0.001$).

3. RESULTS

3.1. Erlin-1 Protein Is a Host Factor Required for Efficient HCV Infection

To investigate the potential role of erlin-1 and erlin-2 proteins in the HCV lifecycle we took advantage of siRNAs that have being previously validated and used to characterize the cellular functions of erlin proteins [24,27]. We transfected Huh-7 cells with siRNAs targeting specifically erlin-1 (i.e., siErlin 1.5) or erlin-2 (i.e., siErlin 2.3) individually, or siRNAs targeting both erlin proteins simultaneously (siErlin 1&2), siRNAs targeting HCV as a positive control (siHCV) and a non-targeting siRNA as a negative control (siCtrol). siRNA transfected cells were inoculated at low multiplicity of infection (moi = 0.2) with a cell culture-adapted HCV virus (D183v) [34] and their susceptibility to infection was assessed by measuring virus production three and five days later. Analysis of the supernatants of erlin-1 down-regulated HCV-infected cells showed ~ 7–10 fold reduction of progeny virus production compared to non-targeting control siRNA-transfected cells that were infected in parallel (Figure 1A).

Remarkably, erlin-2 protein down-regulation did not reduce HCV virus production. As expected the HCV-targeting siRNA prevented the production of infectious HCV virus. Consistent with the effect on virus production, the accumulation of intracellular HCV proteins (e.g., NS3 and NS5A) was also reduced in erlin-1, but not in erlin-2, down-regulated cells (Figure 1B). Despite a small defect in the cellular proliferation rate observed in siErlin 1.5-transfected cells (Figure 1C,D), the cellular respiration capacity of the different siRNA-transfected cells was comparable to that of control cells (Figure 1E). Collectively, these results suggest that the effects observed in erlin-1 down-regulated cells (in siErlin 1.5 and siErlin 1&2) are not due to a reduction in cell viability and most probably reflect a specific requirement of erlin-1 protein for efficient HCV infection.

3.2. Erlin-1 Protein Is Not Required for HCV Cell-Entry

The multiple cycle infection experiments described above suggested that erlin-1 protein plays an important role in the HCV lifecycle. Next, we set out to determine the step in the virus life cycle in which erlin-1 protein was required. Many aspects of HCV cell entry have been studied using HCV-pseudotyped retroviral vectors. Therefore, we used retroviral particles pseudotyped with the JFH-1 envelope glycoproteins E1 and E2 (HCVpp) or the vesicular stomatitis virus protein G (VSVpp), as control, to determine their ability to infect erlin-1 down-regulated cells. siRNA-transfected Huh-7 cells were infected with the corresponding retroviral particles and the intracellular luciferase activity was measured forty-height hours later. Down-regulation efficiency was verified in parallel cultures by WB (Figure 2A). As shown in Figure 2B, transfection of ERLIN1-targeting siRNAs had no effect on viral entry efficiency despite strong reduction in erlin-1 protein expression achieved in those cells. As expected ERLIN2- and HCV-specific siRNAs had no effect on the HCVpp entry process. These results suggest that reduced erlin-1 protein expression does not affect HCV cell entry.

3.3. Erlin-1 Protein Down-Regulation Impairs the Establishment of HCV RNA Replication but Does Not Affect Primary Translation or Maintenance Of Replication

To test whether erlin-1 protein regulates early steps downstream HCV entry such as the primary translation of the incoming RNA or the establishment of RNA replication, we used a highly sensitive approach based on the transfection of an in vitro transcribed JFH-1 subgenomic RNA bearing a luciferase reporter gene (JFH-1 Rluc/SGR RNA). In this system the luciferase activity measured at early time points after transfection (i.e., at six hours post-transfection) derives exclusively from the translation of transfected RNA, while the accumulation of luciferase at later time points (e.g., at 48, 72 or 96 h post-transfection) is the result of both translation and RNA replication processes. Consistent with this, luciferase activity derived from JFH-1 Rluc/SGR wt RNA was equivalent in the presence (white bars) or absence (black bars) of the HCV polymerase inhibitor 2′-C-methyladenosine (2 mAd) six hours after transfection (Figure 3A). However, it was clearly reduced in 2 mAd-treated cells at 96 h post-transfection.

Luciferase activity measured at early time points (6 and 24 h post-transfection) was similar in all siRNA-transfected cell lines, suggesting that primary translation is not affected by the levels of erlin-1 or erlin-2 protein expression (Figure 3B). Interestingly, the luciferase activity in erlin-1 down-regulated cells (siErlin 1.5- and 1&2-transfected cells) was around 50% lower than that of control cells at 48 and 72 h post-transfection. Moreover, the luciferase activity measured at 96 h time point showed greater differences in siErlin 1&2-transfected cells (30% of siCtrol) than in siErlin 1.5-transfected cells (60% of siCtrol), correlating with erlin-1 down-regulation efficiencies achieved by the corresponding siRNAs (Figure 3C). These results suggest that erlin-1 protein regulates the establishment of HCV RNA replication without affecting the primary translation. Supporting this notion, luciferase activity measured six hours after the transfection of a replication deficient JFH-1 RLuc/SGR GND mutant RNA was comparable in erlin-1 down-regulated cells and control cells (Figure 3D). Altogether, these data suggest that erlin-1 protein regulates the initiation of RNA replication without affecting the translation of incoming HCV RNA.

Next, we analyzed the effect of erlin-1 protein down-regulation on the maintenance of HCV RNA replication in the stably replicating JFH-1 subgenomic replicon cells (SGR cells). SGR cells were transfected with the indicated siRNAs and three days later the intracellular erlin protein levels and HCV protein and RNA levels were analyzed by WB and RT-qPCR, respectively. As expected, transfection of the HCV-targeting siRNA produced a significant reduction of HCV proteins (Figure 4A) and RNA (Figure 4B). However, no effect was observed when siRNAs targeting erlin-1 protein were transfected despite the 50–70% reduction in erlin-1 protein expression achieved. Similar results were obtained when JFH-1 full-length replicon cells were used (data not shown). These results suggest that erlin-1 protein does not regulate the ongoing HCV RNA replication process.

3.4. Erlin-1 Protein Down-Regulation Interferes with HCV Protein and Intracellular Infectious Virus Accumulation

Collectively the results from the experiments described above suggested that reduction of erlin-1 protein expression impairs HCV infection by inhibiting the establishment of RNA replication (Figure 3) without affecting HCV entry (Figure 2), HCV IRES dependent primary translation (Figure 3) or ongoing HCV RNA replication (Figure 4). Consistent with those results, quantitation of the intracellular HCV RNA levels 48 h after a high moi (moi = 3) infection showed a modest but statistically significant two to three-fold decrease in erlin-1 down-regulated cells compared to control cells (Figure 5A). Interestingly, the accumulation of intracellular (Figure 5B) and extracellular (Figure 5C) infectious virus was strongly reduced (around ten-fold) in those same cells. These results were confirmed with a different ERLIN1-targeting siRNA (siErlin 1.3). Similarly, the accumulation of HCV core, NS3 and NS5A proteins was reduced by four to ten-fold in erlin-1 down-regulated cells (Figure 5D). The disproportionate effect on the infectivity levels compared to the effect on the intracellular HCV RNA level suggested that erlin-1 protein is required not only for the establishment of RNA replication as shown in Figure 3, but also for a post-replication step in the virus lifecycle. In fact, the assembly rate of infectious virus particles (Figure 5E) but not the secretion rate (Figure 5F) was reduced in erlin-1 down-regulated cells compared the control cells. Supporting the latest, siRNAs that target directly the HCV genome (i.e., siHCV) inhibited HCV RNA accumulation and downstream HCV protein accumulation as well as intra- and extra-cellular infectious virus accumulation to the same extent (around 15–20 fold).

Moreover, dose-dependent and parallel reductions in HCV parameters (i.e., intracellular HCV RNA, protein and infectious virus, and extracellular infectious virus) were observed when cells were infected in the presence of different doses of the HCV polymerase inhibitor 2 mAd (Figure 6). These results imply that the disproportionate effect in virus infectivity levels observed in erlin-1 down-regulated cells could not be explained solely by the modest reduction in intracellular HCV

RNA accumulation suggesting an independent requirement for erlin-1 protein at a post RNA replication step.

To prove unequivocally that erlin-1 protein is a rate limiting factor for a later step in the HCV virus life cycle we took advantage of the possibility of manipulating erlin-1 protein expression after the HCV infection is fully established. This approach allowed us to avoid the effect of erlin-1 protein down-regulation in the establishment of RNA replication described above. siRNAs were transfected into persistently HCV-infected cells and four days later the intracellular HCV RNA (Figure 7A) and the intracellular (Figure 7B) and extracellular (Figure 7C) infectious virus accumulation were quantitated by RT-qPCR and titration assays, respectively. While siRNAs targeting HCV genome significantly suppressed HCV RNA and infectious virus accumulation, ERLIN1-targeting siRNAs reduced infectious virus production by 70–80% with a modest 10–20% reduction on HCV RNA levels. This was reflected in a lower assembly rate (Figure 7E), but not in the secretion rate (Figure 7F) in erlin-1 down-regulated cells. These results were confirmed with a different ERLIN1-targeting siRNA (siErlin 1.3). HCV protein analysis showed stronger defect in NS3 than in NS5A and core protein accumulation in erlin-1 down-regulated cells (Figure 7D). These results strongly suggest that erlin-1 protein regulates later event(s) that lead to infectious virus production.

3.5. Erlin-1 Protein Down-Regulation Increases LD Accumulation in Huh-7 Cells

We have previously described that down-regulation of erlin proteins lead to an intracellular fatty acids and cholesterol increase in Hela cells [27]. Furthermore, it is well known that HCV requires LDs for assembly of infectious virus and that HCV induces LD accumulation in infected cells [10]. Therefore, we analyzed the effect of erlin protein down-regulation on LD accumulation in HCV-infected Huh-7 cells. To do so, erlin down-regulated and control cells were infected with JFH-1 D183 virus at high moi (moi = 3) and fixed forty-eight hours later for analysis. Fixed cells were incubated with LipidTox, a reagent that stains specifically neutral lipids, and they were analyzed by confocal microscopy. As shown in

Figure 8A, reduction in erlin-1 and erlin-2 protein expression produced an increase in LD accumulation compared to control cells. Quantitation of fluorescence intensity in individual cells confirmed those results and revealed a statistically significant increase in LD accumulation in cells where ERLIN1 and ERLIN2 were simultaneously down-regulated (i.e., siErlin 1&2) compared to that of control cells (Figure 8B). These results suggest that erlin protein down-regulation increases the intracellular accumulation of LDs not only in Hela cells, as previously described [27], but also in Huh-7 cells.

3.6. Erlin-1 Protein Deficiency Does Not Impair HCV Core and NS5A Protein Re-Localization to LDs

Late in the infection both core and NS5A proteins are known to be associated to LDs to facilitate virus assembly [10]. Thus, we analyzed the distribution of core and NS5A proteins in ERLIN downregulated cells by confocal microscopy. As shown in Figure 9A, core and NS5A proteins localized to LDs similarly in both control and ERLIN down-regulated cells. A quantitative analysis of the localization of core and NS5A proteins surrounding LDs (Figure 9B) confirmed that the reduced infectious virus accumulation observed in ERLIN1-deficient cells was not due to a mislocalization of core or NS5A proteins during infection.

4. DISCUSSION

The development of an in vitro HCV infection system in 2005 [29,40,41] allowed the identification of viral proteins and cellular factors and pathways that regulate different aspects of the HCV infection (reviewed in [8,42,43]). Using this infection model, we and others have previously identified cellular factors required for initiation of RNA replication [17,32] and HCV assembly and secretion [15,38], and cellular pathways required for the induction [44,45] and evasion [36,46] of the innate immune system. In this study, we have analyzed the consequences of erlin-1 and erlin-2 protein down-regulation on the susceptibility of Huh-7 cells to HCV infection. Our data suggest a role for erlin-1 protein early and late in the HCV life cycle.

Silencing of erlin-1 but not erlin-2 protein expression by siRNAs led to a strong reduction in the production of infectious progeny virus (Figure 1) suggesting that erlin-1 protein is required for efficient HCV infection. Single cycle infection experiments in ERLIN1 down-regulated cells showed a modest 2–3 fold reduction in intracellular HCV RNA accumulation compared to a ~4–10 fold reduction in HCV protein accumulation and ~10-fold reduction in intracellular infectious virus accumulation (Figure 5). Further studies performed in HCV surrogate models that interrogate specific steps in the virus life cycle revealed that erlin-1 protein is required for the initiation of HCV RNA replication (Figure 3) without affecting earlier steps such as HCV cell entry (Figure 2) and HCV IRES dependent translation of the incoming RNA (Figure 3). Interestingly, ERLIN1 silencing in subgenomic replicon cells (Figure 4) or in persistently infected cells (Figure 7) had no effect on the steady-state level of HCV RNA but revealed differences in the steady-state levels of HCV proteins. Collectively, these results suggested that erlin-1 protein plays a role in the initiation of RNA replication but it seems to be dispensable for the maintenance of RNA replication once it has been established. The formation of the first viral replication complexes and the initiation of RNA replication are thought to take place in the presence of very low levels of viral proteins. Therefore, it is conceivable that the initiation step is more likely to be dependent on host factors than later steps when the accumulation of viral factors required for RNA replication is higher and the virus has induced a more favorable environment for replication in the cell by inducing the membranous web [47]. Our results are consistent with this notion and add erlin-1 protein to the list of factors previously reported to show a differential requirement for the initiation and maintenance of the RNA replication processes [17,38,48]. Alternatively, these results might suggest the existence of a different threshold for the requirement of erlin-1 protein in the regulation of the initiation and maintenance of RNA replication i.e., higher threshold for initiation than for maintenance.

Analysis of virus accumulation in acutely (Figure 5) and persistently (Figure 7) infected ERLIN1-deficient cells showed a strong reduction in the production of both intracellular and extracellular infectious virus. Biophysical analysis of infectious virus particles produced in ERLIN1-deficient cells did not reveal any difference in terms of their specific infectivity (FFU per RNA copies) or density profiles compared to the virus produced in control cells (data not shown) suggesting that ERLIN1-deficient cells produce qualitatively normal but quantitatively less infectious virus than ERLIN1-sufficient cells.

The disproportionate reduction in virus production compared to the decrease in HCV RNA levels suggested that erlin-1 protein regulates not only early steps leading to RNA and protein accumulation, but also later steps affecting virus production. In fact, direct targeting of RNA replication by an HCV polymerase inhibitor led to a proportional decrease of HCV RNA, protein and infectious virus, suggesting that the strong inhibition of virus production observed in ERLIN1 down-regulated cells could not be solely explained by the modest reduction in the HCV RNA levels (Figure 6). As mentioned above, the magnitude of HCV NS3, NS5A and core protein reduction was disproportionate to the magnitude of HCV RNA reduction suggesting that ERLIN1 down-regulation affects the steady-state levels of HCV proteins. The HCV-IRES translation results indicated that the translation of HCV RNA itself is not inhibited in ERLIN1-deficient cells (Figure 3) suggesting that the effect is likely to be at the level of protein stability or degradation and not at the translational level. This hypothesis is very appealing since erlin proteins have been implicated in the modulation of the ERAD pathway [23,24,25,26]. In contract to the reduced HCV protein accumulation observed in erlin-1 protein deficient HCV-infected cells (Figure 5), ERLIN1 and ERLIN2 deficiency produces an accumulation of cellular substrates that are typically degraded through the ERAD pathway [23,24,25,26]. These opposite effects on protein homeostasis (HCV vs. cellular proteins) may suggest that cellular proteome changes specifically induced by ERLIN1 down-regulation could indirectly be responsible, at least in part, of the effects observed in the HCV infection. Comparison of cellular proteomes of ERLIN1- and ERLIN2-deficient cells could identify cellular candidate proteins contributing to the effects observed in ERLIN1-deficient HCV-infected cells.

Recruitment of viral NS5A and core proteins to LDs is a prerequisite for virus assembly [10]. Indeed, interfering with viral protein trafficking to LDs reduces virus production [13,49]. However, viral protein localization into LDs is not sufficient for virus particle assembly. For instance, a single mutation in the core protein (Y136A) of J6/JFH-1 virus has been associated with an increased accumulation of core in LD, decreased colocalization with E2 protein and reduced HCV particle assembly [50]. Confocal analysis of infected cells confirmed the reduced expression of HCV NS5A and core proteins and revealed a typical localization of these proteins surrounding LDs late in the infection, suggesting that ERLIN1-deficiency is affecting their abundance (Figure 5 and Figure 7) but not their recruitment to the HCV assembly factories (Figure 9). Collectively these results suggest that the reduction in HCV protein levels could be responsible for the strong inhibition of infectious virus production observed in erlin-1 down-regulated cells. Importantly, given the differential effect in HCV RNA levels and virus production, these results suggest to us the existence of different HCV protein thresholds required to maintain RNA replication and assembly processes.

Another interesting aspect of these studies is that the HCV-related effects are specific to ERLIN1 deficiency but not to ERLIN2 deficiency despite their high similarity and functional cellular redundancy. This difference could be interpreted as a function of protein depletion efficiency achieved by the different siRNAs. ERLIN1-targeting siRNAs (1.3, 1.5 and 1&2) were in general more efficient (stronger reduction and longer-lasting effect) reducing erlin-1 protein expression than ERLIN2-targeting siRNAs (2.3 and 1&2) in silencing erlin-2 protein. The reason for this could be in part the longer half-life of erlin-2 compared to erlin-1 protein and the relative abundance of those proteins in Huh-7 cells, with erlin-2 being ~2–3 times more abundant than erlin-1 protein (data not shown). Alternatively, our results might indeed reflect a specific function of erlin-1 protein that is not shared by erlin-2 protein and that would be the first function exclusively assigned to erlin-1 protein. Importantly, our results identify erlin-1 protein as a new positive modulator of HCV infection and together with the results described by Inoue et al. in the SV40 infection system [28] point to the need of investigating the role of erlin proteins in other viral infections, especially in those in which their life cycle is tightly associated with the ER e.g., dengue virus or zika virus infection.

CONCISE SUMMARY

The creation of cell culture techniques for studying hepatitis C virus (HCV) infection has enabled the discovery of cellular components that control the HCV life cycle. Certain biological variables influence specific stages of the viral life cycle that are closely linked to intracellular membranes originating from the endoplasmic reticulum (ER). In this report, we present the identification of the erlin-1 protein as a crucial cellular component involved in the regulation of HCV infection. Erlin-1 is a protein that binds to cholesterol and is found in certain membranes inside the endoplasmic reticulum (ER) that are resistant to detergents. It plays a role in maintaining cholesterol balance and the pathway for degradation connected with the endoplasmic reticulum. Suppression of erlin-1 protein expression using siRNA resulted in a decline in infection efficiency, as seen by a drop in the concentration of RNA inside cells, reduced expression of HCV proteins, and lower generation of the virus. Studies investigating the mechanisms involved have shown that the erlin-1 protein plays a crucial role in the early stages of infection, after the virus has entered the cell and undergone primary translation. Specifically, erlin-1 is responsible for initiating the replication of RNA. Additionally, erlin-1 is also necessary later in the infection process to facilitate the formation of infectious virus particles. This work identified the erlin-1 protein as a crucial cellular component that controls HCV infection.

REFERENCES

1. Maniloff, J. Identification and classification of viruses that have not been propagated. Arch. Virol. 1995, 140, 1515–1520.

2. Alter, H.J.; Seeff, L.B. Recovery, persistence, and sequelae in hepatitis C virus infection: A perspective on long-term outcome. Semin. Liver Dis. 2000, 20, 17–35.

3. Choo, Q.L.; Richman, K.H.; Han, J.H.; Berger, K.; Lee, C.; Dong, C.; Gallegos, C.; Coit, D.; Medina-Selby, R.; Barr, P.J. Genetic organization and diversity of the hepatitis C virus. Proc. Natl. Acad. Sci. USA 1991, 88, 2451–2455.

4. Friebe, P.; Boudet, J.; Simorre, J.-P.; Bartenschlager, R. Kissing-loop interaction in the 3' end of the hepatitis C virus genome essential for RNA replication. J. Virol. 2005, 79, 380–392.

5. Friebe, P.; Lohmann, V.; Krieger, N.; Bartenschlager, R. Sequences in the 5' nontranslated region of hepatitis C virus required for RNA replication. J. Virol. 2001, 75, 12047–12057.

6. Honda, M.; Beard, M.R.; Ping, L.H.; Lemon, S.M. A phylogenetically conserved stem-loop structure at the 5' border of the internal ribosome entry site of hepatitis C virus is required for cap-independent viral translation. J. Virol. 1999, 73, 1165–1174.

7. Penin, F.; Dubuisson, J.; Rey, F.A.; Moradpour, D.; Pawlotsky, J.-M. Structural biology of hepatitis C virus. Hepatology 2004, 39, 5–19.

8. Alazard-Dany, N.; Denolly, S.; Boson, B.; Cosset, F.-L. Overview of HCV Life Cycle with a Special Focus on Current and Possible Future Antiviral Targets. Viruses 2019, 11, 30.

9. Sharma, N.R.; Mateu, G.; Dreux, M.; Grakoui, A.; Cosset, F.-L.; Melikyan, G.B. Hepatitis C virus is primed by CD81 protein for low pH-dependent fusion. J. Biol. Chem. 2011, 286, 30361–30376.

10. Miyanari, Y.; Atsuzawa, K.; Usuda, N.; Watashi, K.; Hishiki, T.; Zayas, M.; Bartenschlager, R.; Wakita, T.; Hijikata, M.; Shimotohno, K. The lipid droplet is an important organelle for hepatitis C virus production. Nat. Cell Biol. 2007, 9, 1089–1097.

11. Zayas, M.; Long, G.; Madan, V.; Bartenschlager, R. Coordination of Hepatitis C Virus Assembly by Distinct Regulatory Regions in Nonstructural Protein 5A. PLoS Pathog. 2016, 12, e1005376.

12. Popescu, C.-I.; Callens, N.; Trinel, D.; Roingeard, P.; Moradpour, D.; Descamps, V.; Duverlie, G.; Penin, F.; Héliot, L.; Rouille, Y.; et al. NS2 protein of hepatitis C virus interacts with structural and non-structural proteins towards virus assembly. PLoS Pathog. 2011, 7, e1001278.

13. Boson, B.; Granio, O.; Bartenschlager, R.; Cosset, F.-L. A concerted action of hepatitis C virus p7 and nonstructural protein 2 regulates core localization at the endoplasmic reticulum and virus assembly. PLoS Pathog. 2011, 7, e1002144.

14. Gentzsch, J.; Brohm, C.; Steinmann, E.; Friesland, M.; Menzel, N.; Vieyres, G.; Perin, P.M.; Frentzen, A.; Kaderali, L.; Pietschmann, T. Hepatitis C Virus p7 is Critical for Capsid Assembly and Envelopment. PLoS Pathog. 2013, 9, e1003355.

15. Gastaminza, P.; Cheng, G.; Wieland, S.; Zhong, J.; Liao, W.; Chisari, F.V. Cellular determinants of hepatitis C virus assembly, maturation, degradation, and secretion. J. Virol. 2008, 82, 2120–2129.

16. McLauchlan, J.; Lemberg, M.K.; Hope, G.; Martoglio, B. Intramembrane proteolysis promotes trafficking of hepatitis C virus core protein to lipid droplets. EMBO J. 2002, 21, 3980–3988.

17. Friesland, M.; Mingorance, L.; Chung, J.; Chisari, F.V.; Gastaminza, P. Sigma-1 receptor regulates early steps of viral RNA replication at the onset of hepatitis C virus infection. J. Virol. 2013, 87, 6377–6390.

18. Browman, D.T.; Hoegg, M.B.; Robbins, S.M. The SPFH domain-containing proteins: More than lipid raft markers. Trends Cell Biol. 2007, 17, 394–402.

19. Langhorst, M.F.; Reuter, A.; Stuermer, C.A.O. Scaffolding microdomains and beyond: The function of reggie/flotillin proteins. Cell. Mol. Life Sci. 2005, 62, 2228–2240.

20. Browman, D.T.; Resek, M.E.; Zajchowski, L.D.; Robbins, S.M. Erlin-1 and erlin-2 are novel members of the prohibitin family of proteins that define lipid-raft-like domains of the ER. J. Cell Sci. 2006, 119, 3149–3160.

21. Ikegawa, S.; Isomura, M.; Koshizuka, Y.; Nakamura, Y. Cloning and characterization of a novel gene (C8orf2), a human representative of a novel gene family with homology to C. elegans C42.C1.9. Cytogenet. Cell Genet. 1999, 85, 227–231.

22. Hoegg, M.B.; Browman, D.T.; Resek, M.E.; Robbins, S.M. Distinct regions within the erlins are required for oligomerization and association with high molecular weight complexes. J. Biol. Chem. 2009, 284, 7766–7776.

23. Pearce, M.M.P.; Wormer, D.B.; Wilkens, S.; Wojcikiewicz, R.J.H. An endoplasmic reticulum (ER) membrane complex composed of SPFH1 and SPFH2 mediates the ER-associated degradation of inositol 1,4,5-trisphosphate receptors. J. Biol. Chem. 2009, 284, 10433–10445.

24. Pearce, M.M.P.; Wang, Y.; Kelley, G.G.; Wojcikiewicz, R.J.H. SPFH2 mediates the endoplasmic reticulum-associated degradation of inositol 1,4,5-trisphosphate receptors and other substrates in mammalian cells. J. Biol. Chem. 2007, 282, 20104–20115.

25. Jo, Y.; Sguigna, P.V.; DeBose-Boyd, R.A. Membrane-associated ubiquitin ligase complex containing gp78 mediates sterol-accelerated degradation of 3-hydroxy-3-methylglutaryl-coenzyme A reductase. J. Biol. Chem. 2011, 286, 15022–15031.

26. Teranishi, Y.; Hur, J.-Y.; Gu, G.J.; Kihara, T.; Ishikawa, T.; Nishimura, T.; Winblad, B.; Behbahani, H.; Kamali-Moghaddam, M.; Frykman, S.; et al. Erlin-2 is associated with active γ-secretase in brain and affects amyloid β-peptide production. Biochem. Biophys. Res. Commun. 2012, 424, 476–481.

27. Huber, M.D.; Vesely, P.W.; Datta, K.; Gerace, L. Erlins restrict SREBP activation in the ER and regulate cellular cholesterol homeostasis. J. Cell Biol. 2013, 203, 427–436.

28. Inoue, T.; Tsai, B. Regulated Erlin-dependent release of the B12 transmembrane J-protein promotes ER membrane penetration of a non-enveloped virus. PLoS Pathog. 2017, 13, e1006439.

29. Zhong, J.; Gastaminza, P.; Cheng, G.; Kapadia, S.; Kato, T.; Burton, D.R.; Wieland, S.F.; Uprichard, S.L.; Wakita, T.; Chisari, F.V. Robust hepatitis C virus infection in vitro. Proc. Natl. Acad. Sci. USA 2005, 102, 9294–9299.

30. Graham, F.L.; Smiley, J.; Russell, W.C.; Nairn, R. Characteristics of a human cell line transformed by DNA from human adenovirus type 5. J. Gen Virol. 1977, 36, 59–74.

31. Kato, T.; Date, T.; Miyamoto, M.; Furusaka, A.; Tokushige, K.; Mizokami, M.; Wakita, T. Efficient replication of the genotype 2a hepatitis C virus subgenomic replicon. Gastroenterology 2003, 125, 1808–1817.

32. Dreux, M.; Gastaminza, P.; Wieland, S.F.; Chisari, F.V. The autophagy machinery is required to initiate hepatitis C virus replication. Proc. Natl. Acad. Sci. USA 2009, 106, 14046–14051.

33. Gastaminza, P.; Whitten-Bauer, C.; Chisari, F.V. Unbiased probing of the entire hepatitis C virus life cycle identifies clinical compounds that target multiple aspects of the infection. Proc. Natl. Acad. Sci. USA 2010, 107, 291–296.

34. Zhong, J.; Gastaminza, P.; Chung, J.; Stamataki, Z.; Isogawa, M.; Cheng, G.; Mckeating, J.A.; Chisari, F.V. Persistent hepatitis C virus infection in vitro: Coevolution of virus and host. J. Virol. 2006, 80, 11082–11093.

35. Gastaminza, P.; Kapadia, S.B.; Chisari, F.V. Differential biophysical properties of infectious intracellular and secreted hepatitis C virus particles. J. Virol. 2006, 80, 11074–11081.

36. Garaigorta, U.; Chisari, F.V. Hepatitis C virus blocks interferon effector function by inducing protein kinase R phosphorylation. Cell Host Microbe 2009, 6, 513–522.

37. Bartosch, B.; Dubuisson, J.; Cosset, F.-L. Infectious Hepatitis C Virus Pseudo-particles Containing Functional E1–E2 Envelope Protein Complexes. J. Exp. Med. 2003, 197, 633–642.

38. Garaigorta, U.; Heim, M.H.; Boyd, B.; Wieland, S.; Chisari, F.V. Hepatitis C virus (HCV) induces formation of stress granules whose proteins regulate HCV RNA replication and virus assembly and egress. J. Virol. 2012, 86, 11043–11056.

39. Schindelin, J.; Arganda-Carreras, I.; Frise, E.; Kaynig, V.; Longair, M.; Pietzsch, T.; Preibisch, S.; Rueden, C.; Saalfeld, S.; Schmid, B.; et al. Fiji: An open-source platform for biological-image analysis. Nat. Methods 2012, 9, 676–682.

40. Wakita, T.; Pietschmann, T.; Kato, T.; Date, T.; Miyamoto, M.; Zhao, Z.; Murthy, K.; Habermann, A.; Kräusslich, H.G.; Mizokami, M.; et al. Production of infectious hepatitis C virus in tissue culture from a cloned viral genome. Nat. Med. 2005, 11, 791–796.

41. Lindenbach, B.D.; Evans, M.J.; Syder, A.J.; Wolk, B.; Tellinghuisen, T.L.; Liu, C.C.; Maruyama, T.; Hynes, R.O.; Burton, D.R.; Mckeating, J.A.; et al. Complete Replication of Hepatitis C Virus in Cell Culture. Science 2005, 309, 623–626.

42. Alvisi, G.; Madan, V.; Bartenschlager, R. Hepatitis c virus and host cell lipids: An intimate connection. RNA Biol. 2014, 8, 258–269.

43. Zhou, L.-Y.; Zhang, L.-L. Host restriction factors for hepatitis C virus. World J. Gastroenterol. 2016, 22, 1477–1486.

44. Takahashi, K.; Asabe, S.; Wieland, S.; Garaigorta, U.; Gastaminza, P.; Isogawa, M.; Chisari, F.V. Plasmacytoid dendritic cells sense hepatitis C virus-infected cells, produce interferon, and inhibit infection. Proc. Natl. Acad. Sci. USA 2010, 107, 7431–7436.

45. Dreux, M.; Garaigorta, U.; Boyd, B.; Décembre, E.; Chung, J.; Whitten-Bauer, C.; Wieland, S.; Chisari, F.V. Short-Range Exosomal Transfer of Viral RNA from Infected Cells to Plasmacytoid Dendritic Cells Triggers Innate Immunity. Cell Host Microbe 2012, 12, 558–570.

46. Padmanabhan, P.; Garaigorta, U.; Dixit, N.M. Emergent properties of the interferon-signalling network may underlie the success of hepatitis C treatment. Nat. Commun. 2014, 5, 3872.

47. Egger, D.; Wolk, B.; Gosert, R.; Bianchi, L.; Blum, H.E.; Moradpour, D.; Bienz, K. Expression of Hepatitis C Virus Proteins Induces Distinct Membrane Alterations Including a Candidate Viral Replication Complex. J. Virol. 2002, 76, 5974–5984.

48. Mingorance, L.; Castro, V.; Ávila-Pérez, G.; Calvo, G.; Rodriguez, M.J.; Carrascosa, J.L.; Pérez-del-Pulgar, S.; Forns, X.; Gastaminza, P. Host phosphatidic acid phosphatase lipin1 is rate limiting for functional hepatitis C virus replicase complex formation. PLoS Pathog. 2018, 14, e1007284.

49. Herker, E.; Harris, C.; Hernandez, C.; Carpentier, A.; Kaehlcke, K.; Rosenberg, A.R.; Farese, R.V.; Ott, M. Efficient hepatitis C virus particle formation requires diacylglycerol acyltransferase-1. Nat. Med. 2010, 16, 1295–1298.

50. Neveu, G.; Barouch-Bentov, R.; Ziv-Av, A.; Gerber, D.; Jacob, Y.; Einav, S. Identification and targeting of an interaction between a tyrosine motif within hepatitis C virus core protein and AP2M1 essential for viral assembly. PLoS Pathog. 2012, 8, e1002845.

16 EXAMINING THE ROLE OF CALCIUM CHANNELS AND PUMPS IN VIRAL INFECTIONS

1. INTRODUCTION: AN OVERVIEW

Viruses utilize the host cell's environment to reproduce, resulting in the impairment of the host cell's normal functioning. The interaction between a virus and its host is the fundamental basis of pathogenesis and is intimately linked to the intensity and occurrence of a disease. Virus infections are difficult to prevent and treat due to the rapid mutation rates that allow viruses to evade antiviral therapies aimed at targeting viral proteins. Manipulating the conditions within cells has emerged as a crucial approach in the search for and creation of antiviral medications. Ca2+ serves as a crucial second messenger in mammalian cells, facilitating the transmission of sensor input and the generation of output responses for various cellular processes, including stress responses, synaptic plasticity, immunodefenses, protein transport, and endosome formation [1,2]. Research has shown that when a virus infects a host cell, it causes the cell to malfunction and leads to an abnormal level of calcium ions within the cell [3]. A virus can manipulate the host's intracellular Ca2+ system in numerous ways to successfully replicate. For example, viral proteins can directly bind to Ca2+ or disrupt the membrane's ability to allow Ca2+ to pass through by modifying the Ca2+ apparatus. The plasma membrane of the host cell serves as the initial defense mechanism against viral invasion. Multiple Ca2+ channels and pumps are dispersed across the cell plasma membrane. Thus, these membrane proteins serve as the primary target for viral infection. The primary mechanism by which viruses disrupt the host cell's calcium signaling pathway is through their interaction with these membrane proteins. This interaction has the potential to either hinder or enhance the entry of calcium ions and regulate the levels of free calcium ions within the cytoplasm. Upon entering the host cell, viruses modulate the release of calcium from internal storage by influencing calcium-permeable channels, transporters, and exchangers located on the membranes of cellular compartments. Subsequently, the alteration in cytosolic calcium concentration may initiate additional disruption of the host cell system, so promoting virus survival and multiplication. This review focuses on the role of calcium channels and pumps in host cell membranes during viral infection. Inhibiting the activity of these membrane proteins or preventing viruses from interacting with these host calcium-signaling components could reduce the likelihood of virus stability, replication, and release, as well as infection-induced host-cell death and the production of reactive oxygen species, neurotoxicity, and enterotoxins. Therefore, these membrane proteins have the potential to be targeted by antiviral drugs.

2. CALCIUM CHANNELS AND PUMPS IN HOST CA2+ HOMEOSTASIS

Cellular Ca2+ is from two major sources: the internal Ca2+ store (mainly endoplasmic reticulum (ER) or sarcoplasmic reticulum (SR)) and the extracellular medium. Calcium channels on cell plasma membrane mediate the entry of Ca2+ from the extracellular medium. These channels are activated by specific stimuli, such as voltage-gated calcium channels (VGCCs), which are stimulated by membrane depolarization, specific receptor-operated channels (ROC), which are stimulated by external agonists, or intracellular messengers and store-operated calcium channel (SOC), which are stimulated by the depletion of internal Ca2+ stores. The IP3 receptor (IP3R) and the ryanodine receptors (RyR) are the main players in mediating the release of Ca2+ from the internal stores. Inositol-1,4,5-triphosphate (IP3) activates IP3R, triggers Ca2+ release from stores, and further increases IP3R's sensitivity to Ca2+. Calcium pumps (the plasma membrane Ca2+-ATPase (PMCA), sarcoplasmic/endoplasmic reticulum Ca2+-ATPase (SERCA)) and the Na+/Ca2+ exchanger (NCX) are responsible for transporting Ca2+ from the cytosol to external medium or into cellular calcium stores (Figure 1). The normal function of these calcium channels and pump is important for cells to maintain intracellular Ca2+ homeostasis.

These channels and pumps are activated in a flexible and precise manner to generate specific Ca2+ signaling, satisfying various spatiotemporal requirements. During the viral infections, host cells modulate these calcium-signaling components in response to the infection. On the other hand, viruses utilize these components to create a cellular environment that benefits their own lifecycles [4]. Viruses induce elevated cytosolic calcium concentration to activate Ca2+ dependent/sensitive enzymes and transcriptional factors to promote virus replication. Mitochondria could take the Ca2+ to generate more energy to support continuous virus replication. Moreover, regulating the calcium concentration in ER or Golgi may inhibit host proteins trafficking and promote virus protein maturation. The inhibition of host proteins frustrates host antiviral immune responses, while the promotion of virus protein maturation benefits virus propagation (Figure 1). The journey begins when a virus encounters the host cell and attaches to the cell surface. The virus particle penetrates the cytoplasm via direct membrane fusion or receptor-mediated endocytosis followed by exposing its genome to cellular machinery for replication. When the viral proteins and viral genomes are accumulated, they are assembled to form a progeny virion particle and then released [5]. During the viral lifecycles, they utilize various calcium channels and pumps to obviate the cell membrane barriers, enter the host cell, complete replication, acquire infection ability, and release.

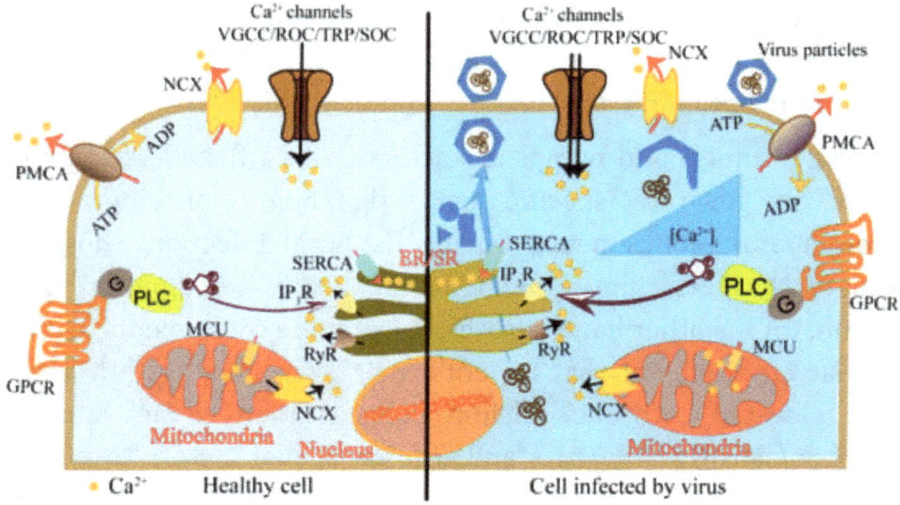

Figure 1. Schematics of host cell elevated cytosolic calcium concentration induced by a virus. Calcium channels (voltage-gated calcium channels (VGCCs), receptor-operated channels (ROC), store-operated Ca2+ (SOC), channels and transient receptor potential (TRP) channels) mediate the entry of Ca2+ from extracellular medium (black arrows). The IP3 receptor (IP3R) and the ryanodine receptors (RyR) on the endoplasmic reticulum (ER) mediate the release of Ca2+ from internal stores (black arrows). Calcium pumps (the plasma membrane Ca2+-ATPase (PMCA), sarcoplasmic/endoplasmic reticulum Ca2+-ATPase (SERCA)) and the Na+/Ca2+ exchanger (NCX) are responsible for transporting Ca2+ from the cytosol to external medium or into cellular calcium stores (red arrows). Viruses utilize these calcium components to elevate cytosolic calcium concentration to activate Ca2+-dependent/sensitive enzymes and transcriptional factors to promote virus replication (right panel).

3. VIRUSES CONTROL HOST VOLTAGE-GATED CALCIUM CHANNELS (VGCCS) AND TWO-PORE CHANNELS (TPCS)

VGCCs are widely found in the membrane of excitable cells [6] and one of the most well studied viral targets because of the availability of specific channel blockers. There are several different types of VGCCs: L-type (Long-conducting) channels are mostly found in skeletal and smooth myocytes, bone (osteoblasts), and ventricular myocytes; N-type (Non-L or Neuronal), P/Q-type (Purkinje and Granular), and R-type (dihydropyridine- Resistant) channels display longer-lasting conductance and are expressed throughout the nervous system; T-type

(Transient) channels produce short-term conductance and are found in neurons and cells that have pacemaker activity [6]. The activation of the channels results in a Ca2+ cascade, which is associated with numerous host physiological functions including excitation–contraction–relaxation coupling of muscles, synaptic transmission, immunoprotection, etc.

Since 1984, it has been known that verapamil, the blocker of VGCCs, inhibits influenza A virus (IAV) infection [7]. Moreover, IAV infection induces Ca2+ influx, and the elevated intracellular Ca2+ promotes endocytic uptake of IAV [8]. Until recently, the underlying mechanism was revealed by Fujioka et al. [9]. They reported that several VGCC blockers and siRNA against the CaV1.2 channel (L-type channel) inhibited H1N1 and H3N2 IAV infection in multiple cell lines. IAV attaches to the target cells by the viral hemagglutinin protein (HA), which is a type I transmembrane protein embedded in the viral envelope, binding to sialic acids [10]. Fujioka et al. showed that the virus HA binds to domain IV of CaV1.2, which contains two potential sialylated asparagine residues (N1436 and N1487) [9,11]. When CaV1.2 mutants in one or both these residues were replaced with glutamine (N1436Q, N1487Q, and N1436Q + N1487Q), the interaction between HA and the Cav1.2 was attenuated compared with the wild-type fragment. They demonstrated that a VGCC blocker, diltiazem, significantly prolonged the survival of IAV-infected mice and allowed the recovery of the survivors. Therefore, CaV1.2 may serve as a host cell surface receptor that binds IAV and is critical for IAV entry (Figure 2A). New world hemorrhagic fever arenaviruses (NWA) are another virus that is reportedly sensitive to the VGCC blocker club member [12,13]. An siRNA screen with Junín virus glycoprotein-pseudotyped viruses identified that VGCCs are involved in NWA entry. Treatment with the channel blocker gabapentin, an FDA (U.S. Food and Drug Administration)-approved analgesic that targets the channel α2δ2 subunit, protects against NWA infection [13]. This research work demonstrated that the interaction between a virus and VGCCs promotes virus entry at the virus-cell fusion step.

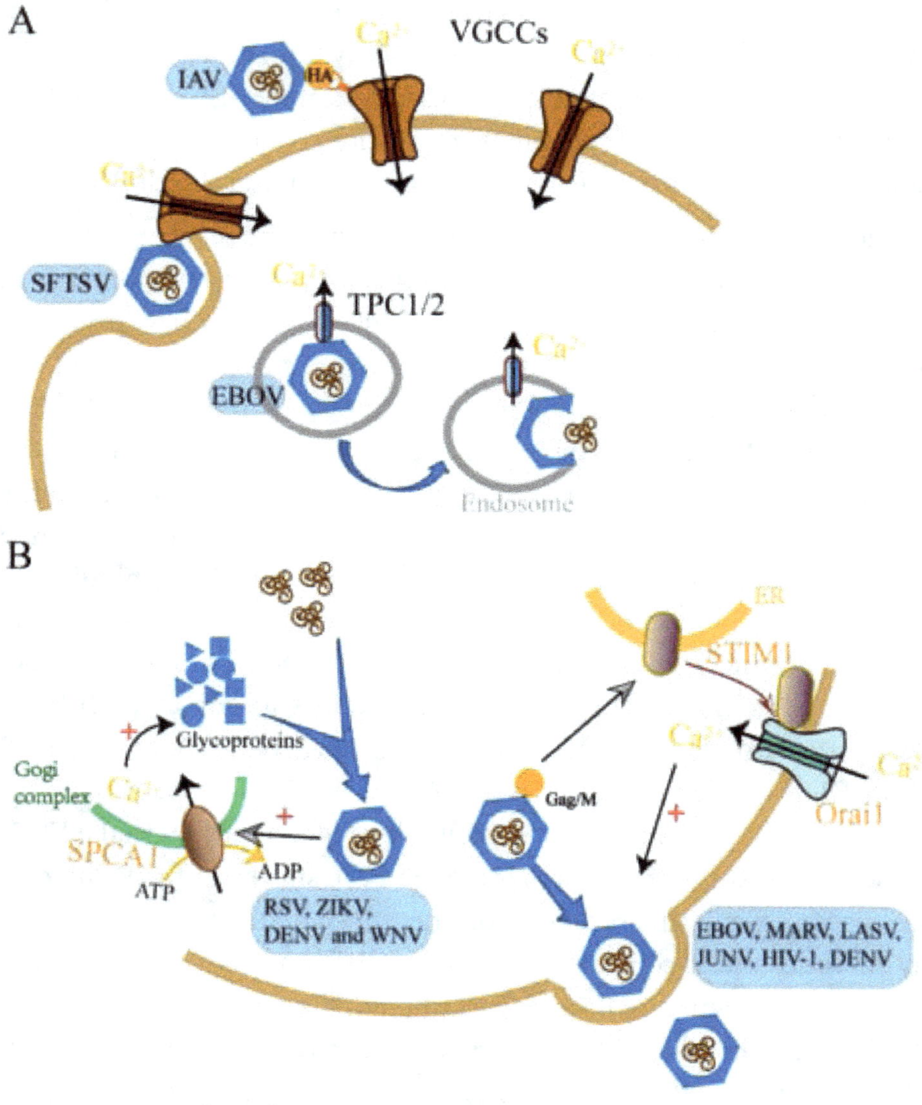

Figure 2. Examples of viruses interplaying with host calcium channels or pumps to achieve viral entry (A) and release (B). VGCCs are important for influenza A virus (IAV) and severe fever with thrombocytopenia syndrome virus (SFTSV) entry into the host cell as well as TPC1/2 for EBOV (A). RSV, Zika virus (ZIKA), dengue virus (DENV) and West Nile virus (WNV) hijack SPCA1 to facilitate their release as well as Ebola virus (EBOV), MARV, LASV, JUNV, HIV-1 and DENV manipulates STIM1/ORAI1 (B). For a complete list of definitions, see Table 1.

Table 1. Calcium channels/pumps utilized by a virus.

Cellular Targets	Virus	Consequences [Ref.]
VGCCs	IAV	$Ca_V1.2$ serves as a host cell surface receptor that binds IAV and is critical for IAV entry [9].
	SFTSV	Benidipine hydrochloride, VGCC blocker, inhibits SFTSV infection via impairing virus internalization and genome replication [12].
	NWV	Virus binds to VGCCs and promotes virus entry at the virus–cell fusion step [13].
	Flavivirus (JEV, ZIKV, DENV, and WNV	VGCCs blockers inhibit flavivirus (JEV, ZIKV, DENV and WNV) infection at the stage of replication [14].
	HIV-1	Tat/gp120 overactivate VGCCs [15,16,17].
	HSV-1	HSV-1 downregulates the $Ca_V3.2$ channel and diminishes the detection of viral infection by host [19,20].
TPCs	EBOV	Facilitates virus–endosome membrane fusion and releases of virus capsid into the cell cytoplasm [27].
STIM1/ORAI1	EBOV, MARV, LASV, JUNV, HIV-1, DENV, and HBV	Promote virion assembly and budding [35,36,37,38,44].
TRPV4	ZIKV	Activation of TRPV4- releases DDX3X and promote the viral RNA metabolism [52].
NMDAr	ZIKV, JEV	NMDAr contributes to ZIKA by triggering the neuronal cell death progress [58,60].
	HIV-1	Increases Ca^{2+} influx [62].
IP_3R	HIV-1, HSV, HRV, and HCMV	These viral proteins deplete ER Ca^{2+} store during early stages of viral infection to increase the replication ability of viruses [37,62,63,64,66].
SPCA1	RSV, ZIKV, DENV and WNV	Trigger to produce functional viral glycoproteins that are essential for virus spread [70]

Abbreviations: Influenza A virus (IAV), Severe fever with thrombocytopenia syndrome virus (SFTSV), New world hemorrhagic fever arenaviruses (NWV), Japanese encephalitis virus (JEV), Zika virus (ZIKV), Dengue virus (DENV), West Nile virus (WNV), Herpes simplex virus (HSV)-1, Ebolavirus (EBOV), Marburgvirus (MARV), Lassa Virus

(LASV), Junín Virus (JUNV), Hepatitis B virus (HBV), Respiratory syncytial virus (RSV), Human rhinovirus (HRV), and Human cytomegalovirus (HCMV). Voltage-gated calcium channels (VGCCs), Two-pore channels (TPCs), Stromal interaction molecule 1 (STIM1), Transient receptor potential vanilloid 4 (TRPV4), N-methyl-D-aspartate receptor (NMDAR), IP3 receptor (IP3R), Secretory pathway Ca2+-ATPase (SPCA1). DEAD-box RNA helicase (DDX3X), Endoplasmic reticulum (ER).

The effect on VGCCs is not only restricted to virus entry. A high-throughput screening of an FDA-approved drug library for inhibitors of Japanese encephalitis virus (JEV) identified five hit drugs, three of which are VGCCs blockers (manidipine, cilnidipine, and benidipine hydrochloride) [14]. Recombinant viral particles (RVPs) with the luciferase-reporting replicon were used to quantify the efficiency of JEV replication, which confirmed that these drugs inhibited JEV infection at the stage of replication. These drugs were subsequently validated for their antiviral activities against other flavivirus, such as Zika virus (ZIKV), dengue virus (DENV), and West Nile virus (WNV). Similarly, another research group screened the FDA-approved drug library and found that nifedipine and benidipine hydrochloride inhibited severe fever with thrombocytopenia syndrome virus (SFTSV) replication in vitro [12]. Moreover, the retrospective clinical investigation on SFTS patients showed that nifedipine can significantly inhibit SFTSV infection. These studies indicate that the VGCCs blockers are excellent candidates for broad-spectrum anti-virus treatment. Most of these FDA-approved VGCC blockers are clinically used to treat cardiovascular diseases. Similar to a double-edged sword, the cardiovascular effect of VGCC blockers may limit their antiviral application. Therefore, repurposing these drugs requires more analysis before clinical trials.

Further examples of virus-regulating VGCCs to service their replication can be found in HIV-1 and herpes simplex virus (HSV)-1. Two HIV-1 proteins, the membrane glycoprotein gp120 [15,16] and the transcriptical transactivator Tat [17,18], have been identified to induce the elevation of host intracellular Ca2+ in various cell types, including neuronal, immune, and epithelial cells, via targeting the activity of

VGCCs. The detailed information about the dysregulation of the L-type calcium channel by Tat can be found in the review [16,17].

Taking together, the infections of viruses could increase the host's intracellular calcium to facilitate viral entry and replication by manipulating host VGCCs. In addition, herpes simplex virus (HSV)-1 could downregulate the VGCCs on infected neuronal cells to escape detection by host cells. T-type Ca2+ channels were reported as the targets of HSV-1 in sensory-like ND7-23 cells [19,20]. HSV-1 infection of differentiated ND7-23 cells causes a significant loss of T-type Ca2+ channels from the membrane, which depends on viral replication and protein synthesis. This downregulation of T-type Ca2+ channel expression may alter the ability of sensory neurons to transmit pain information. Thus, the lower expression of VGCCs may diminish the detection for viral infection by the host, which benefits virus survival and further infection.

Ebola virus (EBOV) used to be considered a VGCC blocker-sensitive virus, and several research groups independently reported that compounds blocking L-type channels (such as verapamil, nimodipine, and diltiazem) inhibited EBOV infection in vivo [21,22,23]. However, gabapentin, representing a fifth distinct class of L-type channel inhibitor, had no effect even at high concentrations. It has been shown that verapamil, nimodipine, and diltiazem also inhibit the activity of two-pore channels (TPCs) [24]. TPCs are intracellular voltage-gated and receptor-operated calcium permeable channels, playing an integral role in membrane trafficking pathways [25,26]. Mouse embryonic fibroblasts (MEFs) lacking TPC1 or TPC2 expression (Tpcn1-/-, Tpcn2-/-) resisted EBOV infection [27]. It turns out that the target of EBOV may not be classical L-type calcium channels but rather endosomal calcium channels termed TPCs (Figure 2A). The calcium channels inhibitors prevented virus–endosome membrane fusion and virus capsid releasing into the cell cytoplasm, which is a late entry step. EBOV acts on TPCs, which control the movement of endosomes containing virus particles, and thereby facilitate its intracellular trafficking [27,28]. The channel inhibitor, tetrandrine, significantly enhanced the survival of mice challenged with mouse-adapted EBOV without any detectable side effects. This indicates that tetrandrine is highly effective against EBOV disease in mice.

4. STORE-OPERATED CALCIUM (SOC) CHANNEL IN VIRAL ASSEMBLY AND EGRESS

SOC channels are the major Ca2+ entry pathways in non-excitable cells. The protein Orai1 on the plasma membrane and STIM1 (stromal interaction molecule) on ER are the molecular identities that are responsible for SOC channels activation. The depletion of ER Ca2+ stores promotes STIM1 proteins aggregation and interaction with Orai1 to open the channel, mediating Ca2+ entry [29,30].

Most enveloped viruses are released extracellularly via exocytosis, also called budding, as an analogy of buds in plants [31,32]. The budding process of the enveloped viruses is triggered by a peptide motif (termed late (L) domains), which was discovered in the Gag polyproteins of retroviruses and M (matrix) proteins of rhabdoviruses [33]. These L domains interact with cellular proteins to promote the formation of virus vesicles that bud away from the cytoplasm [32,34]. Research works established that this essential final step of related viruses depends on the host Ca2+ signal mediated by SOC channels (STIM1/Orai1).

The research work done on four distinct hemorrhagic fever viruses (Ebolavirus (EBOV), Marburgvirus (MARV), Lassa Virus (LASV), and Junín Virus (JUNV)) demonstrated that EBOV, MARV VP40, and JUNV Z proteins trigger host cell Ca2+ signals by activating the ER Ca2+ sensing protein STIM1 and the plasma membrane ORAI1 channel. The STIM1/Orai1-mediated Ca2+ signal is critical for EBOV and related viruses budding from host cells [35]. The suppression of STIM1 expression and genetic inactivation or the pharmacological blockade of ORAI1 inhibits infections of EBOV, MARV, and JUNV in cultured cells (Figure 2B). Obviously, the matrix proteins or live virus activates STIM1 and the ORAI channel. Similarly to hemorrhagic fever viruses, the HIV-1 matrix protein Gag, directing HIV-1 budding, and mediating VLP (Virus-like Particle) formation also exhibit dependence on Ca2+ regulation [36,37].

Therefore, further study is needed to validate the role for STIM1 and the ORAI channel in HIV- 1 budding.

Infections of other enveloped RNA viruses that buds in similar mechanisms may also be inhibited by STIM1 and ORAI1 inhibitors. Indeed, SOC channel antagonists significantly reduced DENV yield [38]. When the human hepatic HepG2 and Huh-7 cells are infected by DENV, STIM1 and ORAI1 were shown to be co-localized in infected cells, indicating activation of the SOC channel [38,39]. Therefore, DENV infection alters cell Ca2+ homeostasis possibly via promoting the interaction between STIM1 and ORAI1.

It is possible that the viral proteins trigger the Ca2+ depletion in ER or prevent ER refilled with Ca2+ to maintain resting ER Ca2+ levels [40,41,42]. Alternately, these viral proteins may directly modify STIM1 and uncouple the activation of STIM1 from Ca2+ levels [35,43]. Thus, STIM1/ORAI might represent a conserved target to regulating the budding of enveloped RNA viruses and possibly DNA viruses that rely on similar host cellular proteins for budding. The example is that the hepatitis B virus X (HBx) protein upregulates the activity of the STIM1–ORAI1 channel complex [44]. The mechanisms by which these viruses activate STIM1 and ORAI1 represent novel therapeutic targets for controlling budding.

Calcium influx through the SOC channel also contributes to the elevated cytosolic calcium concentration induced by rotavirus (RV) infection. It has been well established that the dramatically increased cellular Ca2+ is the hallmark of rotavirus (RV) infection [45,46]. Rotavirus nonstructural protein 4 (NSP4) is an ER-localized viroporin that functionally depletes ER calcium. Thus, in rotavirus-infected cells, STIM1 is constitutively active and colocalizes with the ORAI1 channel. The knockdown of STIM1 or the pharmacological inhibition of SOC channels significantly reduced rotavirus yield, indicating that the SOC channel plays a critical role in the RV lifecycle [47,48].

5. HOST TRANSIENT RECEPTOR POTENTIAL (TRP) CHANNELS AND RECEPTOR-OPERATED CALCIUM (ROC) CHANNELS

5.1. TRP Channels

The TRP channel is a non-selective cation channel predominately permeable for Ca2+ [49,50]. It is divided into six sub-families according to their amino acid sequence: TRP canonical or classical (TRPC), TRP vanilloid (TRPV), TRP melastatin (TRPM), TRP ankyrin (TRPA), TRP polycystin (TRPP), and TRP mucolipin (TRPML) [50]. They are ubiquitously expressed in different tissues and cell types and are a key player in the regulation of intracellular calcium. Nevertheless, reports about virus control TRP channels are not numerous.

TRPV4 mediates intracellular Ca2+ signals in response to several stimuli, including hypotonic cell swelling, mechanical forces, moderating heat, and UVB radiation [51]. When cells are exposed to the Zika virus or the purified viral envelope protein, TRPV4 mediates Ca2+ influx and drives the nuclear translocation of DEAD-box RNA helicase (DDX3X) [52]. DDX3X is an ATP-dependent RNA helicase from the DEAD-box helicase family and is involved in multiple stages of the RNA metabolism, from transcription to translation [53,54]. Moreover, diverse RNA viruses hijack DDX3X to multiply efficiently. Targeting TRPV4 reduces the infectivity of dengue, hepatitis C, and Zika viruses [52]. Overall, the research work demonstrated the role of TRPV4 in the regulation of DDX3X-dependent control of the RNA metabolism and viral infectivity [52].

The TRPV1, TRPA1, and TRPM8 channels are directly activated by chemical, thermal, and mechanical stimuli. They are potentially associated with respiratory virus-induced airway hypersensitivity [55]. The infection of respiratory viruses, such as respiratory syncytial virus (RSV), measles virus (MV), and human rhinovirus (HRV), was reported to increase the expression of these channels in sensory neurons and human bronchial epithelium cells [56,57]. The increase in TRPA1 and TRPV1 levels can be mediated by soluble factors induced by infection, whereas TRPM8 requires virus replication [56,57]. These reports may explain the possible mechanism by which respiratory viruses induce cough. Alternatively, the up-regulated TRP channels may be utilized by the respiratory viruses to create a favorable calcium environment. Further investigation is required to determine the possible pathway by which this happens.

5.2. N-Methyl-D-Aspartate (NMDA) Receptors

NMDArs and IP3Rs are the two receptor-operated calcium (ROC) channels reported to be involved in virus–host interactions. NMDAr, which is activated by the endogenously synthesized excitant amino acid, glutamate, is widely expressed throughout the mammalian central nervous system and is particularly enriched in the cerebral cortex and hippocampus.

The Zika virus infection is a "neurodegenerative" disease. Costa and colleagues showed that blockage of the NMDAr channel activity with FDA-approved memantine or other antagonists prevents neuronal cell death and microgliosis induced by Zika in vitro and in vivo, without affecting the ability of Zika to replicate in the host [58,59]. It seems that NMDAr mainly contributes to Zika, triggering the neuronal cell death progress. Anyway, the blockade of NMDAr may be a viable treatment for patients at risk for Zika infection-induced neurodegeneration [59]. NMDAr blockade also significantly abrogated neuronal cell death and inflammatory response triggered by JEV infection [60]. Therefore, NMDAr are probably common attack targets of flavivirus, inducing host neuron cell death.

5.3. IP3 Receptors

The engagement of receptors or agonist binding to the cell surface receptors activate phospholipase C (PLC), which hydrolyses phosphatidylinositol-4,5-biphosphate (PIP2) to produce IP3. The activation of IP3R triggers Ca2+ release from stores and further increases IP3R's sensitivity to Ca2+. Targeting IP3Rs is more universal among different virus types. In turn, this process generates a more profound effect on host cell physiology, including metabolic stress, neurotoxicity, and enterotoxicity. Modifying IP3 and directly interfering with IP3R are the main ways in which viruses affect calcium release through IP3R. For example, human cytomegalovirus (HCMV) UL37x1 protein interacts with the host P2Y2 purinergic receptor to increase intracellular Ca2+ levels via the PLC–IP3 pathway, and this activity is important to viral replication [61]. HIV-1 Gp120 and Tat upregulate intracellular IP3 [37,62], while HIV Nef [63,64] and p12(I) human T-cell lymphotropic virus type 1 (HTLV-1) [65] activate IP3R directly as an agonist. Glycoprotein E of HSV redistributes the density of IP3Rs within infected cells [66]. Some virus proteins activate IP3R through both ways. Enterotoxin NSP4 of RV and HCMV UL37x1-encoded protein increases the basal permeability of the ER as IP3 diffuses and binds to IP3Rs to stimulate Ca2+ release. These virus proteins deplete ER Ca2+ storage during early stages of viral infection to increase the replication ability of viruses. On the other hand, the depletion of ER calcium storage triggers the calcium influx through SOCs (STIM1/ORAI1).

6. CALCIUM PUMPS

The calcium pumps (Ca2+ ATPase) and the Na+–Ca2+ exchanger are the main regulators of intracellular Ca2+ concentrations [67]. Three Ca2+ ATPase types (pumps) have been described in animal cells located in the membranes of the sarcoplasmic reticulum (SERCA pump), in those of the Golgi network (Secretory Pathway Ca2+ ATPase, SPCA pump), and in the plasma membrane (PMCA pump) [68]. At the end of a stimulus, these pumps participate in returning the cell to its resting state by the decrease in the cytosolic Ca2+ concentration. For example, SERCA transports Ca2+ from the cytoplasm into the lumen of the ER. PMCA pumps function to remove Ca2+ from the cell. They serve as the main regulators of intracellular and extracellular Ca2+ concentrations. All three calcium pumps are reported to be involved in different viral lifecycles.

Maturation is the last assembly step of a virus particle. In this step, the virus acquires the infectivity, which is essential for its lifecycle. The results obtained from a genome-wide knockout screen indicated that SPCA1 in the Golgi network is critical for human respiratory syncytial virus (RSV) infection [69,70,71]. Further study demonstrated that similar to RSV, other viruses from the Paramyxoviridae, Flaviviridae, and Togaviridae families also failed to spread in SPCA1-deficient cells. Studies on this mechanism revealed that Ca2+ pumped into Golgi by SPCA1 is the trigger to produce normal functional viral glycoproteins that are essential for virus spread. Therefore, SPCA1 may be a promising target for therapeutic intervention against a diverse set of viruses (Figure 2B). However, in this study, herpes simplex 1 (HSV-1), vesicular stomatitis (VSV), bunyamwera (BUNV), and LCMV were unaffected by the loss of SPCA1. That is probably because the maturation of these viral glycoproteins requires different triggers.

Cui et al. [72] screened the natural product library and found that cyclopiazonic acid, an inhibitor of SERCA, was shown to have activity in the low micromolar range against RSV strains at the step of virus genome replication. Further study found that another SERCA inhibitor BHQ had a similar effect [72]. It is probable that SERCA inhibitors prevent cytosolic calcium returning to the ER from the cytosol via SERCA, resulting in an increase in intracellular calcium concentration. A continuous higher concentration of intracellular calcium may impair viral genome replication and/or transcription, thereby reducing virus yield.

7. CONCISE SUMMARY

Viral infections include the virus hijacking the host's intracellular calcium signaling system and homeostasis in order to support its own lifecycle. Viral infection can inhibit T-cell responsiveness, antiapoptotic activity, and other potential roles by disrupting the host's Ca2+ system. This paper provides a summary of the calcium permeable channels and calcium pumps present on host cell membranes, which have the potential to be targeted therapeutically for viral infections (Table 1). Nevertheless, the calcium channel family is constantly updated as new members are discovered, and the impact of viruses on the host's intracellular calcium environment is typically not accomplished through a single pathway; it is really complex. Pharmacological intervention can target both upstream messengers and downstream Ca2+ binding proteins. Hence, unraveling the whole understanding of the calcium-virus interaction on a molecular level will pose a significant challenge in the coming years. The prevention and treatment of virus infections are frequently hindered by the elevated mutation rates of viral proteins. Nevertheless, the calcium apparatus proteins that are necessary for viral replication and transmission are extremely conserved, meaning they do not change, and so they could be a vulnerable point for virus infection. As previously mentioned, viruses can target and affect calcium permeable channels and calcium pumps, which play a crucial role in intracellular calcium signaling. One prospective treatment option with therapeutic potential is to target the host cellular proteins that are necessary for virus infection. This approach may help prevent or delay the development of resistance. The aforementioned calcium-permeable proteins serve as appropriate target proteins for the inhibition of viral infection. VGCCs and NMDA receptors are extensively researched targets due to the presence of particular channel antagonists. Furthermore, a portion of these chemicals are medications that have been approved by the FDA and are used in clinical settings. Verapamil, amlodipine, verapamil, and diltiazem, commonly prescribed for cardiovascular illness, have demonstrated inhibitory effects against many viral infections. Memantine, an Alzheimer's medicine licensed by the FDA, mitigates neurological problems linked to Zika virus infection. Clearly, the success of this approach relies on the discovery of specific and powerful chemical compounds that may modify calcium channels. Currently, with the exception of some blockers for VGCCs and the NMDA receptor, there are very few additional channel modulators that can be utilized in clinical settings. However, significant advancements have been made in recent years towards developing high-quality and readily accessible channel modulators. Moreover, it is necessary to produce modulators that possess specific features that render them appropriate for usage within a living organism. This review highlights the encouraging advancements in utilizing a host cell calcium channels modulator as a potential antiviral medication. We consider modulators for host calcium channels to be highly valuable for the development of antiviral drugs.

Ca2+ is crucial for virus entry, viral gene replication, virion maturation, and release. Viruses manipulate host cells' signal transduction processes by changing the balance of Ca2+ in the cells. Calcium-permeable channels and pumps, including as voltage-gated calcium channels, store-operated channels, receptor-operated channels, transient receptor potential ion channels, and Ca2+-ATPase, are responsible for transporting calcium across the plasma membrane or subcellular organelles. This process helps regulate the levels of free calcium inside the cell. Thus, these Ca2+ channels or pumps play a crucial role in viral pathogenicity and the interaction between the virus and its host. According to reports, viruses take control of host calcium channels or pumps, disrupting the balanced levels of Ca2+ in the cells. This disruption enhances the lifecycles of viruses while causing harm to the host cells. New evidence suggests that pharmacologically targeting the calcium channel or calcium release from the endoplasmic reticulum (ER) can disrupt virus lifecycles. Preventing virus-induced disruptions in the balance of intracellular Ca2+ is an effective approach in creating powerful antiviral medications. This review focuses on the identification of cellular calcium channels and pumps as targets for viral attacks.

REFERENCES

1. Martinez de Victoria, E. Calcium, essential for health. Nutr. Hosp. 2016, 33 (Suppl. S4), 341.
2. Berridge, M.J.; Bootman, M.D.; Lipp, P. Calcium—A life and death signal. Nature 1998, 395, 645–648.
3. Olivier, M. Modulation of host cell intracellular Ca2+. Parasitol. Today 1996, 12, 145–150.
4. Clark, K.B.; Eisenstein, E.M. Targeting host store-operated Ca2+ release to attenuate viral infections. Curr. Top. Med. Chem. 2013, 13, 1916–1932.
5. Gonzales-van Horn, S.R.; Sarnow, P. Making the Mark: The Role of Adenosine Modifications in the Life Cycle of RNA Viruses. Cell Host Microbe 2017, 21, 661–669.

6. Atlas, D. Voltage-gated calcium channels function as Ca2+-activated signaling receptors. Trends Biochem. Sci. 2014, 39, 45–52.

7. Nugent, K.M.; Shanley, J.D. Verapamil inhibits influenza A virus replication. Arch. Virol. 1984, 81, 163–170.

8. Fujioka, Y.; Tsuda, M.; Nanbo, A.; Hattori, T.; Sasaki, J.; Sasaki, T.; Miyazaki, T.; Ohba, Y. A Ca2+-dependent signalling circuit regulates influenza a virus internalization and infection. Nat. Commun. 2013, 4, 2763.

9. Fujioka, Y.; Nishide, S.; Ose, T.; Suzuki, T.; Kato, I.; Fukuhara, H.; Fujioka, M.; Horiuchi, K.; Satoh, A.O.; Nepal, P. A Sialylated Voltage-Dependent Ca2+ Channel Binds Hemagglutinin and Mediates Influenza A Virus Entry into Mammalian Cells. Cell Host Microbe 2018, 23, 809–818.

10. Chandrasekaran, A.; Srinivasan, A.; Raman, R.; Viswanathan, K.; Raguram, S.; Tumpey, T.M.; Sasisekharan, V.; Sasisekharan, R. Glycan topology determines human adaptation of avian H5N1 virus hemagglutinin. Nat. Biotechnol. 2008, 26, 107–113.

11. Lazniewska, J.; Weiss, N. Glycosylation of voltage-gated calcium channels in health and disease. Biochim. Biophys. Acta Biomembr. 2017, 1859, 662–668.

12. Li, H.; Zhang, L.K.; Li, S.F.; Zhang, S.F.; Wan, W.W.; Zhang, Y.L.; Xin, Q.L.; Dai, K.; Hu, Y.Y.; Wang, Z.B. Calcium channel blockers reduce severe fever with thrombocytopenia syndrome virus (SFTSV) related fatality. Cell Res. 2019, 29, 739–753.

13. Lavanya, M.; Cuevas, C.D.; Thomas, M.; Cherry, S.; Ross, S.R. siRNA screen for genes that affect Junin virus entry uncovers voltage-gated calcium channels as a therapeutic target. Sci. Transl. Med. 2013, 5, 204ra131.

14. Wang, S.; Liu, Y.; Guo, J.; Wang, P.; Zhang, L.; Xiao, G.; Wang, W. Screening of FDA-Approved Drugs for Inhibitors of Japanese Encephalitis Virus Infection. J. Virol. 2017, 91, e01055-17.

15. Dreyer, E.B.; Kaiser, P.K.; Offermann, J.T.; Lipton, S.A. HIV-1 coat protein neurotoxicity prevented by calcium channel antagonists. Science 1990, 248, 364–367.

16. Haughey, N.J.; Mattson, M.P. Calcium dysregulation and neuronal apoptosis by the HIV-1 proteins Tat and gp120. J. Acquir. Immune. Defic. Syndr. 2002, 31 (Suppl. S2), S55–S61.

17. Hu, X.T. HIV-1 Tat-Mediated Calcium Dysregulation and Neuronal Dysfunction in Vulnerable Brain Regions. Curr. Drug Targets 2016, 17, 4–14.

18. Kruman, I.I.; Nath, A.; Mattson, M.P. HIV-1 protein Tat induces apoptosis of hippocampal neurons by a mechanism involving caspase activation, calcium overload, and oxidative stress. Exp. Neurol. 1998, 154, 276–288.

19. Zhang, Q.; Hsia, S.C.; Martin-Caraballo, M. Regulation of T-type Ca2+ channel expression by herpes simplex virus-1 infection in sensory-like ND7 cells. J. Neurovirol. 2017, 23, 657–670.

20. Zhang, Q.; Hsia, S.C.; Martin-Caraballo, M. Regulation of T-type Ca2+ channel expression by interleukin-6 in sensory-like ND7/23 cells post-herpes simplex virus (HSV-1) infection. J. Neurochem. 2019, 151, 238–254.

21. Kolokoltsov, A.A.; Saeed, M.F.; Freiberg, A.N.; Holbrook, M.R.; Davey, R.A. Identification of novel cellular targets for therapeutic intervention against Ebola virus infection by siRNA screening. Drug Dev. Res. 2009, 70, 255–265.

22. Gehring, G.; Rohrmann, K.; Atenchong, N.; Mittler, E.; Becker, S.; Dahlmann, F.; Pohlmann, S.; Vondran, F.W.; David, S.; Manns, M.P. The clinically approved drugs amiodarone, dronedarone and verapamil inhibit filovirus cell entry. J. Antimicrob. Chemother. 2014, 69, 2123–2131.

23. Madrid, P.B.; Chopra, S.; Manger, I.D.; Gilfillan, L.; Keepers, T.R.; Shurtleff, A.C.; Green, C.E.; Iyer, L.V.; Dilks, H.H.; Davey, R.A. A systematic screen of FDA-approved drugs for inhibitors of biological threat agents. PLoS ONE 2013, 8, e60579.

24. Calcraft, P.J.; Ruas, M.; Pan, Z.; Cheng, X.; Arredouani, A.; Hao, X.; Tang, J.; Rietdorf, K.; Teboul, L.; Chuang, K.T. NAADP mobilizes calcium from acidic organelles through two-pore channels. Nature 2009, 459, 596–600.

25. Zhu, M.X.; Ma, J.; Parrington, J.; Calcraft, P.J.; Galione, A.; Evans, A.M. Calcium signaling via two-pore channels: Local or global, that is the question. Am. J. Physiol. Cell Physiol. 2010, 298, C430–C441.

26. Morgan, A.J. Ca2+ dialogue between acidic vesicles and ER. Biochem. Soc. Trans. 2016, 44, 546–553.

27. Sakurai, Y.; Kolokoltsov, A.A.; Chen, C.C.; Tidwell, M.W.; Bauta, W.E.; Klugbauer, N.; Grimm, C.; Wahl-Schott, C.; Biel, M.; Davey, R.A. Two-pore channels control Ebola virus host cell entry and are drug targets for disease treatment. Science 2015, 347, 995–998.

28. Kintzer, A.F.; Stroud, R.M. Structure, inhibition and regulation of two-pore channel TPC1 from Arabidopsis thaliana. Nature 2016, 531, 258–262.

29. Hogan, P.G.; Lewis, R.S.; Rao, A. Molecular basis of calcium signaling in lymphocytes: STIM and ORAI. Annu. Rev. Immunol. 2010, 28, 491–533.

30. Nwokonko, R.M.; Cai, X.; Loktionova, N.A.; Wang, Y.; Zhou, Y.; Gill, D.L. The STIM-Orai Pathway: Conformational Coupling between STIM and Orai in the Activation of Store-Operated Ca2+ Entry. Adv. Exp. Med. Biol. 2017, 993, 83–98.

31. Sepulveda, C.S.; Garcia, C.C.; Damonte, E.B. Determining the Virus Life-Cycle Stage Blocked by an Antiviral. Methods Mol. Biol. 2018, 1604, 371–392.

32. Hartlieb, B.; Weissenhorn, W. Filovirus assembly and budding. Virology 2006, 344, 64–70.

33. Chen, B.J.; Lamb, R.A. Mechanisms for enveloped virus budding: Can some viruses do without an ESCRT? Virology 2008, 372, 221–232.

34. Licata, J.M.; Simpson-Holley, M.; Wright, N.T.; Han, Z.; Paragas, J.; Harty, R.N. Overlapping motifs (PTAP and PPEY) within the Ebola virus VP40 protein function independently as late budding domains: Involvement of host proteins TSG101 and VPS-4. J. Virol. 2003, 77, 1812–1819.

35. Han, Z.; Madara, J.J.; Herbert, A.; Prugar, L.I.; Ruthel, G.; Lu, J.; Liu, Y.; Liu, W.; Liu, X.; Wrobel, J.E. Calcium Regulation of Hemorrhagic Fever Virus Budding: Mechanistic Implications for Host-Oriented Therapeutic Intervention. PLoS Pathog. 2015, 11, e1005220.

36. Ehrlich, L.S.; Carter, C.A. HIV Assembly and Budding: Ca2+ Signaling and Non-ESCRT Proteins Set the Stage. Mol. Biol. Int. 2012, 2012, 851670.

37. Ehrlich, L.S.; Medina, G.N.; Carter, C.A. ESCRT machinery potentiates HIV-1 utilization of the PI (4,5) P (2)-PLC-IP3R-Ca2+ signaling cascade. J. Mol. Biol. 2011, 413, 347–358.

38. Dionicio, C.L.; Pena, F.; Constantino-Jonapa, L.A.; Vazquez, C.; Yocupicio-Monroy, M.; Rosales, R.; Zambrano, J.L.; Ruiz, M.C.; Del Angel, R.M.; Ludert, J.E. Dengue virus induced changes in Ca2+ homeostasis in human hepatic cells that favor the viral replicative cycle. Virus Res. 2018, 245, 17–28.

39. Cheshenko, N.; Liu, W.; Satlin, L.M.; Herold, B.C. Multiple receptor interactions trigger release of membrane and intracellular calcium stores critical for herpes simplex virus entry. Mol. Biol. Cell 2007, 18, 3119–3130.

40. Robinson, L.C.; Marchant, J.S. Enhanced Ca2+ leak from ER Ca2+ stores induced by hepatitis C NS5A protein. Biochem. Biophys. Res. Commun. 2008, 368, 593–599.

41. Zhadina, M.; Bieniasz, P.D. Functional interchangeability of late domains, late domain cofactors and ubiquitin in viral budding. PLoS Pathog. 2010, 6, e1001153.

42. Ruiz, M.C.; Aristimuno, O.C.; Diaz, Y.; Pena, F.; Chemello, M.E.; Rojas, H.; Ludert, J.E.; Michelangeli, F. Intracellular disassembly

of infectious rotavirus particles by depletion of Ca2+ sequestered in the endoplasmic reticulum at the end of virus cycle. Virus Res. 2007, 130, 140–150.

43. Freedman, B.D.; Harty, R.N. Calcium and filoviruses: A budding relationship. Future Microbiol. 2016, 11, 713–715.

44. Yao, J.H.; Liu, Z.J.; Yi, J.H.; Wang, J.; Liu, Y.N. Hepatitis B Virus X Protein Upregulates Intracellular Calcium Signaling by Binding C-terminal of Orail Protein. Curr. Med. Sci. 2018, 38, 26–34.

45. Michelangeli, F.; Ruiz, M.C.; del Castillo, J.R.; Ludert, J.E.; Liprandi, F. Effect of rotavirus infection on intracellular calcium homeostasis in cultured cells. Virology 1991, 181, 520–527.

46. Pham, T.; Perry, J.L.; Dosey, T.L.; Delcour, A.H.; Hyser, J.M. The Rotavirus NSP4 Viroporin Domain is a Calcium-conducting Ion Channel. Sci. Rep. 2017, 7, 43487.

47. Hyser, J.M.; Utama, B.; Crawford, S.E.; Broughman, J.R.; Estes, M.K. Activation of the endoplasmic reticulum calcium sensor STIM1 and store-operated calcium entry by rotavirus requires NSP4 viroporin activity. J. Virol. 2013, 87, 13579–13588.

48. Chang-Graham, A.L.; Perry, J.L.; Strtak, A.C.; Ramachandran, N.K.; Criglar, J.M.; Philip, A.A.; Patton, J.T.; Estes, M.K.; Hyser, J.M. Rotavirus Calcium Dysregulation Manifests as Dynamic Calcium Signaling in the Cytoplasm and Endoplasmic Reticulum. Sci. Rep. 2019, 9, 10822.

49. Li, J.; Zhang, X.; Song, X.; Liu, R.; Zhang, J.; Li, Z. The structure of TRPC ion channels. Cell Calcium. 2019, 80, 25–28.

50. Hicks, G.A. TRP channels as therapeutic targets: Hot property, or time to cool down? Neurogastroenterol. Motil. 2006, 18, 590–594.

51. Garcia-Elias, A.; Mrkonjic, S.; Jung, C.; Pardo-Pastor, C.; Vicente, R.; Valverde, M.A. The TRPV4 channel. Handb. Exp. Pharm. 2014, 222, 293–319.

52. Donate-Macian, P.; Jungfleisch, J.; Perez-Vilaro, G.; Rubio-Moscardo, F.; Peralvarez-Marin, A.; Diez, J.; Valverde, M.A. The

TRPV4 channel links calcium influx to DDX3X activity and viral infectivity. Nat. Commun. 2018, 9, 2307.

53. Ariumi, Y. Multiple functions of DDX3 RNA helicase in gene regulation, tumorigenesis, and viral infection. Front. Genet. 2014, 5, 423.

54. Valiente-Echeverria, F.; Hermoso, M.A.; Soto-Rifo, R. RNA helicase DDX3: At the crossroad of viral replication and antiviral immunity. Rev. Med. Virol. 2015, 25, 286–299.

55. Fujita, T.; Liu, Y.; Higashitsuji, H.; Itoh, K.; Shibasaki, K.; Fujita, J.; Nishiyama, H. Involvement of TRPV3 and TRPM8 ion channel proteins in induction of mammalian cold-inducible proteins. Biochem. Biophys. Res. Commun. 2018, 495, 935–940.

56. Omar, S.; Clarke, R.; Abdullah, H.; Brady, C.; Corry, J.; Winter, H.; Touzelet, O.; Power, U.F.; Lundy, F.; McGarvey, L.P. Respiratory virus infection up-regulates TRPV1, TRPA1 and ASICS3 receptors on airway cells. PLoS ONE 2017, 12, e0171681.

57. Abdullah, H.; Heaney, L.G.; Cosby, S.L.; McGarvey, L.P. Rhinovirus upregulates transient receptor potential channels in a human neuronal cell line: Implications for respiratory virus-induced cough reflex sensitivity. Thorax 2014, 69, 46–54.

58. Costa, V.V.; Del Sarto, J.L.; Rocha, R.F.; Silva, F.R.; Doria, J.G.; Olmo, I.G.; Marques, R.E.; Queiroz-Junior, C.M.; Foureaux, G.; Araujo, J.M.S. N-Methyl-d-Aspartate (NMDA) Receptor Blockade Prevents Neuronal Death Induced by Zika Virus Infection. MBio 2017, 8, e00543-17.

59. Sirohi, D.; Kuhn, R.J. Can an FDA-Approved Alzheimer's Drug Be Repurposed for Alleviating Neuronal Symptoms of Zika Virus? MBio 2017, 8, e00916-17.

60. Chen, Z.; Wang, X.; Ashraf, U.; Zheng, B.; Ye, J.; Zhou, D.; Zhang, H.; Song, Y.; Chen, H.; Zhao, S. Activation of neuronal N-methyl-D-aspartate receptor plays a pivotal role in Japanese encephalitis virus-induced neuronal cell damage. J. Neuroinflammation 2018, 15, 238.

61. Chen, S.; Shenk, T.; Nogalski, M.T. P2Y2 purinergic receptor modulates virus yield, calcium homeostasis, and cell motility in human cytomegalovirus-infected cells. Proc. Natl. Acad. Sci. USA 2019, 116, 18971–18982.
62. Ehrlich, L.S.; Medina, G.N.; Photiadis, S.; Whittredge, P.B.; Watanabe, S.; Taraska, J.W.; Carter, C.A. Tsg101 regulates PI (4,5) P2/Ca2+ signaling for HIV-1 Gag assembly. Front. Microbiol. 2014, 5, 234.
63. Ehrlich, L.S.; Medina, G.N.; Khan, M.B.; Powell, M.D.; Mikoshiba, K.; Carter, C.A. Activation of the inositol (1,4,5)-triphosphate calcium gate receptor is required for HIV-1 Gag release. J. Virol. 2010, 84, 6438–6451.
64. Manninen, A.; Saksela, K. HIV-1 Nef interacts with inositol trisphosphate receptor to activate calcium signaling in T cells. J. Exp. Med. 2002, 195, 1023–1032.
65. Ding, W.; Albrecht, B.; Kelley, R.E.; Muthusamy, N.; Kim, S.J.; Altschuld, R.A.; Lairmore, M.D. Human T-cell lymphotropic virus type 1 p12 (I) expression increases cytoplasmic calcium to enhance the activation of nuclear factor of activated T cells. J. Virol. 2002, 76, 10374–10382.
66. Cheshenko, N.; Del Rosario, B.; Woda, C.; Marcellino, D.; Satlin, L.M.; Herold, B.C. Herpes simplex virus triggers activation of calcium-signaling pathways. J. Cell Biol. 2003, 163, 283–293.
67. Strehler, E.E.; Zacharias, D.A. Role of alternative splicing in generating isoform diversity among plasma membrane calcium pumps. Physiol. Rev. 2001, 81, 21–50.
68. Brini, M.; Cali, T.; Ottolini, D.; Carafoli, E. Calcium pumps: Why so many? Compr. Physiol. 2012, 2, 1045–1060.
69. Griffiths, C.; Drews, S.J.; Marchant, D.J. Respiratory Syncytial Virus: Infection, Detection, and New Options for Prevention and Treatment. Clin. Microbiol. Rev. 2017, 30, 277–319.

70. Hoffmann, H.H.; Schneider, W.M.; Blomen, V.A.; Scull, M.A.; Hovnanian, A.; Brummelkamp, T.R.; Rice, C.M. Diverse Viruses Require the Calcium Transporter SPCA1 for Maturation and Spread. Cell Host Microbe 2017, 22, 460–470.

71. Cervantes-Ortiz, S.L.; Zamorano Cuervo, N.; Grandvaux, N. Respiratory Syncytial Virus and Cellular Stress Responses: Impact on Replication and Physiopathology. Viruses 2016, 8, 124.

72. Cui, R.; Wang, Y.; Wang, L.; Li, G.; Lan, K.; Altmeyer, R.; Zou, G. Cyclopiazonic acid, an inhibitor of calcium-dependent ATPases with antiviral activity against human respiratory syncytial virus. Antivir. Res. 2016, 132, 38–45.

17 ON THE HOST SIDE OF THE HEPATITIS E VIRUS (HEV) LIFE CYCLE

1. INTRODUCTION: AN OVERVIEW

The Hepatitis E virus (HEV) was determined to be the source of waterborne hepatitis epidemics in the early 1980s [1,2]. In 1990, the genetic material of the virus was replicated and its sequence was determined, enabling the creation of serological assays for investigating its spread and patterns of occurrence [3,4]. The virus has been categorized within the Hepeviridae family, with the majority of strains that are harmful to humans belonging to the Orthohepevirus A species [5]. This species can be categorized into 8 genotypes (gt). Genotypes 1 and 2 are exclusive to humans and are transmitted through the fecal-oral pathway, primarily through drinking water that has been contaminated. Genotypes 3 and 4 are responsible for zoonotic illnesses, which are primarily transmitted by the ingestion of inadequately cooked pig, wild boar, or deer meat. Genotypes 5 and 6 are present in wild boar, while genotypes 7 and 8 infect dromedary and Bactrian camels, respectively. Genotype 7 (gt 7) has been detected in an individual with a weakened immune system following the consumption of camel milk and meat [6] (as discussed in [7]).

However, there have been no documented cases of transmission to humans for genotypes 5, 6, and 8. In recent times, it has been discovered that rabbit HEV, which is closely related to gt 3 within the genus Orthohepevirus A, and rat HEV, which belongs to the species Orthohepevirus C, can also infect people [8,9,10,11]. HEV is a compact, non-enveloped virus that has a spherical shape with 20 triangular faces and a diameter ranging from 27 to 34 nanometers [2]. The genome is composed of a single-stranded RNA molecule with a size of 7.2 kilobases. It has a positive-sense orientation and has a m7G cap at its 5' end and a poly-A tail at its 3' end (Figure 1A). The HEV genome contains three open reading frames (ORF). The ORF1 gene is responsible for encoding the viral replicase, whereas the ORF2 gene is responsible for encoding the capsid. Additionally, the ORF3 gene encodes a tiny protein that plays a role in the release of virions through its possible ion channel function [12]. The initial interaction between HEV and host cells takes place through the binding of entry factor(s) that have not been fully defined yet. Following endocytosis, the viral genome is liberated into the cytoplasm where the host translational machinery synthesizes the ORF1 replicase, which facilitates viral RNA replication (Figure 1B). In this stage, a negative-strand RNA intermediate gives rise to two distinct RNA species: a full-length genomic RNA and a subgenomic RNA measuring 2.2 kb [13,14]. The subgenomic RNA is translated to produce the ORF2 and ORF3 proteins. Subsequent stages of the HEV life cycle involve the formation of viral particles and the liberation of freshly generated virions. HEV, a positive-strand RNA virus that affects the liver, is identical to HAV. It exists in two forms: a 'quasi-enveloped' virion (eHEV) surrounded by exosomal membranes in the blood, and a bare particle in bile and feces. This information is reviewed in reference [15] and may be shown in Figure 1B.

Viruses, being obligate intracellular pathogens, have evolved tactics to seize control of and influence host cell circuits to guarantee successful infection. In addition, RNA viruses, particularly those with smaller genomes and coding capability, like HEV, rely heavily on the cellular machinery of the host. Due to recent limitations in studying HEV, our understanding of the host variables involved in different stages of the viral life cycle is still restricted. Experiments conducted in other organisms, such the yeast two hybrid system, have discovered biological components that interact with HEV proteins [16,17]. Nevertheless, the majority of these possibilities still require validation and further investigation through the use of infectious cell culture systems, in vivo models, and liver biopsies obtained from individuals with hepatitis E. This review will specifically examine host variables that have been confirmed to play a role in the viral life cycle during HEV infection.

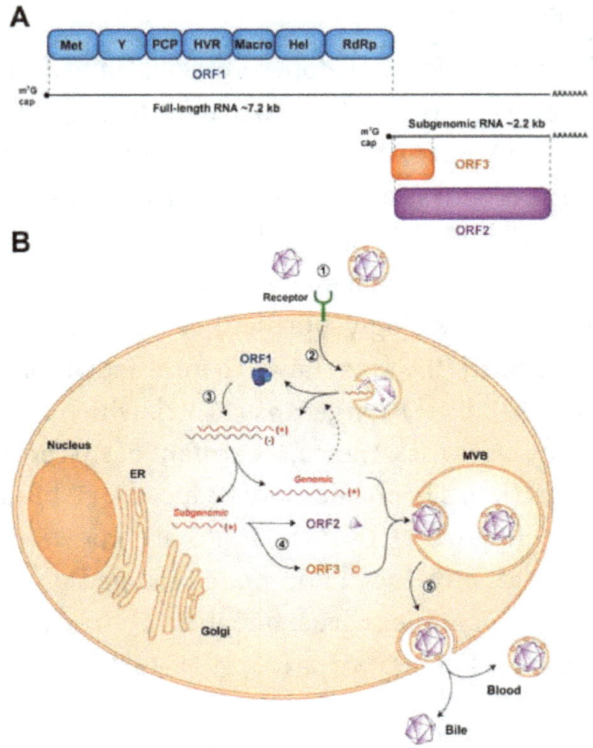

Figure 1. Genome organization and life cycle of hepatitis E virus (HEV). (A) The 7.2 kb positive-strand RNA genome has a 5′ 7-methylguanylate cap (m7G cap) and a 3′ polyadenylated tail (poly-A). It harbors 3 open reading frames (ORFs). ORF1 encodes a replicase of about 190 kDa comprising different functional domains, including a methyltransferase (Met), an RNA helicase (Hel) and an RNA-dependent RNA polymerase (RdRp), as well as less well-characterized domains, such as the Y domain, a putative papain-like cysteine protease (PCP), a hypervariable region (HVR) and the Macro domain. ORF2 and ORF3 encode the viral capsid and a small protein involved in virus secretion respectively, which are translated from a 2.2 kb subgenomic RNA generated during viral replication. (B) The HEV life cycle can be dissected into the following steps: (1) viral entry by as yet unidentified receptor(s), (2) endocytosis and release of the viral positive-strand RNA genome (+) into the cytosol, (3) translation of the ORF1 protein to allow replication of the full-length and generation of the subgenomic RNA through a negative-strand RNA intermediate (-), (4) translation of the subgenomic RNA to produce the ORF2 and ORF3 proteins and (5) genome packaging, virion assembly and release of the virus into the bloodstream and the bile from the basolateral and apical sides, respectively. ER, endoplasmic reticulum; MVB, multivesicular body.

2. HEV ENTRY

As HEV is present in a non-enveloped ("naked") and a quasi-enveloped form (eHEV), the entry pathway of the virus may differ for these two forms. Our knowledge of HEV entry remains scarce but studies using virus-like particles (VLP) as a model system have highlighted possible host factors involved in the initial attachment to the cell and virus internalization, including the 78-kDa glucose-regulated protein (GRP78), ATP synthase subunit β (ATPB5) and asialoglycoprotein receptor (ASGPR) [18,19,20]. Notably, non-enveloped HEV was shown to interact with heparan sulfate proteoglycans (HSPG), likely syndecans [21,22], which are expressed on the surface of many cell types. Hence, treatment of susceptible hepatoma cell lines with heparinase I considerably reduced VLP binding as well as HEV infection [21] (Figure 2). Of note, HSPG are known to mediate cell attachment of several enveloped and nonenveloped viruses, including, among others, herpes simplex virus, hepatitis C virus (HCV), norovirus and human immunodeficiency virus [23,24,25,26,27].

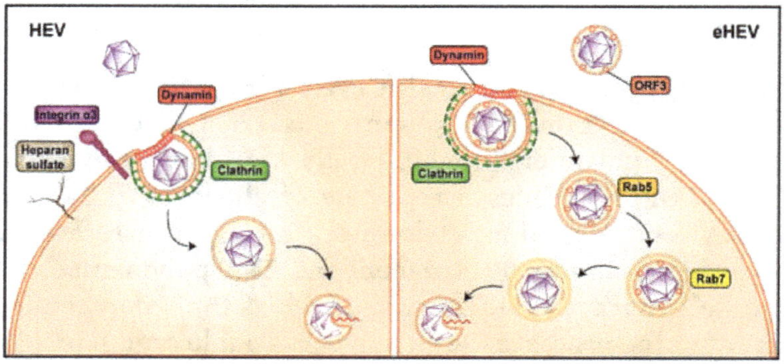

Figure 2. Non-enveloped versus quasi-enveloped hepatitis E virus entry. Non-enveloped hepatitis E virus (HEV) is believed to first bind to heparan sulfate proteoglycans and integrin α3 at the cell surface (left panel). The virus is then internalized via clathrin- and dynamin 2-dependent endocytosis, followed by release of the viral genome into the cytoplasm by an as yet unknown mechanism, possibly involving a conformational change of the capsid. The cofactor(s) and receptor(s) allowing entry of quasi-enveloped HEV (eHEV) are unknown (right panel). Internalization requires clathrin- and dynamin 2-dependent endocytosis as well as trafficking through Rab5- (early) as well as Rab7-positive (late) endosomes and eventually lysosomes to allow release of the viral genome into the cytoplasm, likely by a process similar to that of naked HEV.

A recent microarray analysis comparing gene expression in permissive versus non-permissive cells identified integrin α3 as an entry factor for HEV [28]. Integrins belong to a family of transmembrane proteins localized at the cell surface which can bind to extracellular matrix (ECM) as well as cell surface and intracellular ligands. These interactions trigger a myriad of intracellular signals modulating cell behavior through effects on actin microfilaments, thereby connecting the extracellular space and the cytoskeleton (reviewed in [29]). Integrins were shown to function as receptors for Kaposi's sarcoma-associated herpesvirus [30] and, interestingly, also for HAV [31]. In the case of HEV, overexpression of integrin α3 in non-permissive cells allowed non-enveloped HEV infection only and, conversely, knockout of integrin α3 gene in permissive cells prevented entry of non-enveloped HEV but not of eHEV [28]. Together with the physical interaction between integrin α3 and HEV, the data strongly suggests that this ECM receptor is an entry factor for the non-enveloped viral particle (Figure 2). Integrins are expressed in a broad array of tissues, including in the intestine [26,32], where HEV is known to replicate and which may represent the initial site of infection [33,34]. However, in some organs, such as the liver, integrin α3 expression is rather low [35,36,37], raising the possibility that integrin α3 may act as a cofactor for HEV entry rather than as the key receptor. Further studies are needed to clarify the entry pathway used by non-enveloped HEV.

Internalization of non-enveloped HEV is believed to occur through clathrin- and dynamin 2-dependent endocytosis [22,38,39] (Figure 2). In this context, membrane cholesterol was also shown to be important for HEV entry, as treatment of cells with cholesterol sequestering agents significantly reduced VLP uptake [39]. Following endocytosis, the HEV genome needs to be uncoated to be released into the cytoplasm. It is likely that the capsid protein itself plays a crucial role in this poorly understood process, possibly by undergoing conformational changes induced by interaction with a host protein, as it is seen in other non-enveloped viruses, such as human papilloma virus, murine polyomavirus and poliovirus [40,41,42].

Since the viral capsid is not exposed at the surface of eHEV (see Section 4), the quasi-enveloped particle may use different, as yet unknown entry factor(s). However, as for the naked virus, eHEV is internalized through clathrin- and dynamin 2-dependent endocytosis [38,39]. Unlike non-enveloped HEV, eHEV entry is dependent on the small GTPases, Rab5 and Rab7, both of which are involved in endosomal trafficking. In fact, depletion of one or the other has been shown to considerably reduce virus infection [22] (Figure 2). In addition, eHEV entry depends on endosomal acidification [22], indicating that transition of the endosome to a more acidic cell compartment is necessary (Figure 2). Lysosomal lipid degradation appears to be a required step in eHEV entry. Indeed, depletion of the late endosomal and lysosomal Niemann-Pick disease type C1 protein, involved in cholesterol extraction, significantly reduced eHEV infection [22]. Moreover, treatment with an inhibitor of lysosomal acid lipase, responsible for lipid degradation, resulted in a dose-dependent reduction of eHEV cell entry [22].

These observations suggest that the quasi-envelope of eHEV is removed in endolysosomes. The viral capsid may subsequently interact with an as yet unidentified host factor and undergo the conformational changes required for genome release into the cytoplasm. A similar mechanism was recently proposed for HAV [31]. Since both quasi-enveloped and non-enveloped HEV are internalized in vesicles belonging to the endosomal pathway, it is plausible that they use a common host factor to allow uncoating and release of the genome into the cytoplasm (detailed in [43]).

The ORF3 protein is present within eHEV and interacts with the capsid [44]. However, its role in eHEV entry, uncoating and genome release remains to be explored. Taken together, some HEV entry factors and pathways have been identified, but key elements are still missing. The capacity of HEV to infect a broad array of cell types [45,46,47,48,49,50,51], possibly related to the diverse extrahepatic manifestations of hepatitis E [52,53,54,55], strongly suggests that the entry receptor(s) is (are) not specific to hepatocytes but ubiquitously expressed.

3. VIRAL RNA REPLICATION

Upon release of the viral positive-strand RNA genome into the cytoplasm, the host translational machinery, including ribosomal subunits and elongation factors, seeds on the 5′ untranslated region to start translation of the ORF1 protein. Among the different components, the eukaryotic translation initiation factor 4F (eIF4F) complex has been identified as being important for HEV replication. This complex is known to be involved in the cap-dependent translation and replication of several viruses (reviewed in [56]). In this context, an RNA interference-based loss-of-function study showed that components of the eIF4F complex are required for efficient HEV replication, while known negative regulatory factors of this pathway limit viral RNA synthesis [57]. Along these lines, silvestrol, a natural compound inhibiting part of the eIF4F complex, efficiently inhibits HEV replication in vitro, and in a mouse model in vivo, ORF2 protein production and virus spread [58,59]. Interestingly, silvestrol was reported to also inhibit corona-, picorna-, alpha-, flavi- and filo-viruses and, therefore, displays broad antiviral activity [60,61,62,63].

ORF1 encodes the functional domains required for viral RNA synthesis, including an RNA-dependent RNA polymerase (RdRp) located at the C-terminal end of the polyprotein. In addition to the RdRp, the replicase encoded by ORF1 comprises an RNA helicase, a methyltransferase as well as less well-characterized domains, including the Macro and Y domains, a hypervariable region and a putative papain-like cysteine protease (PCP) (Figure 1A). Of note, the PCP has been associated with deubiquitination and the Macro domain with deribosylation activities likely involved in the posttranslational modifications of host proteins [64,65]. Positive-strand RNA viruses commonly process their polyproteins into individual functional proteins by viral, and in some instances, cellular proteases. In the case of HEV, however, it is unsettled whether, and if so, by which mechanism, the ORF1 protein is processed. While some studies reported processing of the ORF1 polyprotein, including by the PCP or the cellular proteases thrombin and factor Xa [66], others have found that the major form of ORF1 protein is unprocessed (reviewed in [53,67]). Further studies in complete cell culture systems and using more sensitive techniques for the detection of ORF1 protein are required [68].

As for all positive-strand RNA viruses, replication includes synthesis of a complementary negative-strand RNA which serves as a template for the production of positive-strand genomic and, in the case of HEV, an additional subgenomic RNA. The key enzyme responsible for these steps is the RdRp. The mechanisms regulating the production of these RNA species from the negative-strand RNA are still unknown but imply cis-acting elements within the genome [69]. Regulation of transcription may also involve host factor(s), such as heterogeneous nuclear ribonucleoproteins (hnRNP) [70,71]. hnRNPs play a role in nuclear RNA metabolism and have been reported to re-localize to the cytoplasm of HEV-infected cells [71], similarly to what had been observed for infection with other RNA viruses [72,73,74,75,76], suggesting a potential role of hnRNPs in genomic and subgenomic HEV RNA transcription.

Replication of positive-strand RNA viruses takes place in membrane-associated replication complexes composed of viral proteins, the replicating viral RNA, rearranged cellular membranes and other host factors [77]. The subcellular site of HEV RNA replication has not been identified to date but recent advances such as the development of tagged functional HEV genomes may facilitate progress in this area [68]. Insertion of a hemagglutinin epitope tag in the ORF1 polyprotein allowed us to visualize the HEV replicase together with viral RNA in cytoplasmic dot-like structures, likely indicating the site of active replication. ORF1 protein was found to colocalize best with exosomal markers as well as with the ORF2 and ORF3 proteins, suggesting that HEV RNA replication takes place in close proximity to virion assembly sites [68]. Virus-induced membrane rearrangements may serve several purposes, including the physical support and organization of the replication complex, the compartmentalization and local concentration of viral and host factors required for RNA replication, tethering of the viral RNA during unwinding and coordination of its translation, replication and packaging, provision of lipid constituents important for replication, and physical protection from host antiviral defenses (reviewed in [77,78]). Of note, HCV as well as picorna- and coronaviruses require the guanine nucleotide-exchange factor Golgi brefeldin A resistance factor 1 (GBF1) for the induction of membrane alterations making up their viral replication complexes [79,80,81,82,83]. Interestingly, similar observations have been made for HEV [84]. However, no colocalization of GBF1 with HEV ORF1 and ORF2 proteins or relocalization of GBF1 upon HEV infection have been observed. One may thus hypothesize that GBF1 is not recruited to HEV replication sites and is not involved in replication complex formation but rather plays an indirect role in HEV replication [84], as also proposed for HCV [79,80] and mouse hepatitis virus [85].

4. VIRION ASSEMBLY AND INFECTIOUS PARTICLE RELEASE

Virion assembly consists of the packaging of genomic viral RNA in the capsid. While the subcellular site of HEV assembly has not been identified yet, it is likely tightly connected to replication complexes. The mechanisms driving HEV assembly are poorly understood but early observations showed that RNA and the ORF2 protein can spontaneously assemble into virus-like particles in insect cells [86] (reviewed in [87]). These findings argue in favor of a self-assembly process involving a limited number of host factors.

The ORF2 protein exists in several forms, of which a non-glycosylated form is involved in virion formation [88,89,90,91]. It is currently debated whether this form, harboring a truncated and non-functional signal peptide, results from translation at an alternative start site [90] or from proteolytic cleavage, as suggested by mass spectrometry analyses [89]. The non-glycosylated ORF2 protein may represent the major intracellular form detected in the cytosol as well as the cell nucleus [89,91,92]. While the role of nuclear ORF2 protein is unknown, its nucleocytoplasmic shuttling likely involves the host nuclear import/export machineries. In contrast to the non-glycosylated form packaging the viral genome, the glycosylated form of ORF2 protein is rapidly released extracellularly through the secretory pathway [88,89,90,91,93,94]. This implies recognition of its signal peptide by the translocon, translocation into the ER lumen, cleavage by signal peptidase, followed by sialylation as well as N- and O-glycosylation by yet to be defined glycosyltransferases in the ER and Golgi [89], and secretion. Glycosylated ORF2 protein is present in at least two forms, of which the smaller may result from cleavage by furin-like proteases at an RRR motif [91]. Overall, these studies suggest that ORF2 protein likely has more than one function in genome packaging and point toward the implication of different host factors.

HEV assembly involves the non-glycosylated ORF2 protein and the viral RNA but does not require the ORF3 protein, as genomes harboring a mutated ORF3 start codon yield infectious particles; however, these are not secreted from the cell [95,96]. Infectious virus is believed to be released as quasi-enveloped particles from both the basolateral and apical sides of hepatocytes, facing the liver sinusoids and the bile canaliculi, respectively [33,97,98]. The pseudo-envelope of virions secreted into the bile is likely delipidated by bile acids and/or pancreatic enzymes (Figure 3). Interestingly, HEV is preferentially secreted from the apical side, explaining the high viral load detected in feces. Based on these observations, different host factors may be involved in virus secretion from the apical and basolateral sides of hepatocytes. Moreover, HEV ORF3 protein is essential for the secretion of infectious particles, possibly by connecting the capsid with the host factors required for egress. Phosphorylation of ORF3 protein at a serine residue (Ser 70 and Ser 71 in gt 3 and gt 1, respectively) may trigger the interaction with assembled non-glycosylated ORF2 protein [44]. Based on sequence information, phosphorylation may involve the p34cdc2 kinase and mitogen-activated protein kinase [99]. Furthermore, ORF3 protein recruits Tsg101, a member of the endosomal sorting complex required for transport-I (ESCRT-I), via a highly conserved PSAP motif [100,101,102]. The presence of this motif in the C-terminal region of the ORF3 protein is required for virus release [101]. It is known that proline-rich motifs P(S/T)AP and PPXY (X being any amino acid), which are also called "late domains", are essential for budding of enveloped viruses such as human immunodeficiency virus (HIV)-1 or Ebola [103,104].

The ESCRT machinery is composed of 4 complexes, ESCRT-0 to ESCRT-III, which are involved in membrane remodeling, leading to budding reactions or membrane involution. These complexes can further recruit accessory proteins, such as vacuolar protein sorting 4 (Vps4) and apoptosis-linked gene-2 interacting protein X (Alix), both interacting with the ESCRT-III complex, to close newly formed vesicles (reviewed in [105]). Further confirming the requirement for the ESCRT machinery, hepatocyte growth factor-regulated tyrosine kinase substrate (Hrs), a member of ESCRT-0, and Vps4 were shown to be required for HEV particle release [102,106]. In addition, the envelope wrapping the virion has been shown to be derived from exosomes, small vesicles generated from multivesicular bodies (MVB) by the ESCRT pathway [102,106]. These MVB, which fuse with the plasma membrane to release their content into the extracellular milieu, require the Rab27a protein, which has been shown to colocalize with ORF3 protein [102,106] (Figure 3).

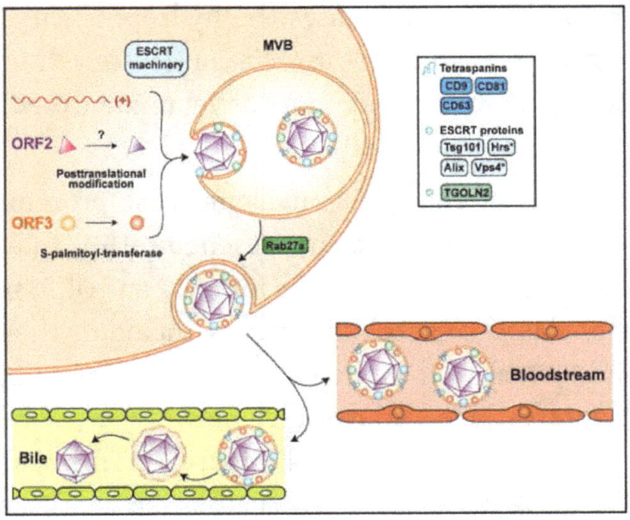

Figure 3. Assembly and release of infectious hepatitis E virus. Packaging of the viral genome into the capsid is believed to occur by spontaneous self-assembly of the non-glycosylated ORF2 protein. The non-glycosylated ORF2 protein may undergo post-translational modification by yet unknown enzyme(s). Formation of the quasi-enveloped particle involves phosphorylation and palmitoylation of the ORF3 protein and the ESCRT machinery, of which the components Tsg101, Hrs, Vps4 and Alix were shown to be required (see text for abbreviations). The virion is wrapped in

an exosomal membrane harboring the tetraspanins CD9, CD63 and CD81, as well as the trans-Golgi network protein 2 (TGOLN2), Alix and Tsg101. Release of quasi-enveloped HEV (eHEV) involves Rab27a-dependent trafficking of multivesicular bodies (MVB) and fusion with the plasma membrane. Secreted particles remain associated with the lipid membrane in the bloodstream while they are delipidated in the bile. Asterisks indicate host factors that were not found on the quasi-envelope of eHEV.

Current evidence indicates that none of the viral proteins are present on the surface of eHEV but identified ORF3 protein beneath the quasi-envelope [107,108,109]. Palmitoylation at N-terminal cysteine residues mediates membrane association as well as stability of the ORF3 protein and is required for virus secretion [110]. A membrane topology where ORF3 protein is located entirely on the cytosolic side, corresponding to the exosome lumen, has therefore been proposed [110] and is further supported by the interaction of its PSAP motif with Tsg101. Of note, palmitoylation is a reversible protein modification taking place in the cytosol, which increases the hydrophobicity of a protein and contributes to its membrane association (reviewed in [111]). This implies the requirement for one or more as yet unidentified S-palmitoyl-transferase(s) of the host cell [110].

Given the origin of the quasi-envelope, host proteins, and more specifically, exosomal proteins, can be present on eHEV. Indeed, particles released from infected cells and positive for ORF2 and ORF3 proteins were shown to harbor classical exosomal markers, including the tetraspanins CD81, CD63 and CD9, as well as the ESCRT components Alix and Tsg101 [106,108,112]. Interestingly, trans-Golgi network protein 2 (TGOLN2) has also been detected on eHEV [107] (Figure 3). This observation and the fact that TGOLN2 is located in the cytoplasm confirms that quasi-enveloped particles are produced within the cell and not at the plasma membrane.

The quasi-enveloped nature of virions circulating in blood may provide several advantages to HEV, including protection from neutralizing antibodies. However, the presence of eHEV in blood may also favor virus dissemination to organs other than the liver, as exosomes containing viral genetic material were shown to lead to productive infection by other viruses and to modulate cellular responses (reviewed in [113]). Hence, it is plausible that the production of quasi-enveloped virions has additional functions beyond the spread of HEV within the liver.

5. CONCISE SUMMARY

Despite the fast growth of HEV research, our understanding of the virus life cycle is still constrained by significant gaps in information. We have insufficient data regarding the cellular receptor(s) responsible for virus entrance, as well as the process of uncoating the viral RNA. Furthermore, it is recommended that future endeavors focus on investigating the molecular mechanisms and subcellular compartments that are involved in the processes of RNA replication and assembly. Although the release of viruses is a well-researched stage in the viral life cycle, there are still many issues that require clarification, particularly regarding the involvement of host factors and the composition and function of quasi-enveloped particles. Given HEV's capacity for replication in many tissues and its ability to infect a broad spectrum of animals, it is plausible to predict that HEV does not exhibit strong selectivity in terms of its host dependency. This lack of selectivity may facilitate the transmission of the virus across species barriers. Currently, research on the life cycle of HEV relies on the use of in vitro model systems. However, these systems have limitations, particularly when it comes to the presence of host factors seen in differentiated hepatocytes. For instance, the initial findings indicated that cyclophilin inhibitors have a stimulating effect on HEV replication in hepatoma cells [114]. Nevertheless, this finding was validated only using a cell culture-adapted infectious clone in stem cell-derived hepatocyte-like cells, but not with natural HEV isolates [115]. Future advancements in in vitro models should incorporate the utilization of authentic HEV isolates in conjunction with hepatocyte-like cells produced from primary and stem cells, polarized cell culture models, as well as ex vivo and in vivo infection model systems to validate the findings obtained in vitro. In the end, it will be necessary to confirm the main discoveries by examining liver samples obtained from people diagnosed with hepatitis E.

To gain a more comprehensive and impartial understanding of the host components involved in the life cycle of HEV, it is probable that we will need to rely on innovative technologies like CRISPR/Cas9-based genome-wide screening and proteomic proximity tagging methods. CRISPR/Cas9-based screens have enhanced our comprehension of various aspects, such as virus entry [116,117] and replication [118,119], and have also aided the discovery of novel targets for antiviral treatments [120]. Proximity labeling has effectively been utilized, as demonstrated by its application in characterizing the microenvironment of coronavirus replication complexes [121]. In the future, the integration of enhanced HEV model systems and innovative technologies is expected to enhance our understanding of the host variables necessary for successful HEV infection. Hepatitis E virus (HEV) infection is a prevalent cause of acute hepatitis globally. HEV is a positive-strand RNA virus that is transmitted through the gastrointestinal tract. It is present as a non-enveloped particle in bile and stool, and as a quasi-enveloped particle in blood. Our current knowledge of the molecular mechanisms and host variables involved in successful HEV infection is limited. However, the development of new model systems has greatly accelerated work in this field. In this article, we present a comprehensive analysis of the life cycle of HEV, with particular emphasis on the essential host components necessary for viral entry, RNA replication, assembly, and release. In the future, the expansion of HEV model systems and the introduction of new technologies are expected to provide a more comprehensive understanding.

Viruses, being obligate intracellular pathogens, have evolved tactics to seize control of and influence host cell circuits to guarantee successful infection. In addition, RNA viruses, particularly those with smaller genomes and coding capability, like HEV, rely heavily on the cellular machinery of the host. Due to recent limitations in studying HEV, our understanding of the host variables involved in different stages of the viral life cycle is still restricted. Experiments conducted in other organisms, such the yeast two hybrid system, have discovered biological components that interact with HEV proteins [16,17]. Nevertheless, the majority of these possibilities still require validation and further investigation through the use of infectious cell culture systems, in vivo models, and liver biopsies obtained from individuals with hepatitis E. This review will specifically examine host variables that have been confirmed to play a role in the viral life cycle during HEV infection.

REFERENCES

1. Khuroo, M.S. Study of an epidemic of non-A, non-B hepatitis. Possibility of another human hepatitis virus distinct from post-transfusion non-A, non-B type. Am. J. Med. 1980, 68, 818–824.
2. Balayan, M.S.; Andjaparidze, A.G.; Savinskaya, S.S.; Ketiladze, E.S.; Braginsky, D.M.; Savinov, A.P.; Poleschuk, V.F. Evidence for a virus in non-A, non-B hepatitis transmitted via the fecal-oral route. Intervirology 1983, 20, 23–31.
3. Reyes, G.R.; Purdy, M.A.; Kim, J.P.; Luk, K.C.; Young, L.M.; Fry, K.E.; Bradley, D.W. Isolation of a cDNA from the virus responsible for enterically transmitted non-A, non-B hepatitis. Science 1990, 247, 1335–1339.
4. Tam, A.W.; Smith, M.M.; Guerra, M.E.; Huang, C.C.; Bradley, D.W.; Fry, K.E.; Reyes, G.R. Hepatitis E virus (HEV): Molecular cloning and sequencing of the full-length viral genome. Virology 1991, 185, 120–131.
5. Smith, D.B.; Simmonds, P.; Jameel, S.; Emerson, S.U.; Harrison, T.J.; Meng, X.J.; Okamoto, H.; Van der Poel, W.H.; Purdy, M.A. Consensus proposals for classification of the family Hepeviridae. J. Gen. Virol. 2015, 96, 1191–1192.
6. Lee, G.H.; Tan, B.H.; Teo, E.C.; Lim, S.G.; Dan, Y.Y.; Wee, A.; Aw, P.P.; Zhu, Y.; Hibberd, M.L.; Tan, C.K.; et al. Chronic infection with camelid hepatitis E virus in a liver transplant recipient who regularly consumes camel meat and milk. Gastroenterology 2016, 150, 355.e353–357.e353.
7. Nimgaonkar, I.; Ding, Q.; Schwartz, R.E.; Ploss, A. Hepatitis E virus: Advances and challenges. Nat. Rev. Gastroenterol. Hepatol. 2018, 15, 96–110.

8. Abravanel, F.; Lhomme, S.; El Costa, H.; Schvartz, B.; Peron, J.M.; Kamar, N.; Izopet, J. Rabbit hepatitis E virus infections in humans, France. Emerg. Infect. Dis. 2017, 23, 1191–1193.
9. Sahli, R.; Fraga, M.; Semela, D.; Moradpour, D.; Gouttenoire, J. Rabbit HEV in immunosuppressed patients with hepatitis E acquired in Switzerland. J. Hepatol. 2019, 70, 1023–1025.
10. Sridhar, S.; Yip, C.C.Y.; Wu, S.; Cai, J.; Zhang, A.J.; Leung, K.H.; Chung, T.W.H.; Chan, J.F.W.; Chan, W.M.; Teng, J.L.L.; et al. Rat hepatitis E virus as cause of persistent hepatitis after liver transplant. Emerg. Infect. Dis. 2018, 24, 2241–2250.
11. Sridhar, S.; Yip, C.C.; Wu, S.; Chew, N.F.; Leung, K.H.; Chan, J.F.; Zhao, P.S.; Chan, W.M.; Poon, R.W.; Tsoi, H.W.; et al. Transmission of rat hepatitis E virus infection to humans in Hong Kong: A clinical and epidemiological analysis. Hepatology 2020.
12. Ding, Q.; Heller, B.; Capuccino, J.M.; Song, B.; Nimgaonkar, I.; Hrebikova, G.; Contreras, J.E.; Ploss, A. Hepatitis E virus ORF3 is a functional ion channel required for release of infectious particles. Proc. Natl. Acad. Sci. USA 2017, 114, 1147–1152.
13. Graff, J.; Torian, U.; Nguyen, H.; Emerson, S.U. A bicistronic subgenomic mRNA encodes both the ORF2 and ORF3 proteins of hepatitis E virus. J. Virol. 2006, 80, 5919–5926.
14. Ichiyama, K.; Yamada, K.; Tanaka, T.; Nagashima, S.; Jirintai; Takahashi, M.; Okamoto, H. Determination of the 5′-terminal sequence of subgenomic RNA of hepatitis E virus strains in cultured cells. Arch. Virol. 2009, 154, 1945–1951.
15. Feng, Z.; Hensley, L.; McKnight, K.L.; Hu, F.; Madden, V.; Ping, L.; Jeong, S.H.; Walker, C.; Lanford, R.E.; Lemon, S.M. A pathogenic picornavirus acquires an envelope by hijacking cellular membranes. Nature 2013, 496, 367–371.
16. Geng, Y.; Yang, J.; Huang, W.; Harrison, T.J.; Zhou, Y.; Wen, Z.; Wang, Y. Virus host protein interaction network analysis reveals that the HEV ORF3 protein may interrupt the blood coagulation process. PLoS ONE 2013, 8, e56320.

17. Subramani, C.; Nair, V.P.; Anang, S.; Mandal, S.D.; Pareek, M.; Kaushik, N.; Srivastava, A.; Saha, S.; Shalimar; Nayak, B.; et al. Host-virus protein interaction network reveals the involvement of multiple host processes in the life cycle of hepatitis E virus. mSystems 2018, 3, e00135-17.

18. Yu, H.; Li, S.; Yang, C.; Wei, M.; Song, C.; Zheng, Z.; Gu, Y.; Du, H.; Zhang, J.; Xia, N. Homology model and potential virus-capsid binding site of a putative HEV receptor Grp78. J. Mol. Model. 2011, 17, 987–995.

19. Ahmed, Z.; Holla, P.; Ahmad, I.; Jameel, S. The ATP synthase subunit β (ATP5B) is an entry factor for the hepatitis E virus. bioRxiv 2016.

20. Zhang, L.; Tian, Y.; Wen, Z.; Zhang, F.; Qi, Y.; Huang, W.; Zhang, H.; Wang, Y. Asialoglycoprotein receptor facilitates infection of PLC/PRF/5 cells by HEV through interaction with ORF2. J. Med. Virol. 2016, 88, 2186–2195.

21. Kalia, M.; Chandra, V.; Rahman, S.A.; Sehgal, D.; Jameel, S. Heparan sulfate proteoglycans are required for cellular binding of the hepatitis E virus ORF2 capsid protein and for viral infection. J. Virol. 2009, 83, 12714–12724.

22. Yin, X.; Ambardekar, C.; Lu, Y.; Feng, Z. Distinct entry mechanisms for nonenveloped and quasi-enveloped hepatitis E viruses. J. Virol. 2016, 90, 4232–4242.

23. WuDunn, D.; Spear, P.G. Initial interaction of herpes simplex virus with cells is binding to heparan sulfate. J. Virol. 1989, 63, 52–58.

24. Shieh, M.T.; WuDunn, D.; Montgomery, R.I.; Esko, J.D.; Spear, P.G. Cell surface receptors for herpes simplex virus are heparan sulfate proteoglycans. J. Cell Biol. 1992, 116, 1273–1281.

25. Mondor, I.; Ugolini, S.; Sattentau, Q.J. Human immunodeficiency virus type 1 attachment to HeLa CD4 cells is CD4 independent and gp120 dependent and requires cell surface heparans. J. Virol. 1998, 72, 3623–3634.

26. Barth, H.; Schafer, C.; Adah, M.I.; Zhang, F.; Linhardt, R.J.; Toyoda, H.; Kinoshita-Toyoda, A.; Toida, T.; Van Kuppevelt, T.H.; Depla, E.; et al. Cellular binding of hepatitis C virus envelope glycoprotein E2 requires cell surface heparan sulfate. J. Biol. Chem. 2003, 278, 41003–41012.

27. Tamura, M.; Natori, K.; Kobayashi, M.; Miyamura, T.; Takeda, N. Genogroup II noroviruses efficiently bind to heparan sulfate proteoglycan associated with the cellular membrane. J. Virol. 2004, 78, 3817–3826.

28. Shiota, T.; Li, T.C.; Nishimura, Y.; Yoshizaki, S.; Sugiyama, R.; Shimojima, M.; Saijo, M.; Shimizu, H.; Suzuki, R.; Wakita, T.; et al. Integrin alpha3 is involved in non-enveloped hepatitis E virus infection. Virology 2019, 536, 119–124.

29. Takada, Y.; Ye, X.; Simon, S. The integrins. Genome Biol. 2007, 8, 215.

30. Akula, S.M.; Pramod, N.P.; Wang, F.Z.; Chandran, B. Integrin alpha3beta1 (CD 49c/29) is a cellular receptor for Kaposi's sarcoma-associated herpesvirus (KSHV/HHV-8) entry into the target cells. Cell 2002, 108, 407–419.

31. Rivera-Serrano, E.E.; González-López, O.; Das, A.; Lemon, S.M. Cellular entry and uncoating of naked and quasi-enveloped human hepatoviruses. Elife 2019, 8, e43983.

32. Choy, M.Y.; Richman, P.I.; Horton, M.A.; MacDonald, T.T. Expression of the VLA family of integrins in human intestine. J. Pathol. 1990, 160, 35–40.

33. Marion, O.; Lhomme, S.; Nayrac, M.; Dubois, M.; Pucelle, M.; Requena, M.; Migueres, M.; Abravanel, F.; Peron, J.M.; Carrere, N.; et al. Hepatitis E virus replication in human intestinal cells. Gut 2020, 69, 901–910.

34. Oechslin, N.; Moradpour, D.; Gouttenoire, J. Hepatitis E virus finds its path through the gut. Gut 2020, 69, 796–798.

35. de Melker, A.A.; Sterk, L.M.; Delwel, G.O.; Fles, D.L.; Daams, H.; Weening, J.J.; Sonnenberg, A. The A and B variants of the alpha 3 integrin subunit: Tissue distribution and functional characterization. Lab. Invest. 1997, 76, 547–563.

36. Volpes, R.; van den Oord, J.J.; Desmet, V.J. Distribution of the VLA family of integrins in normal and pathological human liver tissue. Gastroenterology 1991, 101, 200–206.

37. Volpes, R.; van den Oord, J.J.; Desmet, V.J. Integrins as differential cell lineage markers of primary liver tumors. Am. J. Pathol. 1993, 142, 1483–1492.

38. Kapur, N.; Thakral, D.; Durgapal, H.; Panda, S.K. Hepatitis E virus enters liver cells through receptor-dependent clathrin-mediated endocytosis. J. Viral Hepat. 2012, 19, 436–448.

39. Holla, P.; Ahmad, I.; Ahmed, Z.; Jameel, S. Hepatitis E virus enters liver cells through a dynamin-2, clathrin and membrane cholesterol-dependent pathway. Traffic 2015, 16, 398–416.

40. Bubeck, D.; Filman, D.J.; Cheng, N.; Steven, A.C.; Hogle, J.M.; Belnap, D.M. The structure of the poliovirus 135S cell entry intermediate at 10-angstrom resolution reveals the location of an externalized polypeptide that binds to membranes. J. Virol. 2005, 79, 7745–7755.

41. Cavaldesi, M.; Caruso, M.; Sthandier, O.; Amati, P.; Garcia, M.I. Conformational changes of murine polyomavirus capsid proteins induced by sialic acid binding. J. Biol. Chem. 2004, 279, 41573–41579.

42. Sapp, M.; Bienkowska-Haba, M. Viral entry mechanisms: human papillomavirus and a long journey from extracellular matrix to the nucleus. FEBS J. 2009, 276, 7206–7216.

43. Yin, X.; Feng, Z. Hepatitis E virus entry. Viruses 2019, 11, 883.

44. Tyagi, S.; Korkaya, H.; Zafrullah, M.; Jameel, S.; Lal, S.K. The phosphorylated form of the ORF3 protein of hepatitis E virus interacts with its non-glycosylated form of the major capsid protein, ORF2. J. Biol. Chem. 2002, 277, 22759–22767.

45. Drave, S.A.; Debing, Y.; Walter, S.; Todt, D.; Engelmann, M.; Friesland, M.; Wedemeyer, H.; Neyts, J.; Behrendt, P.; Steinmann, E. Extra-hepatic replication and infection of hepatitis E virus in neuronal-derived cells. J. Viral Hepat. 2016, 23, 512–521.

46. Fu, R.M.; Decker, C.C.; Dao Thi, V.L. Cell culture models for hepatitis E virus. Viruses 2019, 11, 608.

47. Gouilly, J.; Chen, Q.; Siewiera, J.; Cartron, G.; Levy, C.; Dubois, M.; Al-Daccak, R.; Izopet, J.; Jabrane-Ferrat, N.; El Costa, H. Genotype specific pathogenicity of hepatitis E virus at the human maternal-fetal interface. Nat. Commun. 2018, 9, 4748.

48. Knegendorf, L.; Drave, S.A.; Dao Thi, V.L.; Debing, Y.; Brown, R.J.P.; Vondran, F.W.R.; Resner, K.; Friesland, M.; Khera, T.; Engelmann, M.; et al. Hepatitis E virus replication and interferon responses in human placental cells. Hepatol. Commun. 2018, 2, 173–187.

49. Meister, T.L.; Bruening, J.; Todt, D.; Steinmann, E. Cell culture systems for the study of hepatitis E virus. Antiviral Res. 2019, 163, 34–49.

50. Shukla, P.; Nguyen, H.T.; Torian, U.; Engle, R.E.; Faulk, K.; Dalton, H.R.; Bendall, R.P.; Keane, F.E.; Purcell, R.H.; Emerson, S.U. Cross-species infections of cultured cells by hepatitis E virus and discovery of an infectious virus-host recombinant. Proc/ Natl. Acad. Sci. USA 2011, 108, 2438–2443.

51. Zhou, X.; Huang, F.; Xu, L.; Lin, Z.; de Vrij, F.M.S.; Ayo-Martin, A.C.; van der Kroeg, M.; Zhao, M.; Yin, Y.; Wang, W.; et al. Hepatitis E virus infects neurons and brains. J. Infect. Dis. 2017, 215, 1197–1206.

52. Kamar, N.; Bendall, R.; Legrand-Abravanel, F.; Xia, N.S.; Ijaz, S.; Izopet, J.; Dalton, H.R. Hepatitis E. Lancet 2012, 379, 2477–2488.

53. Debing, Y.; Moradpour, D.; Neyts, J.; Gouttenoire, J. Update on hepatitis E virology: Implications for clinical practice. J. Hepatol. 2016, 65, 200–212.

54. Geng, Y.; Zhao, C.; Huang, W.; Harrison, T.J.; Zhang, H.; Geng, K.; Wang, Y. Detection and assessment of infectivity of hepatitis E virus in urine. J. Hepatol. 2016, 64, 37–43.

55. Kamar, N.; Abravanel, F.; Lhomme, S.; Rostaing, L.; Izopet, J. Hepatitis E virus: Chronic infection, extra-hepatic manifestations, and treatment. Clin. Res. Hepatol. Gastroenterol. 2015, 39, 20–27.

56. Montero, H.; Perez-Gil, G.; Sampieri, C.L. Eukaryotic initiation factor 4A (eIF4A) during viral infections. Virus Genes 2019, 55, 267–273.

57. Zhou, X.; Xu, L.; Wang, Y.; Wang, W.; Sprengers, D.; Metselaar, H.J.; Peppelenbosch, M.P.; Pan, Q. Requirement of the eukaryotic translation initiation factor 4F complex in hepatitis E virus replication. Antiviral Res. 2015, 124, 11–19.

58. Glitscher, M.; Himmelsbach, K.; Woytinek, K.; Johne, R.; Reuter, A.; Spiric, J.; Schwaben, L.; Grunweller, A.; Hildt, E. Inhibition of hepatitis E virus spread by the natural compound silvestrol. Viruses 2018, 10, 301.

59. Todt, D.; Moeller, N.; Praditya, D.; Kinast, V.; Friesland, M.; Engelmann, M.; Verhoye, L.; Sayed, I.M.; Behrendt, P.; Dao Thi, V.L.; et al. The natural compound silvestrol inhibits hepatitis E virus (HEV) replication in vitro and in vivo. Antiviral Res. 2018, 157, 151–158.

60. Biedenkopf, N.; Lange-Grunweller, K.; Schulte, F.W.; Weisser, A.; Muller, C.; Becker, D.; Becker, S.; Hartmann, R.K.; Grunweller, A. The natural compound silvestrol is a potent inhibitor of Ebola virus replication. Antiviral Res. 2017, 137, 76–81.

61. Elgner, F.; Sabino, C.; Basic, M.; Ploen, D.; Grunweller, A.; Hildt, E. Inhibition of Zika virus replication by silvestrol. Viruses 2018, 10, 149.

62. Henss, L.; Scholz, T.; Grunweller, A.; Schnierle, B.S. Silvestrol inhibits Chikungunya virus replication. Viruses 2018, 10, 592.

63. Muller, C.; Schulte, F.W.; Lange-Grunweller, K.; Obermann, W.; Madhugiri, R.; Pleschka, S.; Ziebuhr, J.; Hartmann, R.K.; Grunweller, A. Broad-spectrum antiviral activity of the eIF4A inhibitor silvestrol against corona- and picornaviruses. Antiviral Res. 2018, 150, 123–129.

64. Karpe, Y.A.; Lole, K.S. Deubiquitination activity associated with hepatitis E virus putative papain-like cysteine protease. J. Gen. Virol. 2011, 92, 2088–2092.

65. Li, C.; Debing, Y.; Jankevicius, G.; Neyts, J.; Ahel, I.; Coutard, B.; Canard, B. Viral macro domains reverse protein ADP-ribosylation. J. Virol. 2016, 90, 8478–8486.

66. Kanade, G.D.; Pingale, K.D.; Karpe, Y.A. Activities of thrombin and factor Xa are essential for replication of hepatitis E virus and are possibly implicated in ORF1 polyprotein processing. J. Virol. 2018, 92, e01853-17.

67. LeDesma, R.; Nimgaonkar, I.; Ploss, A. Hepatitis E virus replication. Viruses 2019, 11, 719.

68. Szkolnicka, D.; Pollan, A.; Da Silva, N.; Oechslin, N.; Gouttenoire, J.; Moradpour, D. Recombinant hepatitis E viruses harboring tags in the ORF1 protein. J. Virol. 2019, 93, e00459-19.

69. Ding, Q.; Nimgaonkar, I.; Archer, N.F.; Bram, Y.; Heller, B.; Schwartz, R.E.; Ploss, A. Identification of the intragenomic promoter controlling hepatitis E virus subgenomic RNA transcription. mBio 2018, 9, e00769-18.

70. Kanade, G.D.; Pingale, K.D.; Karpe, Y.A. Protein interactions network of hepatitis E virus RNA and polymerase with host proteins. Front. Microbiol. 2019, 10, 2501.

71. Pingale, K.D.; Kanade, G.D.; Karpe, Y.A. Heterogeneous nuclear ribonucleoproteins participate in hepatitis E virus (HEV) replication. J. Mol. Biol. 2020, 432, 2369–2387.

72. Burnham, A.J.; Gong, L.; Hardy, R.W. Heterogeneous nuclear ribonuclear protein K interacts with Sindbis virus nonstructural proteins and viral subgenomic mRNA. Virology 2007, 367, 212–221.

73. Pettit Kneller, E.L.; Connor, J.H.; Lyles, D.S. hnRNPs relocalize to the cytoplasm following infection with vesicular stomatitis virus. J. Virol. 2009, 83, 770–780.

74. Bourai, M.; Lucas-Hourani, M.; Gad, H.H.; Drosten, C.; Jacob, Y.; Tafforeau, L.; Cassonnet, P.; Jones, L.M.; Judith, D.; Couderc, T.; et al. Mapping of Chikungunya virus interactions with host proteins identified nsP2 as a highly connected viral component. J. Virol. 2012, 86, 3121–3134.

75. Brunetti, J.E.; Scolaro, L.A.; Castilla, V. The heterogeneous nuclear ribonucleoprotein K (hnRNP K) is a host factor required for dengue virus and Junin virus multiplication. Virus Res. 2015, 203, 84–91.

76. Poenisch, M.; Metz, P.; Blankenburg, H.; Ruggieri, A.; Lee, J.Y.; Rupp, D.; Rebhan, I.; Diederich, K.; Kaderali, L.; Domingues, F.S.; et al. Identification of hnRNPK as regulator of hepatitis C virus particle production. PLoS Pathog. 2015, 11, e1004573.

77. Paul, D.; Madan, V.; Bartenschlager, R. Hepatitis C virus RNA replication and assembly: Living on the fat of the land. Cell Host Microbe 2014, 16, 569–579.

78. Moradpour, D.; Penin, F.; Rice, C.M. Replication of hepatitis C virus. Nature Rev. Microbiol. 2007, 5, 453–463.

79. Farhat, R.; Seron, K.; Ferlin, J.; Feneant, L.; Belouzard, S.; Goueslain, L.; Jackson, C.L.; Dubuisson, J.; Rouillé, Y. Identification of class II ADP-ribosylation factors as cellular factors required for hepatitis C virus replication. Cell. Microbiol. 2016, 18, 1121–1133.

80. Goueslain, L.; Alsaleh, K.; Horellou, P.; Roingeard, P.; Descamps, V.; Duverlie, G.; Ciczora, Y.; Wychowski, C.; Dubuisson, J.; Rouille, Y. Identification of GBF1 as a cellular factor required for hepatitis C virus RNA replication. J. Virol. 2010, 84, 773–787.

81. Belov, G.A.; Feng, Q.; Nikovics, K.; Jackson, C.L.; Ehrenfeld, E. A critical role of a cellular membrane traffic protein in poliovirus RNA replication. PLoS Pathog. 2008, 4, e1000216.

82. Lanke, K.H.; van der Schaar, H.M.; Belov, G.A.; Feng, Q.; Duijsings, D.; Jackson, C.L.; Ehrenfeld, E.; van Kuppeveld, F.J. GBF1, a guanine nucleotide exchange factor for Arf, is crucial for coxsackievirus B3 RNA replication. J. Virol. 2009, 83, 11940–11949.

83. Wang, J.; Du, J.; Jin, Q. Class I ADP-ribosylation factors are involved in enterovirus 71 replication. PLoS ONE 2014, 9, e99768.

84. Farhat, R.; Ankavay, M.; Lebsir, N.; Gouttenoire, J.; Jackson, C.L.; Wychowski, C.; Moradpour, D.; Dubuisson, J.; Rouille, Y.; Cocquerel, L. Identification of GBF1 as a cellular factor required for hepatitis E virus RNA replication. Cell. Microbiol. 2018, 20, e12804.

85. Verheije, M.H.; Raaben, M.; Mari, M.; Te Lintelo, E.G.; Reggiori, F.; van Kuppeveld, F.J.; Rottier, P.J.; de Haan, C.A. Mouse hepatitis coronavirus RNA replication depends on GBF1-mediated ARF1 activation. PLoS Pathog. 2008, 4, e1000088.

86. Xing, L.; Li, T.C.; Mayazaki, N.; Simon, M.N.; Wall, J.S.; Moore, M.; Wang, C.Y.; Takeda, N.; Wakita, T.; Miyamura, T.; et al. Structure of hepatitis E virion-sized particle reveals an RNA-dependent viral assembly pathway. J. Biol. Chem. 2010, 285, 33175–33183.

87. Mori, Y.; Matsuura, Y. Structure of hepatitis E viral particle. Virus Res. 2011, 161, 59–64.

88. Graff, J.; Zhou, Y.H.; Torian, U.; Nguyen, H.; St Claire, M.; Yu, C.; Purcell, R.H.; Emerson, S.U. Mutations within potential glycosylation sites in the capsid protein of hepatitis E virus prevent the formation of infectious virus particles. J. Virol. 2008, 82, 1185–1194.

89. Montpellier, C.; Wychowski, C.; Sayed, I.M.; Meunier, J.C.; Saliou, J.M.; Ankavay, M.; Bull, A.; Pillez, A.; Abravanel, F.; Helle, F.; et al. Hepatitis E virus lifecycle and identification of 3 forms of the ORF2 capsid protein. Gastroenterology 2018, 154, 211.e218–223.e218.

90. Yin, X.; Ying, D.; Lhomme, S.; Tang, Z.; Walker, C.M.; Xia, N.; Zheng, Z.; Feng, Z. Origin, antigenicity, and function of a secreted form of ORF2 in hepatitis E virus infection. Proc. Natl. Acad. Sci. USA 2018, 115, 4773–4778.

91. Ankavay, M.; Montpellier, C.; Sayed, I.M.; Saliou, J.M.; Wychowski, C.; Saas, L.; Duvet, S.; Aliouat-Denis, C.M.; Farhat, R.; de Masson d'Autume, V.; et al. New insights into the ORF2 capsid protein, a key player of the hepatitis E virus lifecycle. Sci. Rep. 2019, 9, 6243.

92. Lenggenhager, D.; Gouttenoire, J.; Malehmir, M.; Bawohl, M.; Honcharova-Biletska, H.; Kreutzer, S.; Semela, D.; Neuweiler, J.; Hurlimann, S.; Aepli, P.; et al. Visualization of hepatitis E virus RNA and proteins in the human liver. J. Hepatol. 2017, 67, 471–479.

93. Jameel, S.; Zafrullah, M.; Ozdener, M.H.; Panda, S.K. Expression in animal cells and characterization of the hepatitis E virus structural proteins. J. Virol. 1996, 70, 207–216.

94. Zafrullah, M.; Ozdener, M.H.; Kumar, R.; Panda, S.K.; Jameel, S. Mutational analysis of glycosylation, membrane translocation, and cell surface expression of the hepatitis E virus ORF2 protein. J. Virol. 1999, 73, 4074–4082.

95. Emerson, S.U.; Nguyen, H.; Torian, U.; Purcell, R.H. ORF3 protein of hepatitis E virus is not required for replication, virion assembly, or infection of hepatoma cells in vitro. J. Virol. 2006, 80, 10457–10464.

96. Yamada, K.; Takahashi, M.; Hoshino, Y.; Takahashi, H.; Ichiyama, K.; Nagashima, S.; Tanaka, T.; Okamoto, H. ORF3 protein of hepatitis E virus is essential for virion release from infected cells. J. Gen. Virol. 2009, 90, 1880–1891.

97. Capelli, N.; Marion, O.; Dubois, M.; Allart, S.; Bertrand-Michel, J.; Lhomme, S.; Abravanel, F.; Izopet, J.; Chapuy-Regaud, S. Vectorial release of hepatitis E virus in polarized human hepatocytes. J. Virol. 2019, 93.

98. Dao Thi, V.L.; Wu, X.; Belote, R.L.; Andreo, U.; Takacs, C.N.; Fernandez, J.P.; Vale-Silva, L.A.; Prallet, S.; Decker, C.C.; Fu, R.M.;

et al. Stem cell-derived polarized hepatocytes. Nat. Commun. 2020, 11, 1677.

99. Zafrullah, M.; Ozdener, M.H.; Panda, S.K.; Jameel, S. The ORF3 protein of hepatitis E virus is a phosphoprotein that associates with the cytoskeleton. J. Virol. 1997, 71, 9045–9053.

100. Surjit, M.; Oberoi, R.; Kumar, R.; Lal, S.K. Enhanced alpha1 microglobulin secretion from hepatitis E virus ORF3-expressing human hepatoma cells is mediated by the tumor susceptibility gene 101. J. Biol. Chem. 2006, 281, 8135–8142.

101. Nagashima, S.; Takahashi, M.; Jirintai; Tanaka, T.; Yamada, K.; Nishizawa, T.; Okamoto, H. A PSAP motif in the ORF3 protein of hepatitis E virus is necessary for virion release from infected cells. J. Gen. Virol. 2011, 92, 269–278.

102. Nagashima, S.; Takahashi, M.; Jirintai, S.; Tanaka, T.; Nishizawa, T.; Yasuda, J.; Okamoto, H. Tumour susceptibility gene 101 and the vacuolar protein sorting pathway are required for the release of hepatitis E virions. J. Gen. Virol. 2011, 92, 2838–2848.

103. Garrus, J.E.; von Schwedler, U.K.; Pornillos, O.W.; Morham, S.G.; Zavitz, K.H.; Wang, H.E.; Wettstein, D.A.; Stray, K.M.; Cote, M.; Rich, R.L.; et al. Tsg101 and the vacuolar protein sorting pathway are essential for HIV-1 budding. Cell 2001, 107, 55–65.

104. Martin-Serrano, J.; Zang, T.; Bieniasz, P.D. HIV-1 and Ebola virus encode small peptide motifs that recruit Tsg101 to sites of particle assembly to facilitate egress. Nat. Med. 2001, 7, 1313–1319.

105. Vietri, M.; Radulovic, M.; Stenmark, H. The many functions of ESCRTs. Nat. Rev. Mol. Cell Biol. 2020, 21, 25–42.

106. Nagashima, S.; Jirintai, S.; Takahashi, M.; Kobayashi, T.; Tanggis; Nishizawa, T.; Kouki, T.; Yashiro, T.; Okamoto, H. Hepatitis E virus egress depends on the exosomal pathway, with secretory exosomes derived from multivesicular bodies. J. Gen. Virol. 2014, 95, 2166–2175.

107. Nagashima, S.; Takahashi, M.; Jirintai, S.; Tanggis; Kobayashi, T.; Nishizawa, T.; Okamoto, H. The membrane on the surface of hepatitis E virus particles is derived from the intracellular membrane and contains trans-Golgi network protein 2. Arch. Virol. 2014, 159, 979–991.

108. Nagashima, S.; Takahashi, M.; Kobayashi, T.; Nishizawa, T.; Nishiyama, T.; Primadharsini, P.P.; Okamoto, H. Characterization of the quasi-enveloped hepatitis E virus particles released by the cellular exosomal pathway. J. Virol. 2017, 91, e00822-17.

109. Qi, Y.; Zhang, F.; Zhang, L.; Harrison, T.J.; Huang, W.; Zhao, C.; Kong, W.; Jiang, C.; Wang, Y. Hepatitis E virus produced from cell culture has a lipid envelope. PLoS ONE 2015, 10, e0132503.

110. Gouttenoire, J.; Pollan, A.; Abrami, L.; Oechslin, N.; Mauron, J.; Matter, M.; Oppliger, J.; Szkolnicka, D.; Dao Thi, V.L.; van der Goot, F.G.; et al. Palmitoylation mediates membrane association of hepatitis E virus ORF3 protein and is required for infectious particle secretion. PLoS Pathog. 2018, 14, e1007471.

111. Blaskovic, S.; Blanc, M.; van der Goot, F.G. What does S-palmitoylation do to membrane proteins? FEBS J. 2013, 280, 2766–2774.

112. Chapuy-Regaud, S.; Dubois, M.; Plisson-Chastang, C.; Bonnefois, T.; Lhomme, S.; Bertrand-Michel, J.; You, B.; Simoneau, S.; Gleizes, P.E.; Flan, B.; et al. Characterization of the lipid envelope of exosome encapsulated HEV particles protected from the immune response. Biochimie 2017, 141, 70–79.

113. Chahar, H.S.; Bao, X.; Casola, A. Exosomes and their role in the life cycle and pathogenesis of RNA viruses. Viruses 2015, 7, 3204–3225.

114. Wang, Y.; Zhou, X.; Debing, Y.; Chen, K.; Van Der Laan, L.J.; Neyts, J.; Janssen, H.L.; Metselaar, H.J.; Peppelenbosch, M.P.; Pan, Q. Calcineurin inhibitors stimulate and mycophenolic acid inhibits replication of hepatitis E virus. Gastroenterology 2014, 146, 1775–1783.

115. Wu, X.; Dao Thi, V.L.; Liu, P.; Takacs, C.N.; Xiang, K.; Andrus, L.; Gouttenoire, J.; Moradpour, D.; Rice, C.M. Pan-genotype hepatitis E virus replication in stem cell-derived hepatocellular systems. Gastroenterology 2018, 154, 663.e7–674.e7.

116. Meertens, L.; Hafirassou, M.L.; Couderc, T.; Bonnet-Madin, L.; Kril, V.; Kummerer, B.M.; Labeau, A.; Brugier, A.; Simon-Loriere, E.; Burlaud-Gaillard, J.; et al. FHL1 is a major host factor for Chikungunya virus infection. Nature 2019, 574, 259–263.

117. Zhang, R.; Kim, A.S.; Fox, J.M.; Nair, S.; Basore, K.; Klimstra, W.B.; Rimkunas, R.; Fong, R.H.; Lin, H.; Poddar, S.; et al. Mxra8 is a receptor for multiple arthritogenic alphaviruses. Nature 2018, 557, 570–574.

118. Han, J.; Perez, J.T.; Chen, C.; Li, Y.; Benitez, A.; Kandasamy, M.; Lee, Y.; Andrade, J.; tenOever, B.; Manicassamy, B. Genome-wide CRISPR/Cas9 screen identifies host factors essential for influenza virus replication. Cell Rep. 2018, 23, 596–607.

119. Richardson, R.B.; Ohlson, M.B.; Eitson, J.L.; Kumar, A.; McDougal, M.B.; Boys, I.N.; Mar, K.B.; De La Cruz-Rivera, P.C.; Douglas, C.; Konopka, G.; et al. A CRISPR screen identifies IFI6 as an ER-resident interferon effector that blocks flavivirus replication. Nat. Microbiol. 2018, 3, 1214–1223.

120. Flint, M.; Chatterjee, P.; Lin, D.L.; McMullan, L.K.; Shrivastava-Ranjan, P.; Bergeron, E.; Lo, M.K.; Welch, S.R.; Nichol, S.T.; Tai, A.W.; et al. A genome-wide CRISPR screen identifies N-acetylglucosamine-1-phosphate transferase as a potential antiviral target for Ebola virus. Nat. Commun. 2019, 10, 285.

121. V'Kovski, P.; Gerber, M.; Kelly, J.; Pfaender, S.; Ebert, N.; Braga Lagache, S.; Simillion, C.; Portmann, J.; Stalder, H.; Gaschen, V.; et al. Determination of host proteins composing the microenvironment of coronavirus replicase complexes by proximity-labeling. Elife 2019, 8, e42037.

18 EXAMINING THE SPREAD OF HIV-1 IN TISSUE-LIKE 3D CULTURES: A NOVEL APPROACH TO ENVIRONMENTAL RESTRICTIONS

1. INTRODUCTION: AN OVERVIEW

Viruses, being obligate intracellular parasites, rely on infecting host cells that facilitate their reproduction and the generation of new viral offspring. To successfully initiate virus replication in a new host, the virus must effectively disseminate once the initial infection occurs at the entrance point. The replication of infectious offspring and subsequent infection of additional host cells is the primary mechanism by which viruses propagate. The release of virus particles into the extracellular space allows for the possibility of infecting new target cells through cell-free infection (Figure 1a). Furthermore, viruses can be transmitted from infected donor cells to uninfected target cells by direct physical contact between the cells (referred to as cell-cell transmission) (as seen in Figure 1b-d). Cell-associated modes of virus transmission involve the transfer of virus particles between cells in close proximity. This can occur through the release of virus particles at points of contact between cells, the movement of virus particles along or within cellular protrusions that connect donor and target cells, and the fusion of cells. These modes of transmission are generally considered to be more effective than the transmission of free virus particles. Although cell-free infection has been extensively studied, there is growing evidence supporting the use of cell-associated modes of virus transmission. This transmission mode has been observed in various viruses, including Vaccinia virus, Hepatitis C virus, Herpes Simplex virus, Epstein–Barr Virus, Dengue

Virus, Human Immunodeficiency Virus type 1 (HIV-1), and Human T-cell Lymphotropic Virus type 1 (HTLV-1). The majority of these viruses are capable of spreading through both cell-free and cell-associated modes of transmission. However, several viruses, like HTLV-I, have a specific preference for cell-to-cell transmission and rely primarily on this method. It is rare to isolate cell-free infectious virus from these viruses [12]. Assessing the relative contribution of cell-free and cell-associated transmission modes to the overall spread of viruses is challenging, and the significance of cell-associated transmission in terms of its impact on the body's functioning is sometimes uncertain. It is unsurprising that traditional virology notions have mostly concentrated on infections that occur without the involvement of cells. This emphasis is still evident in most experimental research done.

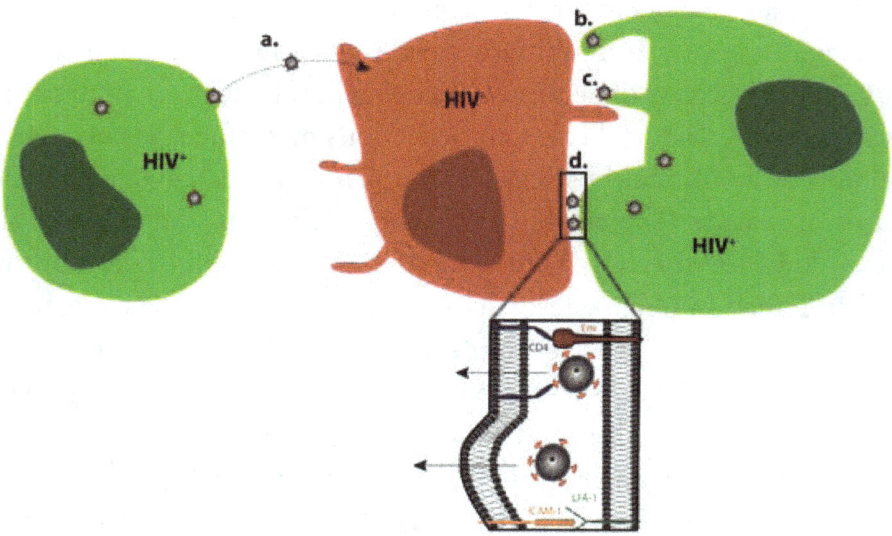

Figure 1. Transmission modes of HIV-1. Viral particles infect target cells via cell-free (a) or cell-associated (b-d) modes of transmission. (a) Viral particles bud at the surface of infected donor cells, mature, diffuse, and infect non-adjacent target cells. (b,c) Virions can bud at the tip (b) and surf along (c) filopodia to enter in adjacent target cells. In addition, infected and non-infected cells establish close contact, forming a virological synapse (d). Whether HIV-1 enters the target cell via fusion at the plasma membrane or following prior internalization [30,31] remains a matter of debate, and may depend on the nature of the target cell (reviewed in Reference [32]).

HIV-1 is a virus that has been extensively researched in terms of its methods of transmission. Originally believed to only spread by cell-free virus, early research demonstrated that infected cells are a more effective means of transmitting the virus to a new culture than cell-free virus [13]. The experiment showed that continuous stirring of infected CD4+ T cells or physically separating infected cells from uninfected cells using transwells prevents the formation of cell-to-cell contacts and efficient spread of the virus. This indicates that modes of transmission involving direct cell-to-cell contact are crucial for efficient spread of HIV-1 in CD4+ T-cell cultures. Extensive imaging experiments have confirmed that HIV-1 can spread effectively through cell-cell interactions, in addition to being transmitted through free-floating viral particles. HIV-1 cell-cell transmission occurs between CD4+ T cells, dendritic cells and CD4+ T cells, CD4+ T cells and macrophages, and myeloid cells. Although the exact process may differ significantly, this transfer is documented. The interactions between infected donor cells and target cells are known as virological synapses (VSs). In other cases, when donor cells such dendritic cells carry the virus for transmission without being infected themselves, these interactions are referred to as infectious synapses [17,18,19,20,21,22]. The contact-dependent transmission method has been discovered to be significantly more effective than the uptake of cell-free virus, with an estimated efficacy in mediating viral spread that is 10 to 18,000 times higher [14,23,24,25]. Although there is substantial evidence showing that HIV-1 effectively utilizes cell-associated methods of transmission in experimental HIV-1 infection, the ability to research this element of HIV-1 biology in living organisms is hindered by technological obstacles. As a result, there is limited evidence available regarding the utilization of this transmission mode in the infected host. Only a limited number of studies have documented the observation of cell-cell interactions resembling a virological synapse (VS) in living organisms [26,27,28], or confirmed that the movement of cells carrying infectious HIV-1 is crucial for the spread of HIV-1 in infected humanized mice [27,29]. The transmission route of HIV-1 in living organisms and its potential dependence on the unique tissue environment have yet to be determined.

2. 3D CULTURE SYSTEMS

Questions such as the relevance of individual virus transmission modes for viral spread and disease progression would be best addressed in in vivo infection models, as these reflect the complex tissue organization and pathogen–host interactions of infected patients [33]. In the case of HIV-1 infection, various humanized mouse models that mirror a range of but not all pathological events of AIDS are available for such studies [34]. However, experimental parameters are very difficult to control in these complex in vivo systems, and the number of experiments that can be conducted is limited by logistic, financial, and/or ethical concerns. In turn, simple monotypic, two-dimensional (2D) cultures do not reflect the complex architecture and cell heterogeneity of HIV-1 target tissue and, even for suspension cells, such as CD4+ T cells, quickly organize into a cell monolayer with dense cell packing. In vivo, the physical distance between donor and target cells often is the main barrier for efficient virus transmission, be it via diffusion of cell-free particles or the motility of the cell towards forming cell–cell contacts for virus transmission. Thus, densely packed 2D cultures are not suitable to study such processes.

As these constraints apply to all areas of biology, major efforts are being undertaken to establish and exploit experimental 3D systems that faithfully reflect specific aspects of tissue organization and physiology. One major area of development involves establishing organ or organoid cultures, either by culturing original tissues obtained from surgery or by reconstituting organ-like tissue ex vivo by differentiation from induced pluripotent stem cells [35,36]. In the case of HIV-1, such organotypic cell systems include cultures of human tonsil tissue as surrogates for lymph node tissues or mucosa explants as mimics of sites of virus transmission during primary HIV-1 infection [33,37]. While these models are of great value for advancing our understanding of pathological mechanisms, they do not offer tight control over critical experimental parameters such as cell density and composition, are subject to significant donor-to-donor variability, and their availability is limited. These limitations have precluded their use for defining the relative contribution of individual virus transmission modes and generated the need for the development of alternative experimental approaches. In, for example, immunology, significant efforts have been made to establish synthetic 3D systems to complement our portfolio of experimental systems with approaches that offer tight experimental control and direct access to visualization and genetic modification of individual cells. Particularly for lymphocytes and dendritic cells, 3D matrices made of collagen have emerged as the gold standard for such studies [38,39,40]. Collagen has the advantage of being a major constituent of extracellular matrix, supporting cell viability for up to several weeks, and allowing the easy removal of cells by collagenase digestion. Moreover, the good optical properties of the matrix provide access to live cell microscopy and the architecture of the matrix can be modified easily by adjusting polymerization conditions (collagen concentration, polymerization temperature, and medium used) [41,42,43]. The bovine and rat tail collagen classically used in 3D culture systems share high amino-acid similarities with the human type I collagen protein (91.1% identity for the alpha I chains and 87.4% for the alpha 2 chains), suggesting that they likely mirror the properties of human collagen. Consistent with this idea, CD4+ cells embedded in bovine and rat collagen matrices adopt migrating behaviors reminiscent of in vivo situations [44]. The xenogeneic effects associated with culturing human cells in this type of culture thus seem to be minimal with respect to cell migration, although species differences in molecular regulation between different collagens cannot be entirely excluded. These aspects allow quantitative insight to be gained into, for example, CD4+ T-cell function, such as their interaction and communication with DCs, or the ability to incorporate specific small RNAs into exosomes that match their behavior in vivo [45,46]. We therefore embarked on an attempt to reconstitute key aspects of HIV-1 spread in target tissue using 3D collagen cultures [47].

3. NEW INSIGHTS FROM COMBINING THE STUDY OF HIV-1 SPREAD IN 3D COLLAGEN CULTURES WITH COMPUTATION

In our recent proof-of-concept study [47], 3D cultures used to study HIV-1 spread included exclusively lymphocytes, considering HIV-1-infected and uninfected CD4+ T cells. Using a reporter virus expressing a sortable cell surface tag enabled the generation of pure HIV-1-infected donor cell populations that were mixed with uninfected target cells at a defined ratio prior to being embedded in 3D collagen. Cells were viable over several weeks in this system and cultures allowed the quantification of viral titers, quantification of expansion or depletion of specific cell populations, and the determination of cell motility and cell–cell contact formation/duration over time. Finally, the system allows for the use of collagen types with different density, with rat and bovine collagen resulting in denser or looser meshworks, respectively [48]. Recording the dynamics of HIV-1 spread and CD4+ T-cell depletion in parallel suspension and 3D cultures immediately revealed that the accurate interpretation of this plethora of quantitative kinetic data required computational approaches. Moreover, addressing the potential relationship between cell motility and virus spread, as well as disentangling the relative contribution of cell-free vs. cell-associated virus transmission to the infection dynamics, was impossible to assess experimentally but required the development of customized computational models. Our analysis relied on several iterative cycles between experimental quantification and computation and resulted in an Integrative method to Study Pathogen spread by Experiment and Computation within Tissue-like 3D cultures(INSPECT-3D) workflow depicted in Figure 2.

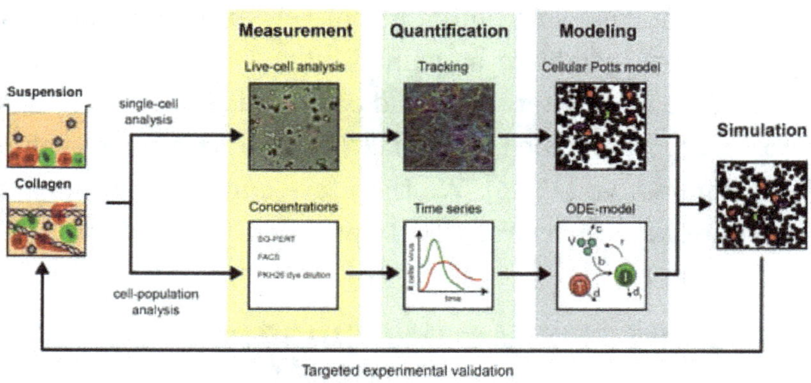

Figure 2. Schematic of the INSPECT-3D workflow that combined single-cell and cell-population measurements by mathematical modeling to reveal the dynamics of infection spread. Infected and non-infected Peripheral Blood Mononuclear Cells (PBMCs) were cultured in suspension or collagen

cultures at a 1:20 ratio. In suspension cultures, cells rapidly settled at the bottom of the culture dish to form dense 2D monolayers. In collagen, cells were spatially separated from another, migrated along the 3D scaffold, and eventually made contact with other cells. Measurements from live-cell imaging of single cells and bulk dynamics allowed quantification of individual cell motilities and interactions, as well as long-term kinetics of viral load and cell concentrations. Appropriate mathematical model systems that either rely on spatially resolved cellular Potts models describing single-cell behavior or systems of ordinary differential equations (ODE) enabled us to disentangle processes of cell motility, cellular turnover, and viral transmission and replication. Theses analyses led to an integrative computational model for HIV-1 spread that provides a mechanistic and quantitative understanding of viral transmission within these multicellular systems. Recurrent experimental validation of model predictions by in silico manipulation of experimental conditions was used to refine the model system and reveal key processes governing the dynamics of HIV-1 spread.

Together, the development and first application of INSPECT-3D revealed that HIV-1 spread is delayed relative to suspension cultures when cells are placed in a dense matrix but, after an initial phase characterized by low virus replication, follows kinetics that are similar to those seen in suspension cultures. In contrast, low-density collagen matrices did not support a marked spread of HIV-1. Several findings allowed these surprising observations to be explained. First, 3D collagen, irrespective of its density, potently suppresses the infectivity of cell-free HIV-1 particles (approx. 20-fold). Moreover, virion diffusion rates are too slow to efficiently deliver virus particles to new target cells in the spacing of 3D collagen (i.e., diffusion of virions to the nineteen nearest neighboring cells requires ~22.6 h, in comparison to half-life HIV-1 particle infectivity of t1/2 = 17.9 h). Together, these results revealed that 3D tissue-like environments potently restrict cell-free HIV-1 infection (see discussion of environmental restriction below). Analysis of the experimental data by mathematical modeling allowed us to estimate the relative contribution of cell-free and cell-associated HIV transmission under these various conditions. Hereby, consistent with the experimental observations, we found that HIV-1 spread in 3D collagen predominantly relies on cell–cell transmission (~78% (73–100%) and ~63% (55.8–100%) of all infections for loose and dense collagen, respectively). In contrast, in suspension cultures, there was no dominance of cell–cell transmission found and its contribution ranged widely between 0 and 100% (Table 1). Viral spread could thus be explained by the exclusive use of cell-free or cell–cell transmission. This revealed that in suspension cultures, no selection pressures for any of the transmission modes exist. This flexibility is consistent with the observation that physical separation of cells in suspension by agitation or transwells, which create conditions in which (i) cell-free infection depends on long-range diffusion of virus particles which compromises their infectivity and (ii) cell–cell contact is limited, markedly reduces virus spread [14,15]. Moreover, our analyses revealed that transmission dynamics substantially differ between suspension cultures and collagen environments.

Since 3D collagen suppressed cell-free HIV-1 infection and virus spread depended on cell-associated virus transmission, we investigated this scenario in more detail and found that in dense 3D collagen, contacts of extended duration between donor and target cells were more frequent in comparison to in loose collagen. These findings suggested that the dense 3D environment specifically promotes cell-associated HIV-1 transmission, while the low virus replication rates observed in loose collagen stem from the combination of suppression of cell-free infectivity by the 3D matrix and the lack of support for cell-associated virus transmission due to a lack of long-lasting cell–cell contacts. Importantly, the computational model was able to predict the extent to which cell densities had to be increased experimentally to overcome these barriers to HIV-1 spread in loose 3D collagen, suggesting that adopting a 2D monolayer type of configuration could override adverse effects of a 3D matrix on HIV-1 spread. These findings also imply that in a dense matrix, in which cells are not unphysiologically densely packed, cell motility is a key parameter governing the efficacy of HIV-1 spread, i.e., allowing transport of the infection to other areas as cell-free infection is impaired. Quantitative assessments of the relationship of HIV-1 spread in specific target organs will hence require the individual reconstitution of cell and matrix density as well as of the spacing between donor and target cells.

Table 1. Estimates for HIV-1 infection dynamics obtained from the analysis of experimental infection in cell lines or primary cells and in vivo patient data. Obtained estimates vary over several orders of magnitude due to different mathematical models and data types used. Please note that viral production rates vary in terms of units due to different quantification methods used. Numbers marked with (*) indicate estimates obtained for Simian Immunodeficiency Virus (SIV) infection.

Parameter	Description	Unit	Cell Lines	Primary Cells [47]			In Vivo
				2D Suspension	3D Collagen Loose	3D Collagen Dense	
ρ	Viral production rate	$\times 10^4$ day^{-1}	2.61 (1.55, 3.70) (RNA copies) [60]	1.02 (0.80, 1.38) (RT cell^{-1})	0.48 (0.37, 0.66) (RT cell^{-1})	0.43 (0.32, 0.61) (RT cell^{-1})	5.0 * (1.3, 12.0) (virion cell^{-1}) [61] 0.07–0.34 [56,62]
ρI	Total viral production rate	$\times 10^{10}$ virions day^{-1}					1.03 ± 1.17 [52,53]
δ_I	Death rate of infected cells	day^{-1}	0.50 ± 0.10 [55]	0.42 (0.40, 0.44)	0.48 (0.46, 0.50)	0.52 (0.51, 0.56)	0.48–1.36 ± 0.16 [53,63] 0.7 ± 0.25 [51] 1.0 ± 0.3 [54] 0.88 * ± 0.40 [61]
β_f	Cell-free infection rate	$\times 10^{-5}$ day^{-1}	0.42 ± 0.14 [55] (p24)	2.3 (0, 3.0) (RT)	0.14 × suspension	0.14 × suspension	
β_c	Cell–cell infection rate	$\times 10^{-5}$ (cell × day)$^{-1}$	0.11 ± 0.03 [55]	10^{-6} (0, 7.0)	4.3 (3.6, 5.4)	1.7 (1.4, 2.6)	
P_f	Predicted percentage of infections by cell-free transmission	%	57% ± 7% [55]	99.6% (0.0%, 100%)	22.0% (0.0%, 27.0%)	37% (0.0%, 44.2%)	

Correlating cell–cell contact duration and productive HIV-1 transmission also allowed us to predict the minimal donor–target cell contact time required for productive infection of new target cells, which was estimated to be in the range of ~25 min. Low-resolution inspection of productive donor–target cell contacts in dense 3D collagen revealed a strikingly close and intimate organization of these contacts, and it will be interesting to dissect how this morphology translates into elevated rates of virus transmission and how dense, but not loose, 3D environments induce these events.

4. HOW WELL DO EXPERIMENTAL AND COMPUTATIONAL INSPECT-3D DATA MATCH IN VIVO?

Although ex vivo 3D collagen cultures allow an improved experimental assessment of physiologically relevant cellular behaviour, they still only provide a reduced representation of relevant tissue conditions and do not completely mirror specific target tissues. However, comparison of the kinetics inferred for viral transmission and cellular turnover dynamics within these culture systems can help to assess their relevance for understanding the spread of HIV-1 in vivo. Mathematical modeling has long been used as an essential tool to dissect the multifactorial complexity of viral transmission and replication dynamics (reviewed in References [49,50]). Standard models of viral dynamics describing the turnover of uninfected and infected cells and the progression of the viral load by systems of ordinary differential equations have been applied to various experimental and clinical data assessing the dynamics of HIV-1 viral spread [51,52,53]. In particular, these models allowed the quantification of rates describing the kinetics of viral production (ρ) or the death of infected cells (δ) based on patient data. Using HIV-1 plasma RNA levels as an indicator for viral spread in patients, several studies estimated an average production of ~1010 HIV-1 virions in total per day, with estimates of the death rate of infected cells between δ ≈ 0.46–1.40 day-1, corresponding to average half-lives of infected cells between t1/2 = 0.5 and 1.5 days (t1/2 = log2/ δ) [51,52,53,54]. Studies based on standard in vitro culture systems have produced estimates that are at the upper end of the half-lives determined for infected cells in vivo, with estimates of the corresponding death rates being in the order of δ = 0.5 +/- 0.1 day-1 [55]. The estimates on the death rate of infected cells assessed via the INSPECT-3D workflow are within the same range, with a slight tendency towards higher death rates within 3D collagen environments compared to 2D suspension cultures (δ = 0.46–0.56 day-1 in dense and loose collagen vs. δ = 0.40–0.44 day-1 in suspension; ranges based on 95% confidence intervals of estimates, Table 1). Although the differences are minor, this might indicate that environmental factors, in combination with additional clearance mechanisms such as immune responses, could also contribute to the shorter half-lives of infected cells observed in vivo. While mathematical analyses of HIV-1 viral load kinetics in patients usually allow an appropriate quantification of viral and infected cell decay dynamics, as they work under the assumption that Highly Active Antiretroviral Therapy (HAART) effectively blocks novel infections [50,52], assessing the kinetics of viral transmission and spread in vivo is much more difficult. This requires an approximation of the available number of target cells, which might vary dependent on the time-point of infection and the different compartments of viral replication [56]. In vitro systems, with their associated complete knowledge of the initial experimental conditions, provide an advantage for assessing these dynamics, but have to account for the general lack of potential physiological relevance. As cell-free and cell-associated transmission modes are synergistic, unraveling their relative contributions to viral spread in vitro usually requires the impairment of at least one of them [4,55,57,58,59]. Iwami et al. [55] set out to assess the transmission rates for HIV-1 using shaken suspension cultures where cell–cell transmission was considered to be impaired [14]. Based on their data, they estimated that cell-associated viral transmission contributed to approximately 60% of viral infections. Without the assumption of efficiently blocking one transmission mode, we found similar contributions of cell-associated transmission modes to viral spread within 3D collagen (with only ~22% (0%, 27%) and ~37% (0%, 44.2%) of infections within loose and dense collagen, respectively, due to cell-free transmission). Thus, there was a clear difference in this contribution compared to spread within 2D suspension cultures. The simultaneous analysis of all three environments revealed a ~4–7-fold higher probability of cells becoming infected by cell–cell transmission within dense and loose collagen compared to suspension. The impaired contribution of cell-free transmission within 3D collagen compared to suspension cultures is further indicated by a ~50% reduction in viral production rates, and a measured relative infectivity of virions that was reduced to ~14% within collagen (βf, Table 1). In addition to previous reports that blocking CD4+ T-cell exit from lymph nodes restricted HIV-1 spread [27], these findings obtained for viral spread within 3D collagen environments support the argument that cell–cell transmission is the main contributor to HIV-1 spread in vivo. The importance of cell-contact-dependent transmission modes in tissue-like conditions is further indicated by the increased fraction of CD4+ T cells predicted to be initially refractory to infection in collagen (loose ~50%; dense ~30%) compared to suspension (~17%; fractions at 2.5 days post infection). This could point towards cells that are inaccessible for contact-dependent transmission modes due to physical barriers, as they are blocked by collagen within certain areas of the culture.

Combining live-cell microscopy data with a spatially detailed cellular Potts model in our INSPECT3D workflow also allowed us to estimate a minimal contact duration between infected and uninfected cells required for productive infection of ~25 min. Importantly, this matches very well the time required for the transfer of fluorescent viral material across the VS ex vivo [17,20], indicating that the events observed by these imaging approaches typically result in the productive infection of the target cell.

Based on the relative comparison of the viral kinetics between 2D suspension and 3D culture systems, our study indicated the need to account for physiological conditions when aiming to quantitatively understand viral replication and transmission dynamics in vivo. Arguably, the dense collagen conditions we used [47] might come close to mucosal tissue without reflecting the cellular composition of this tissue environment. Extending the existing culture systems by incorporating additional cell types and molecular factors will be needed to approximate specific tissue conditions. Other relevant physiological parameters that will have to be considered to be implemented to advance the experimental 3D model further include dynamic tissue perfusion, which might impact HIV-1 spread, e.g., by affecting the dynamics of cell-free infections or the frequency and/or stability of cell–cell interactions. It will be interesting to assess how these additional factors might impact the quantification of the viral processes obtained so far.

5. REQUIREMENTS AND LIMITATIONS OF CURRENT COMPUTATIONAL APPROACHES TO SIMULATING HIV-1 SPREAD IN TISSUE

The majority of previous studies combining mathematical models and experimental data have relied on model systems based on ordinary differential equations that use time-resolved measurements of cell or viral concentrations [50]. These analyses have increased our knowledge of various aspects of HIV-1 life-cycle kinetics and transmission dynamics, and are still the method of choice for gathering most types of experimental and clinical data. However, a more detailed analysis, and thus computational representation, of single-cell dynamics within tissues is needed to understand the dynamics of HIV-1 spread within multicellular systems. Population dynamic models have their limitations when it comes to analyzing cell-based transmission modes, as their ability to describe single-cell transmission dynamics appropriately is, by definition, hampered. Several extensions have been developed that adjust standard models for viral dynamics to account for cell-contact requirements dependent on the tissue environment [64,65]. Nevertheless, the need to clearly quantify the kinetics of cell-associated viral transmission modes, and to determine the contributions of cell-free and cell-associated transmission, as well as local immunity to viral spread, provides novel challenges for computational analysis.

Individual-cell-based models that describe single-cell behavior have been used previously to simulate and analyze viral spread [66]. However, these model systems were mainly used for qualitative assessment of dynamics, e.g., for the spread of HIV-1 [67,68], as appropriate data and tools for model parameterization remained limited. Both limitations have now been overcome. Advanced imaging and visualization techniques used in vitro and in vivo [33,69,70] have increased our ability to observe and quantify cellular behavior and infection processes at a single-cell level. In addition, improved parameter inference tools and modeling systems have enabled us to quantitatively adapt individual-cell-based models to these novel types of data [71,72]. However, several challenges remain to appropriately analyze HIV-1 spread within different tissue environments using mathematical and computational models (Figure 3). A major challenge for computational approaches that analyze the spatiotemporal dynamics of HIV-1 spread is the appropriate representation of cell motility and the underlying tissue architecture, e.g., as represented by the collagen matrix within 3D ex vivo culture systems. Determining the influence of the environment on cell and infection dynamics requires a detailed description of these matrices and networks that structure the tissue or culture system. Previous attempts to describe the topology of the fibroblastic reticular network within lymph nodes in order to analyze T-cell motility and activation dynamics have shown the difficulty of such tasks [73,74,75,76]. High-resolution images of collagen matrices and fibroblastic reticular networks help to improve our understanding of the structures shaping and influencing tissue architecture and cell motility [48]. Nevertheless, the identification of appropriate quantities for network description, such as the pore size, fiber length, and connectivity, and how they relate to network density [42,43,77,78,79], is also required in order to provide a reliable computational representation of 3D tissue environments that accounts for the different resolutions of fibers and cells [73,76,80,81,82]. The challenges of describing cell motility based on live-cell microscopy data, and possible ways to refine these analyses, have been already discussed elsewhere [83]. Considering relevant physiological tissue conditions will also have to account for various cell types, such as T cells, dendritic cells, and macrophages, and requires a detailed characterization of their complex morphology and dynamics in 3D. In particular, this also applies to the appropriate representation of the diffusion dynamics of soluble factors, such as virions or chemokines, that are important for cell–cell communication, infection spread, or cell movement. Being able to measure gradients and concentrations of these factors, especially their release dynamics from single cells, would help to mimic their dynamics and to determine the influence of these factors on viral spread and the efficacy of the counteracting immune response [84].

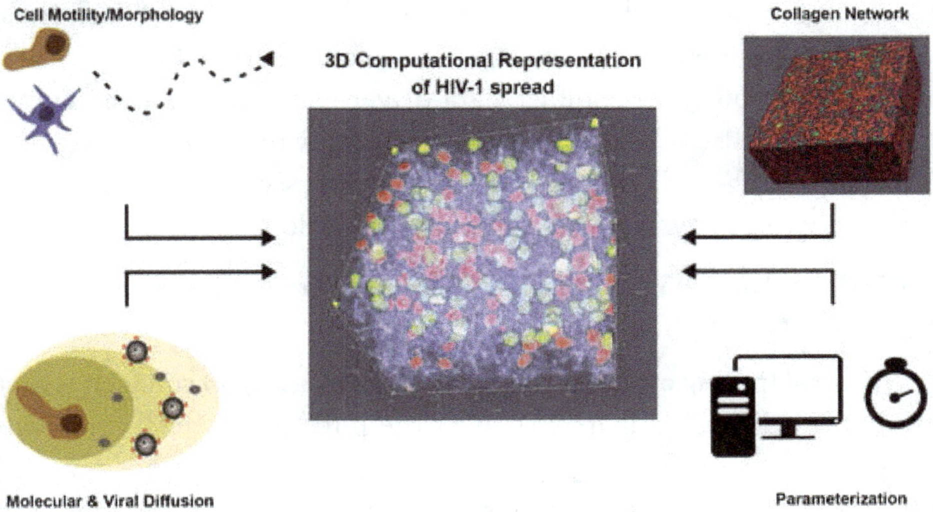

Figure 3. 3D computational representation of HIV-1 spread: Appropriate representations to infer the processes that govern the dynamics of infection require a detailed description and quantification of individual cell motility and morphology, the underlying tissue structure, and the diffusion dynamics of soluble components (chemokines, virions, etc.) within these environments. Data-driven parameterization and subsequent simulation of these 3D model systems requires computationally efficient methods. The sketch shows a computational representation of 100 HIV-1-infected (green) and 100 uninfected (red) cells within a 3D collagen matrix (blue fibers).

Finally, a major challenge for the computational modeling of 3D tissue conditions on a single-cell level is the increased level of detail required for its description, and especially its parameterization. The higher complexity of the model systems and the adaptation of their stochastic dynamics to heterogeneous experimental data requires the use of increased computational resources, including high-performance computing clusters and sophisticated parameter inference tools.

Improved parallelization and advanced parameter estimation methods, such as approximated Bayesian computing [72], could help to reduce computational run time costs. In addition, simulation areas in terms of the size and number of cells considered have to be chosen with care in order to balance the need for sufficient tissue volumes that account for stochastic variability while simultaneously ensuring computational efficiency for the simulation of long-term infection dynamics. Thus, novel advances in imaging techniques and computational methods will improve our ability to conduct data-driven computational modeling of HIV-1 spread within multicellular systems, which will be essential to disentangling and understanding the processes that shape these dynamics within physiologically relevant conditions.

6. ENVIRONMENTAL RESTRICTION TO CELL-FREE HIV-1 INFECTION

A central finding of the studies on HIV-1 spread in 3D collagen is that residing within such a matrix reduces the infectivity of cell-free virus particles approx. 20-fold [47]. Since single-particle tracking has revealed that particles frequently undergo short and transient interactions with the collagen matrix, this negative impact on particle infectivity likely represents the consequence of physical stress imposed on virions in the context of these interactions. It is noteworthy that marked effects of the tissue environment on virus infectivity have been reported. One example is the interaction of viruses with bile acids, which can induce fusogenicity (e.g., norovirus [85]) or restrict infection (cytomegalovirus [86]; Hepatitis Delta virus [87]). Similarly, seminal fluids can impact virion infectivity via peptide-mediated enhancement of virus infectivity (e.g., retroviruses including HIV-1 [88], Ebola virus [89]) or by affecting particle aggregation [90,91]. Since the infectivity reduction in 3D collagen results from negative physical impact by the surrounding matrix and thus represents a direct effect of tissue architecture rather than the consequence of compounds released in the extracellular space, we propose the novel term "environmental restriction" for this phenomenon. Considering that this environmental restriction seems to represent a built-in antiviral activity of the 3D collagen matrix, this is conceptually similar to intrinsic immune barriers such as host cell restriction factors, and could thus be viewed as a novel element of innate immunity (Figure 4). In contrast to cell-autonomous mechanisms exerted by, for example, restriction factors, this activity would reflect a tissue-autonomous restriction. Since the restriction of virus spread by extracellular tissue architecture has only recently begun to become apparent, we can only speculate about the effector function mediating the restriction. Such environmental restrictions could include direct effects on the pathogen, such as the impairment of cell-free infectivity [92] (see below), but could also impact the host cells (e.g., via the promotion of longer-lasting cell contacts with architecture optimized for virus transmission). It can also be envisioned that, for example, transcriptional and epigenetic changes resulting from mechanosensing of the cells in the 3D environment would affect expression and activity of pro- and antiviral host cell factors. Assessing the effect of tissue-like 3D environments on the permissivity of various HIV-1 target cells for infection and their ability to sense the infection will be an important area of future research. Cell-associated HIV-1 transmission is currently considered to be beneficial for the virus, as it allows for more polarized and therefore efficient short-term transfer of particles to target cells, allows host cell restriction factors such as tetherin to be partially overcome, and avoids the recognition of certain neutralizing antibodies [25,93,94,95]. In addition, we propose that cell-associated HIV-1 transmission evolved from the need to bypass the environmental restriction against cell-free virion infectivity exerted by HIV-1 target tissue.

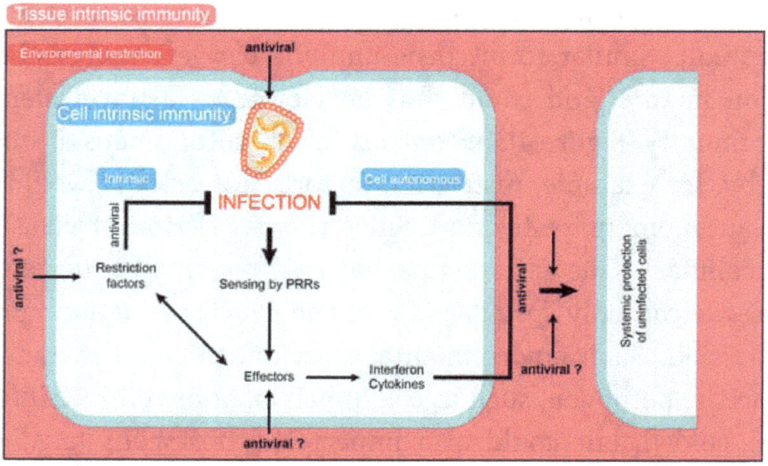

Figure 4. Schematic representation of the environmental restriction concept. Human cells bear cell-intrinsic mechanisms to limit virus spread, such as intrinsic immunity by restriction factors and cell-autonomous immunity by recognition of viral components by pattern-recognition receptors (PRRs). Both pathways can lead to the induction of antiviral effectors such as interferons and cytokines, which can impair virus replication in both the infected and bystander cells. In addition, we propose that the tissue environment can provide tissue-intrinsic immunity by exerting environmental restrictions on virus replication. Such activities could be exerted via direct effects on virus particles, but we also hypothesize that tissue interactions impact the expression and activity of cell-autonomous immune mechanisms.

Following the initial description of the environmental restrictions on cell-free HIV-1 infectivity in 3D collagen cultures, it will be important to dissect the mechanism by which tissue-like 3D matrices impair the infectivity of cell-free virions. Intuitively, physical interactions with the matrix could enhance shedding of the Env glycoprotein. However, overall Env levels in virions were unaffected by interactions of virus particles with the 3D matrix. Alternatively, the 3D matrix could affect virion infectivity by alteration of their aggregation state [90]. However, single-particle tracking of HIV-1 particles in 3D matrices did not reveal virion aggregates in either suspension or 3D collagen, suggesting that HIV-1 particles do not form aggregates under these experimental conditions. Finally, the reduced particle infectivity in 3D collagen may also reflect adaptation of the producer cells to the 3D environment;

however, the reduction of virion infectivity in 3D collagen was similar when particles were produced within the matrix or embedded in a cell-free matrix, indicating that the matrix exerts direct effects on HIV-1 virions. Future studies will be thus required to dissect whether the 3D matrix affects HIV-1 infection at the level of particle entry into target cells, e.g., by affecting the conformation of Env glycoproteins on virions, or at subsequent post entry steps (Figure 5). Assessing how conserved this effect may be among viruses other than HIV-1 will be helpful in dissecting the underlying mechanisms. Another key question is whether this barrier is indeed in place in HIV-1 target tissue in infected individuals. Analyzing this question is complicated by the fact that HIV-1 particles derived from patient samples represent a pool of particles of heterogeneous and undefined origin, and that potential effects of environmental tissue restrictions could not easily be distinguished from effects due to soluble factors present in bodily fluids. Organotypic cultures will likely be instrumental for addressing this question.

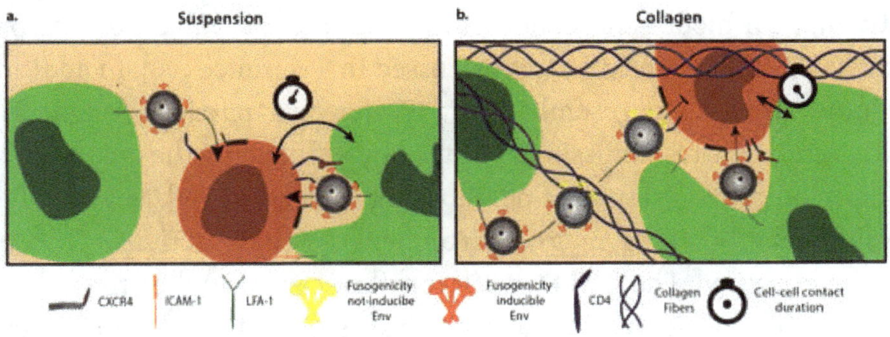

Figure 5. HIV-1 spread in suspension and 3D collagen cultures. (a) In suspension cultures, virions can be transmitted to uninfected target cells via both cell-free and cell-associated modes of transmission. (b) In collagen cultures, cell-associated transmission is the main driving force for HIV-1 spread. This reflects on one hand that the infectivity of cell-free particles is impaired, which coincides with frequent transient interactions with collagen fibres. These interactions could impair the ability of Env proteins to adopt a fusogenic conformation in response to interaction with the HIV-1 receptor/coreceptor complex on target cells, and thus inhibit virus entry. Moreover, cell-associated HIV-1 transmission may benefit from alterations in the architecture of contacts between donor and target cells, which are more stable and involve larger areas of membrane compared to cells in suspension.

7. CONCISE SUMMARY

The effectiveness of virus transmission within a living organism is determined by a complicated set of factors that can only be understood by using both synthetic and organoid 3D models, as well as computer analysis. Our work, which serves as a preliminary demonstration, investigated the spread of HIV-1 in 3D collagen cultures. The results showed that the tissue-like environment had a notable influence on the ways and effectiveness of HIV-1 dissemination. Additionally, the study shows the presence of inherent mechanisms inside the tissue that hinder viral replication (environmental limitation). Due of the ongoing advancements in physiological ex vivo 3D culture methods and enhanced computational analyses, it is expected that this methodology will further uncover significant insights into the mechanics of virus dissemination in living organisms. A crucial element of future endeavors will involve unraveling the role of antibody-mediated neutralization of HIV-1 during cell-cell transmission. This will be accomplished by expanding the mathematical model and conducting experimental validations as part of the INSPECT-3D workflow, as discussed in Reference [96]. In addition, exosomes that have resemblance to retroviral particles might also contribute to the transmission of viruses by influencing the vulnerability to infection or the lifespan of neighboring cells [97,98,99]. Nevertheless, the influence of tissue microenvironment on these activities has not been investigated. The INSPECT-3D workflow can be applied to study the dissemination of other T-cell-tropic viruses, like HTLV-1. The concept of cell-associated transmission through a virological synapse (VS) was initially established for HTLV-1. Infection with HTLV-1 is believed to primarily rely on cell-to-cell transmission, even in suspension cultures.

REFERENCES

1. Zhong, P.; Agosto, L.M.; Munro, J.B.; Mothes, W. Cell-to-cell transmission of viruses. Curr. Opin. Virol. 2013, 3, 44–50.

2. Sattentau, Q. Avoiding the void: Cell-to-cell spread of human viruses. Nat. Rev. Microbiol. 2008, 6, 815–826.

3. Roberts, K.L.; Smith, G.L. Vaccinia virus morphogenesis and dissemination. Trends Microbiol. 2008, 16, 472–479.

4. Graw, F.; Martin, D.N.; Perelson, A.S.; Uprichard, S.L.; Dahari, H. Quantification of Hepatitis C Virus Cell-to-Cell Spread Using a Stochastic Modeling Approach. J. Virol. 2015, 89, 6551–6561.

5. Polcicova, K.; Goldsmith, K.; Rainish, B.L.; Wisner, T.W.; Johnson, D.C. The extracellular domain of herpes simplex virus gE is indispensable for efficient cell-to-cell spread: Evidence for gE/gI receptors. J. Virol. 2005, 79, 11990–12001.

6. Imai, S.; Nishikawa, J.; Takada, K. Cell-to-cell contact as an efficient mode of Epstein-Barr virus infection of diverse human epithelial cells. J. Virol. 1998, 72, 4371–4378.

7. Yang, C.F.; Tu, C.H.; Lo, Y.P.; Cheng, C.C.; Chen, W.J. Involvement of Tetraspanin C189 in Cell-to-Cell Spreading of the Dengue Virus in C6/36 Cells. PLoS Negl. Trop. Dis. 2015, 9, e0003885.

8. Jolly, C.; Kashefi, K.; Hollinshead, M.; Sattentau, Q.J. HIV-1 cell to cell transfer across an Env-induced, actin-dependent synapse. J. Exp. Med. 2004, 199, 283–293.

9. Bracq, L.; Xie, M.; Benichou, S.; Bouchet, J. Mechanisms for Cell-to-Cell Transmission of HIV-1. Front. Immunol. 2018, 9, 260.

10. Igakura, T.; Stinchcombe, J.C.; Goon, P.K.; Taylor, G.P.; Weber, J.N.; Griffiths, G.M.; Tanaka, Y.; Osame, M.; Bangham, C.R. Spread of HTLV-I between lymphocytes by virus-induced polarization of the cytoskeleton. Science 2003, 299, 1713–1716.

11. Pais-Correia, A.M.; Sachse, M.; Guadagnini, S.; Robbiati, V.; Lasserre, R.; Gessain, A.; Gout, O.; Alcover, A.; Thoulouze, M.I. Biofilm-like extracellular viral assemblies mediate HTLV-1 cell-to-cell transmission at virological synapses. Nat. Med. 2010, 16, 83–89.

12. Gross, C.; Thoma-Kress, A.K. Molecular Mechanisms of HTLV-1 Cell-to-Cell Transmission. Viruses 2016, 8, 74.

13. Dimitrov, D.S.; Willey, R.L.; Sato, H.; Chang, L.J.; Blumenthal, R.; Martin, M.A. Quantitation of human immunodeficiency virus type 1 infection kinetics. J. Virol. 1993, 67, 2182–2190.

14. Sourisseau, M.; Sol-Foulon, N.; Porrot, F.; Blanchet, F.; Schwartz, O. Inefficient human immunodeficiency virus replication in mobile lymphocytes. J. Virol. 2007, 81, 1000–1012.

15. Orlandi, C.; Flinko, R.; Lewis, G.K. A new cell line for high throughput HIV-specific antibody-dependent cellular cytotoxicity (ADCC) and cell-to-cell virus transmission studies. J. Immunol. Methods 2016, 433, 51–58.

16. Xie, M.; Leroy, H.; Mascarau, R.; Woottum, M.; Dupont, M.; Ciccone, C.; Schmitt, A.; Raynaud-Messina, B.; Verollet, C.; Bouchet, J.; et al. Cell-to-Cell Spreading of HIV-1 in Myeloid Target Cells Escapes SAMHD1 Restriction. Mbio 2019, 10.

17. Hubner, W.; McNerney, G.P.; Chen, P.; Dale, B.M.; Gordon, R.E.; Chuang, F.Y.; Li, X.D.; Asmuth, D.M.; Huser, T.; Chen, B.K. Quantitative 3D video microscopy of HIV transfer across T cell virological synapses. Science 2009, 323, 1743–1747.

18. Sowinski, S.; Jolly, C.; Berninghausen, O.; Purbhoo, M.A.; Chauveau, A.; Kohler, K.; Oddos, S.; Eissmann, P.; Brodsky, F.M.; Hopkins, C.; et al. Membrane nanotubes physically connect T cells over long distances presenting a novel route for HIV-1 transmission. Nat. Cell Biol. 2008, 10, 211–219.

19. Felts, R.L.; Narayan, K.; Estes, J.D.; Shi, D.; Trubey, C.M.; Fu, J.; Hartnell, L.M.; Ruthel, G.T.; Schneider, D.K.; Nagashima, K.; et al. 3D visualization of HIV transfer at the virological synapse between dendritic cells and T cells. Proc. Natl. Acad. Sci. USA 2010.

20. Real, F.; Sennepin, A.; Ganor, Y.; Schmitt, A.; Bomsel, M. Live Imaging of HIV-1 Transfer across T Cell Virological Synapse to Epithelial Cells that Promotes Stromal Macrophage Infection. Cell Rep. 2018, 23, 1794–1805.

21. McDonald, D.; Wu, L.; Bohks, S.M.; KewalRamani, V.N.; Unutmaz, D.; Hope, T.J. Recruitment of HIV and its receptors to dendritic cell-T cell junctions. Science 2003, 300, 1295–1297.

22. Earl, L.A.; Lifson, J.D.; Subramaniam, S. Catching HIV 'in the act' with 3D electron microscopy. Trends Microbiol. 2013, 21, 397–404.

23. Chen, P.; Hubner, W.; Spinelli, M.A.; Chen, B.K. Predominant mode of human immunodeficiency virus transfer between T cells is mediated by sustained Env-dependent neutralization-resistant virological synapses. J. Virol. 2007, 81, 12582–12595.

24. Martin, N.; Welsch, S.; Jolly, C.; Briggs, J.A.; Vaux, D.; Sattentau, Q.J. Virological synapse-mediated spread of human immunodeficiency virus type 1 between T cells is sensitive to entry inhibition. J. Virol. 2010, 84, 3516–3527.

25. Kolodkin-Gal, D.; Hulot, S.L.; Korioth-Schmitz, B.; Gombos, R.B.; Zheng, Y.; Owuor, J.; Lifton, M.A.; Ayeni, C.; Najarian, R.M.; Yeh, W.W.; et al. Efficiency of cell-free and cell-associated virus in mucosal transmission of human immunodeficiency virus type 1 and simian immunodeficiency virus. J. Virol. 2013, 87, 13589–13597.

26. Sewald, X.; Gonzalez, D.G.; Haberman, A.M.; Mothes, W. In vivo imaging of virological synapses. Nat. Commun. 2012, 3, 1320.

27. Murooka, T.T.; Deruaz, M.; Marangoni, F.; Vrbanac, V.D.; Seung, E.; von Andrian, U.H.; Tager, A.M.; Luster, A.D.; Mempel, T.R. HIV-infected T cells are migratory vehicles for viral dissemination. Nature 2012, 490, 283–287.

28. Ladinsky, M.S.; Khamaikawin, W.; Jung, Y.; Lin, S.; Lam, J.; An, D.S.; Bjorkman, P.J.; Kieffer, C. Mechanisms of virus dissemination in bone marrow of HIV-1-infected humanized BLT mice. ELife 2019, 8.

29. Sewald, X.; Ladinsky, M.S.; Uchil, P.D.; Beloor, J.; Pi, R.; Herrmann, C.; Motamedi, N.; Murooka, T.T.; Brehm, M.A.; Greiner, D.L.; et al. Retroviruses use CD169-mediated trans-infection of permissive lymphocytes to establish infection. Science 2015, 350, 563–567.

30. Miyauchi, K.; Kim, Y.; Latinovic, O.; Morozov, V.; Melikyan, G.B. HIV enters cells via endocytosis and dynamin-dependent fusion with endosomes. Cell 2009, 137, 433–444.

31. Dale, B.M.; McNerney, G.P.; Thompson, D.L.; Hubner, W.; de Los Reyes, K.; Chuang, F.Y.; Huser, T.; Chen, B.K. Cell-to-cell transfer of HIV-1 via virological synapses leads to endosomal virion maturation that activates viral membrane fusion. Cell Host Microbe 2011, 10, 551–562.

32. Melikyan, G.B. HIV entry: A game of hide-and-fuse? Curr. Opin. Virol. 2014, 4, 1–7.

33. Fackler, O.T.; Murooka, T.T.; Imle, A.; Mempel, T.R. Adding new dimensions: Towards an integrative understanding of HIV-1 spread. Nat. Rev. Microbiol. 2014, 12, 563–574.

34. Marsden, M.D.; Zack, J.A. Humanized Mouse Models for Human Immunodeficiency Virus Infection. Annu. Rev. Virol. 2017, 4, 393–412.

35. Rossi, G.; Manfrin, A.; Lutolf, M.P. Progress and potential in organoid research. Nat. Rev. Genet. 2018, 19, 671–687.

36. Bar-Ephraim, Y.E.; Kretzschmar, K.; Clevers, H. Organoids in immunological research. Nat. Rev. Immunol. 2019.

37. Merbah, M.; Introini, A.; Fitzgerald, W.; Grivel, J.C.; Lisco, A.; Vanpouille, C.; Margolis, L. Cervico-vaginal tissue ex vivo as a model to study early events in HIV-1 infection. Am. J. Reprod. Immunol. 2011, 65, 268–278.

38. Sixt, M.; Lammermann, T. In vitro analysis of chemotactic leukocyte migration in 3D environments. Methods Mol. Biol. 2011, 769, 149–165.

39. Wolf, K.; Alexander, S.; Schacht, V.; Coussens, L.M.; von Andrian, U.H.; van Rheenen, J.; Deryugina, E.; Friedl, P. Collagen-based cell migration models in vitro and in vivo. Semin. Cell Dev. Biol. 2009, 20, 931–941.

40. Lammermann, T.; Bader, B.L.; Monkley, S.J.; Worbs, T.; Wedlich-Soldner, R.; Hirsch, K.; Keller, M.; Forster, R.; Critchley, D.R.; Fassler, R.; et al. Rapid leukocyte migration by integrin-independent flowing and squeezing. Nature 2008, 453, 51–55.

41. Miron-Mendoza, M.; Seemann, J.; Grinnell, F. The differential regulation of cell motile activity through matrix stiffness and porosity in three dimensional collagen matrices. Biomaterials 2010, 31, 6425–6435.

42. Olivares, V.; Condor, M.; Del Amo, C.; Asin, J.; Borau, C.; Garcia-Aznar, J.M. Image-based Characterization of 3D Collagen Networks and the Effect of Embedded Cells. Microsc. Microanal. 2019, 25, 971–981.

43. Jansen, K.A.; Licup, A.J.; Sharma, A.; Rens, R.; MacKintosh, F.C.; Koenderink, G.H. The Role of Network Architecture in Collagen Mechanics. Biophys. J. 2018, 114, 2665–2678.

44. Friedl, P.; Brocker, E.B. The biology of cell locomotion within three-dimensional extracellular matrix. Cell. Mol. Life Sci. 2000, 57, 41–64.

45. Thippabhotla, S.; Zhong, C.; He, M. 3D cell culture stimulates the secretion of in vivo like extracellular vesicles. Sci. Rep. 2019, 9, 13012.

46. Abu-Shah, E.; Demetriou, P.; Balint, S.; Mayya, V.; Kutuzov, M.A.; Dushek, O.; Dustin, M.L. A tissue-like platform for studying engineered quiescent human T-cells' interactions with dendritic cells. ELife 2019, 8.

47. Imle, A.; Kumberger, P.; Schnellbacher, N.D.; Fehr, J.; Carrillo-Bustamante, P.; Ales, J.; Schmidt, P.; Ritter, C.; Godinez, W.J.; Muller, B.; et al. Experimental and computational analyses reveal that environmental restrictions shape HIV-1 spread in 3D cultures. Nat. Commun. 2019, 10, 2144.

48. Wolf, K.; Te Lindert, M.; Krause, M.; Alexander, S.; Te Riet, J.; Willis, A.L.; Hoffman, R.M.; Figdor, C.G.; Weiss, S.J.; Friedl, P. Physical limits

of cell migration: Control by ECM space and nuclear deformation and tuning by proteolysis and traction force. J. Cell Biol. 2013, 201, 1069–1084.

49. Graw, F.; Perelson, A.S. Modeling Viral Spread. Annu. Rev. Virol. 2016, 3, 555–572.

50. Perelson, A.S. Modelling viral and immune system dynamics. Nat. Rev. Immunol. 2002, 2, 28–36.

51. Perelson, A.S.; Neumann, A.U.; Markowitz, M.; Leonard, J.M.; Ho, D.D. HIV-1 dynamics in vivo: Virion clearance rate, infected cell life-span, and viral generation time. Science 1996, 271, 1582–1586.

52. Ramratnam, B.; Bonhoeffer, S.; Binley, J.; Hurley, A.; Zhang, L.; Mittler, J.E.; Markowitz, M.; Moore, J.P.; Perelson, A.S.; Ho, D.D. Rapid production and clearance of HIV-1 and hepatitis C virus assessed by large volume plasma apheresis. Lancet 1999, 354, 1782–1785.

53. Perelson, A.S.; Essunger, P.; Cao, Y.; Vesanen, M.; Hurley, A.; Saksela, K.; Markowitz, M.; Ho, D.D. Decay characteristics of HIV-1-infected compartments during combination therapy. Nature 1997, 387, 188–191.

54. Markowitz, M.; Louie, M.; Hurley, A.; Sun, E.; Di Mascio, M.; Perelson, A.S.; Ho, D.D. A novel antiviral intervention results in more accurate assessment of human immunodeficiency virus type 1 replication dynamics and T-cell decay in vivo. J. Virol. 2003, 77, 5037–5038.

55. Iwami, S.; Takeuchi, J.S.; Nakaoka, S.; Mammano, F.; Clavel, F.; Inaba, H.; Kobayashi, T.; Misawa, N.; Aihara, K.; Koyanagi, Y.; et al. Cell-to-cell infection by HIV contributes over half of virus infection. ELife 2015, 4.

56. De Boer, R.J.; Ribeiro, R.M.; Perelson, A.S. Current estimates for HIV-1 production imply rapid viral clearance in lymphoid tissues. PLoS Comput. Biol. 2010, 6, e1000906.

57. Timpe, J.M.; Stamataki, Z.; Jennings, A.; Hu, K.; Farquhar, M.J.; Harris, H.J.; Schwarz, A.; Desombere, I.; Roels, G.L.; Balfe, P.; et al. Hepatitis C virus cell-cell transmission in hepatoma cells in the presence of neutralizing antibodies. Hepatology 2008, 47, 17–24.
58. Barretto, N.; Sainz, B., Jr.; Hussain, S.; Uprichard, S.L. Determining the involvement and therapeutic implications of host cellular factors in hepatitis C virus cell-to-cell spread. J. Virol. 2014, 88, 5050–5061.
59. Barretto, N.; Uprichard, S.L. Hepatitis C virus Cell-to-cell Spread Assay. Bio Protoc. 2014, 4.
60. Iwami, S.; Holder, B.P.; Beauchemin, C.A.; Morita, S.; Tada, T.; Sato, K.; Igarashi, T.; Miura, T. Quantification system for the viral dynamics of a highly pathogenic simian/human immunodeficiency virus based on an in vitro experiment and a mathematical model. Retrovirology 2012, 9, 18.
61. Chen, H.Y.; Di Mascio, M.; Perelson, A.S.; Ho, D.D.; Zhang, L. Determination of virus burst size in vivo using a single-cycle SIV in rhesus macaques. Proc. Natl. Acad. Sci. USA 2007, 104, 19079–19084.
62. Reilly, C.; Wietgrefe, S.; Sedgewick, G.; Haase, A. Determination of simian immunodeficiency virus production by infected activated and resting cells. AIDS (London, England) 2007, 21, 163–168.
63. Althaus, C.L.; De Vos, A.S.; De Boer, R.J. Reassessing the human immunodeficiency virus type 1 life cycle through age-structured modeling: Life span of infected cells, viral generation time, and basic reproductive number, R0. J. Virol. 2009, 83, 7659–7667.
64. Kumberger, P.; Durso-Cain, K.; Uprichard, S.L.; Dahari, H.; Graw, F. Accounting for Space-Quantification of Cell-To-Cell Transmission Kinetics Using Virus Dynamics Models. Viruses 2018, 10, 200.
65. Pilyugin, S.S.; Antia, R. Modeling immune responses with handling time. Bull. Math. Biol. 2000, 62, 869–890.
66. Bauer, A.L.; Beauchemin, C.A.; Perelson, A.S. Agent-based modeling of host-pathogen systems: The successes and challenges. Infor. Sci. 2009, 179, 1379–1389.

67. Strain, M.C.; Richman, D.D.; Wong, J.K.; Levine, H. Spatiotemporal dynamics of HIV propagation. J. Theor. Biol. 2002, 218, 85–96.

68. dos Santos, R.M.Z.; Coutinho, S. Dynamics of HIV Infection: A cellular automata approach. Phys. Rev. Lett. 2001, 87, 168102.

69. Coombes, J.L.; Robey, E.A. Dynamic imaging of host-pathogen interactions in vivo. Nat. Rev. Immunol. 2010, 10, 353–364.

70. Germain, R.N.; Miller, M.J.; Dustin, M.L.; Nussenzweig, M.C. Dynamic imaging of the immune system: Progress, pitfalls and promise. Nat. Rev. Immunol. 2006, 6, 497–507.

71. Starruss, J.; de Back, W.; Brusch, L.; Deutsch, A. Morpheus: A user-friendly modeling environment for multiscale and multicellular systems biology. Bioinformatics 2014, 30, 1331–1332.

72. Klinger, E.; Rickert, D.; Hasenauer, J. pyABC: Distributed, likelihood-free inference. Bioinformatics 2018, 34, 3591–3593.

73. Novkovic, M.; Onder, L.; Cheng, H.W.; Bocharov, G.; Ludewig, B. Integrative Computational Modeling of the Lymph Node Stromal Cell Landscape. Front. Immunol. 2018, 9, 2428.

74. Novkovic, M.; Onder, L.; Bocharov, G.; Ludewig, B. Graph Theory-Based Analysis of the Lymph Node Fibroblastic Reticular Cell Network. Meth. Mol. Biol. 2017, 1591, 43–57.

75. Novkovic, M.; Onder, L.; Cupovic, J.; Abe, J.; Bomze, D.; Cremasco, V.; Scandella, E.; Stein, J.V.; Bocharov, G.; Turley, S.J.; et al. Topological Small-World Organization of the Fibroblastic Reticular Cell Network Determines Lymph Node Functionality. PLoS Biol. 2016, 14, e1002515.

76. Novkovic, M.; Onder, L.; Bocharov, G.; Ludewig, B. Topological Structure and Robustness of the Lymph Node Conduit System. Cell Rep. 2020, 30, 893–904.

77. Licup, A.J.; Munster, S.; Sharma, A.; Sheinman, M.; Jawerth, L.M.; Fabry, B.; Weitz, D.A.; MacKintosh, F.C. Stress controls the mechanics of collagen networks. Proc. Natl. Acad. Sci. USA 2015, 112, 9573–9578.

78. Lindstrom, S.B.; Vader, D.A.; Kulachenko, A.; Weitz, D.A. Biopolymer network geometries: Characterization, regeneration, and elastic properties. Phys. Rev. E. 2010, 82, 051905.

79. Stein, A.M.; Vader, D.A.; Jawerth, L.M.; Weitz, D.A.; Sander, L.M. An algorithm for extracting the network geometry of three-dimensional collagen gels. J. Microsc. 2008, 232, 463–475.

80. Harjanto, D.; Zaman, M.H. Modeling extracellular matrix reorganization in 3D environments. PLoS ONE 2013, 8, e52509.

81. Graw, F.; Balagopal, A.; Kandathil, A.J.; Ray, S.C.; Thomas, D.L.; Ribeiro, R.M.; Perelson, A.S. Inferring viral dynamics in chronically HCV infected patients from the spatial distribution of infected hepatocytes. PLoS Computat. Biol. 2014, 10, e1003934.

82. Beltman, J.B.; Henrickson, S.E.; von Andrian, U.H.; de Boer, R.J.; Maree, A.F. Towards estimating the true duration of dendritic cell interactions with T cells. J. Immunol. Methods 2009, 347, 54–69.

83. Beltman, J.B.; Maree, A.F.; de Boer, R.J. Analysing immune cell migration. Nat. Rev. Immunol. 2009, 9, 789–798.

84. Howat, T.J.; Barreca, C.; O'Hare, P.; Gog, J.R.; Grenfell, B.T. Modelling dynamics of the type I interferon response to in vitro viral infection. J. R. Soc. Interface 2006, 3, 699–709.

85. Nelson, C.A.; Wilen, C.B.; Dai, Y.N.; Orchard, R.C.; Kim, A.S.; Stegeman, R.A.; Hsieh, L.L.; Smith, T.J.; Virgin, H.W.; Fremont, D.H. Structural basis for murine norovirus engagement of bile acids and the CD300lf receptor. Proc. Natl. Acad. Sci. USA 2018, 115, E9201–E9210.

86. Schupp, A.K.; Trilling, M.; Rattay, S.; Le-Trilling, V.T.K.; Haselow, K.; Stindt, J.; Zimmermann, A.; Haussinger, D.; Hengel, H.; Graf, D. Bile Acids Act as Soluble Host Restriction Factors Limiting Cytomegalovirus Replication in Hepatocytes. J. Virol. 2016, 90, 6686–6698.

87. Veloso Alves Pereira, I.; Buchmann, B.; Sandmann, L.; Sprinzl, K.; Schlaphoff, V.; Dohner, K.; Vondran, F.; Sarrazin, C.; Manns, M.P.;

Pinto Marques Souza de Oliveira, C.; et al. Primary biliary acids inhibit hepatitis D virus (HDV) entry into human hepatoma cells expressing the sodium-taurocholate cotransporting polypeptide (NTCP). PLoS ONE 2015, 10, e0117152.

88. Munch, J.; Rucker, E.; Standker, L.; Adermann, K.; Goffinet, C.; Schindler, M.; Wildum, S.; Chinnadurai, R.; Rajan, D.; Specht, A.; et al. Semen-derived amyloid fibrils drastically enhance HIV infection. Cell 2007, 131, 1059–1071.

89. Bart, S.M.; Cohen, C.; Dye, J.M.; Shorter, J.; Bates, P. Enhancement of Ebola virus infection by seminal amyloid fibrils. Proc. Natl. Acad. Sci. USA 2018, 115, 7410–7415.

90. Sanjuan, R. Collective Infectious Units in Viruses. Trends Microbiol. 2017, 25, 402–412.

91. Andreu-Moreno, I.; Sanjuan, R. Collective Infection of Cells by Viral Aggregates Promotes Early Viral Proliferation and Reveals a Cellular-L0evel Allee Effect. Curr. Biol. 2018, 28, 3212–3219.

92. Zotova, A.; Atemasova, A.; Pichugin, A.; Filatov, A.; Mazurov, D. Distinct Requirements for HIV-1 Accessory Proteins during Cell Coculture and Cell-Free Infection. Viruses 2019, 11, 390.

93. Malbec, M.; Porrot, F.; Rua, R.; Horwitz, J.; Klein, F.; Halper-Stromberg, A.; Scheid, J.F.; Eden, C.; Mouquet, H.; Nussenzweig, M.C.; et al. Broadly neutralizing antibodies that inhibit HIV-1 cell to cell transmission. J. Exp. Med. 2013, 210, 2813–2821.

94. Zhong, P.; Agosto, L.M.; Ilinskaya, A.; Dorjbal, B.; Truong, R.; Derse, D.; Uchil, P.D.; Heidecker, G.; Mothes, W. Cell-to-Cell Transmission Can Overcome Multiple Donor and Target Cell Barriers Imposed on Cell-Free HIV. PLoS ONE 2013, 8, e53138.

95. Jolly, C.; Booth, N.J.; Neil, S.J. Cell-cell spread of human immunodeficiency virus type 1 overcomes tetherin/BST-2-mediated restriction in T cells. J. Virol. 2010, 84, 12185–12199.

96. Dufloo, J.; Bruel, T.; Schwartz, O. HIV-1 cell-to-cell transmission and broadly neutralizing antibodies. Retrovirology 2018, 15, 51.

97. Chen, P.; Chen, B.K.; Mosoian, A.; Hays, T.; Ross, M.J.; Klotman, P.E.; Klotman, M.E. Virological synapses allow HIV-1 uptake and gene expression in renal tubular epithelial cells. J. Am. Soc. Nephrol. 2011, 22, 496–507.

98. Lee, J.H.; Schierer, S.; Blume, K.; Dindorf, J.; Wittki, S.; Xiang, W.; Ostalecki, C.; Koliha, N.; Wild, S.; Schuler, G.; et al. HIV-Nef and ADAM17-Containing Plasma Extracellular Vesicles Induce and Correlate with Immune Pathogenesis in Chronic HIV Infection. EBioMedicine 2016, 6, 103–113.

99. Ostalecki, C.; Wittki, S.; Lee, J.H.; Geist, M.M.; Tibroni, N.; Harrer, T.; Schuler, G.; Fackler, O.T.; Baur, A.S. HIV Nef- and Notch1-dependent Endocytosis of ADAM17 Induces Vesicular TNF Secretion in Chronic HIV Infection. EBioMedicine 2016, 13, 294–304.

100. Mazurov, D.; Ilinskaya, A.; Heidecker, G.; Lloyd, P.; Derse, D. Quantitative comparison of HTLV-1 and HIV-1 cell-to-cell infection with new replication dependent vectors. PLoS Pathog. 2010, 6, e1000788.

101. Shunaeva, A.; Potashnikova, D.; Pichugin, A.; Mishina, A.; Filatov, A.; Nikolaitchik, O.; Hu, W.S.; Mazurov, D. Improvement of HIV-1 and Human T Cell Lymphotropic Virus Type 1 Replication-Dependent Vectors via Optimization of Reporter Gene Reconstitution and Modification with Intronic Short Hairpin RNA. J. Virol. 2015, 89, 10591–10601.

19 THE RUBELLA VIRUS, KNOWN FOR ITS TERATOGENIC EFFECTS, MODIFIES THE ABILITY OF HUMAN INDUCED PLURIPOTENT STEM CELLS TO DIFFERENTIATE INTO ENDODERMAL CELLS

1. INTRODUCTION: AN OVERVIEW

Rubella virus (RV), a member of the Rubivirus genus within the Togaviridae family, is an enclosed, single-stranded RNA virus with a positive-sense genome. It is unique among viruses in its ability to produce intrauterine infections. The mechanisms by which these viruses are passed from an infected mother to the fetus and their effects on human development are only poorly understood. RV, a highly efficient teratogen, causes congenital rubella syndrome (CRS) in humans. CRS is characterized by a classical triad of clinical symptoms, which includes sensorineural deafness, congenital heart disease (including cardiovascular and vascular anomalies), and cataracts [1,2]. Cardiorespiratory syndrome (CRS) can involve abnormalities in the heart such as ventricular and atrial septal defects, patent ductus arteriosus, and patent foramen ovale. Congenital rubella can cause several eye conditions, such as cataract, microphthalmia, glaucoma, and pigmentary retinopathy [1,2]. Moreover, in tissue samples obtained from three fatal cases of CRS, the presence of RV was identified in fibroblasts located in the cardiac and adventitia (aorta and pulmonary artery) regions, in conjunction with vascular lesions [3]. The risk of developing congenital abnormalities is particularly high after maternal rubella infection during gestational weeks 11 and 12 [4,5,6]. Therefore, the risk of intrauterine RV infection is only significant in the first trimester of pregnancy. Although congenital abnormalities are frequently observed, there is no significant increase in premature delivery and stillbirths following intrauterine RV infection [1]. The study of human embryogenesis and congenital abnormalities is subject to several ethical limitations, particularly due to the inaccessibility of early implantation stages of human embryos [7]. Embryonic stem cells (ESCs) and induced pluripotent stem cells (iPSCs) are two types of human pluripotent stem cells (PSCs). These innovative cell culture platforms, known as ex vivo, enable the study of human embryonic germ layer segregation and also facilitate developmental toxicity testing [8]. They serve as a cell culture model that mimics the blastocyst stage and can be further developed into stages resembling gastrulation by differentiating into embryonic germ layer derivatives, including ectoderm, mesoderm, and endoderm. Furthermore, their appropriateness as a developmental model has been proven for cardiac commitment throughout embryonic development [9], given that the heart is the initial organ to form and cardiac cell fate determinations take place at an early stage. In addition, the growth of embryonic stem cells (ESCs) can be enhanced by using appropriate 3D matrices or by co-culturing them with trophoblast cells. This allows the development of blastoids, gastruloids, and even embryoids (or embryo-like structures) that can serve as models for studying human embryogenesis in culture dishes [7,10]. PSCs and PSC-based differentiation models, particularly the mouse ESC test, have already been confirmed as reliable methods for assessing teratogenic and embryotoxic chemicals, such as thalidomide (marketed as Contergan®) [11,12]. Nevertheless, the assessment of their potential in studying infections during pregnancy is still in its early stages [13,14]. iPSCs have inherent mechanisms that limit virus infections, which aligns with the small number of viruses that can induce perinatal infection. Furthermore, iPSCs exhibit a greater degree of expression for a specific group of interferon (IFN)-induced genes, as compared to differentiated somatic cells [14]. This seems to compensate for the lack of a type I IFN response in iPSCs, which is a crucial part of the body's natural defense against viruses [15]. The teratogenic rubella virus (RV) may be preserved in induced pluripotent stem cells (iPSCs) through multiple passages. These iPSCs can then be guided to differentiate into cells of the embryonic germ layers, providing valuable insights into the initial processes behind rubella embryopathy. This demonstrates the potential of iPSCs as a model for studying the early mechanisms of rubella-induced birth defects. In continuation of this investigation, our objective was to identify molecular changes generated by RV in these cells both before and after the start of directed differentiation using transcriptomics. The most significant impacts linked to RV infection were observed in endodermal cells produced from RV-infected induced pluripotent stem cells (iPSCs). Markers indicating the presence of definitive endoderm were increased, and this was accompanied by significant changes in the epigenetic profile. Additionally, there was an increase in the expression of proteins involved in the formation of blood vessels, and a decrease in the activity of the TGF-β signaling pathway. In addition, ectodermal cells displayed a modified expression pattern of crucial transcription factors involved in the development of the eye field following respiratory virus infection. Therefore, investigating the impact of RV infection on iPSCs and their derived lineages offers valuable information about how the virus can affect the early stages of development and contribute to the development of congenital disorders.

2. MATERIALS AND METHODS

2.1. Cell Lines and Cultivation

Vero (green monkey kidney epithelial cell line, ATCC CCL-81) and A549 (human lung carcinoma epithelial cells, ATCC, LGC Standards GmbH, Wesel, Germany) were cultured in Dulbecco's modified Eagle's medium (DMEM; Thermo Fisher Scientific, Darmstadt, Germany) with high glucose, GlutaMAX, 10% fetal calf serum (FCS) and 100 U/mL penicillin–streptomycin. If not otherwise indicated, the vector-free human episomal A18945 iPS cell line (alias TMOi001-A), (Thermo Fisher Scientific) was maintained in mTeSR™1 medium (StemCell Technologies,

Cologne, Germany) with 10 µg/mL gentamycin on Matrigel™ (BD Biosciences, dispensed in DMEM/F-12)-coated culture plates with daily medium change. They were passaged enzymatically at a ratio of 1:6 to 1:10 every 3 to 5 days with collagenase type IV (Thermo Fisher Scientific) in DMEM-F12 with the addition of 10 µM Y-27632 ROCK inhibitor.

2.2. Directed and Undirected Differentiation of iPSCs

Directed differentiation was performed as an endpoint differentiation assay through the STEMdiffTM trilineage differentiation kit (StemCell Technologies). The differentiation protocol was performed according to the manufacturer's instructions and required cultivation of A18945 iPSCs in mTeSR™1 medium. Single cells, as obtained after treatment with Accutase (Merck/Sigma-Aldrich Chemie GmbH, Taufkirchen, Germany), were plated on Matrigel. Every 24 h medium change of the respective STEMdiffTM trilineage differentiation medium for ectoderm, mesoderm, and endoderm was performed. Samples were collected after 5 days (mesoderm and endoderm) and 7 days (ectoderm) of cultivation. Undirected differentiation was initiated 24 h after collagenase-passaging of iPSC cultures at a ratio of 1:4 through application of undirected differentiation medium (DMEM-F12, 1x MEM-NEAA, 0.2 mM L-glutamine, 20% FBS, 0.11 mM β-mercaptoethanol, and 100 U/mL penicillin) followed by further cultivation for 5 days.

2.3. Embryoid Body Formation

EB formation as based on a previous publication [16] and (http://www.biolamina.com/media.ashx/instructions-bl010.pdf) was carried out in suspension culture and single cell suspensions were obtained after Accutase (Sigma-Aldrich) treatment. A total of 1 × 106 cells was seeded in 200 µL of EB culture medium (DMEM-F12, 20% KnockOut™ Serum Replacement (Thermo Fisher Scientific), 1× MEM-NEAA, 0.2 mM L-glutamine, 0.11 mM β-mercaptoethanol and 1 mg/mL Gentamicin) medium into one well of a nontreated conical 96-well plate and centrifuged at 600× g for 5 min. After cultivation for 2 days the EBs were transferred according to the protocol to a low attachment flat-bottom six-well plate and medium was changed every third day.

2.4. Virus Infection and Interferon Assays

The supernatant of infected Vero cells was collected and cleared from cellular debris by centrifugation at 350× g for 10 min at 4 °C and filtration through a 0.45 μm syringe filter. Thereafter ultracentrifugation with a 20% sucrose cushion (w/v in PBS) was performed for 2 h at 25,000 rpm and 4 °C. The obtained pellets were resuspended in mTeSR1. Viral titers were determined by standard plaque assay. As described previously [13], iPSC cultures with a 40–50% confluency were acutely infected with 7.5 × 105 plaque forming units (PFU) of RV per well of a 24-well plate. This corresponds approximately to an MOI of 20. The applied MOI can only be estimated as iPSCs were passaged enzymatically in clumps. The inoculum was replaced with fresh mTesR1 medium after 2 h of incubation [13]. After 4 to 5 days of cultivation, RV-infected iPSCs were passaged.

For exogenous (or paracrine) IFN treatment human recombinant IFN lambda 1 (IL-29, #300-02L) and 2 (IL28A, #300-02K), were purchased from Peprotech (Hamburg, Germany), and 3 (IL-28B, #CS26) from Novoprotein (Novoprotein, PELOBIOTECH GmbH, Planegg/Martinsried, Germany). The Accuri C6 flow cytometer (BD Bioscience, Heidelberg, Germany) was used for IFN measurement by the LEGENDplex human type 1/2/3 IFN panel (BioLegend, San Diego, CA, USA). The double-stranded (ds) RNA analogue polyinosinic-polycytidylic acid (poly I:C; Santa Cruz Biotechnology, Heidelberg, Germany) was added either directly to the cell culture or transfected at a concentration of 1 μg using Lipofectamine 2000 (Thermo Fisher Scientific) as transfection reagent.

2.5. Calcein Live Cell Staining

For live cell staining, EBs were incubated with mTeSR1 plus calcein FM (Sigma-Aldrich) at 1 μM. After an incubation period for 30 min, EBs were washed twice with PBS and analyzed on an inverted fluorescence microscope.

2.6. RNA Isolation

Total RNA was extracted from mock- and RV-infected cells by Trizol reagent (Thermo Fisher Scientific). The purification was performed with the Direct-zol RNA kit (Zymo Research, Freiburg, Germany) according to manufacturer's instructions. The integrity of the RNA samples was confirmed through analysis on a fragment analyzer (Advanced Analytical). Only samples with a RIN (RNA integrity number as a means of quality assessment) equal to 7 or greater were subjected to further analysis.

2.7. Microarray Gene Expression Analysis and SOM Portrayal

Isolated RNA was processed and hybridized to Illumina HT-12 v4 Expression BeadChips (Illumina, San Diego, CA, USA) and measured on the Illumina HiScan. Raw intensity data of 47,323 gene probes was extracted by Illumina GenomeStudio and subsequently background corrected, transformed into log10-scale, quantile normalized, and centralized to obtain gene expression estimates. Two independent samples per condition and cell type were processed.

Expression data were then further processed using self-organizing map (SOM) machine learning. The method distributes the gene-centered expression values among 2500 microclusters called meta-genes, which were arranged in a two-dimensional 50 × 50 lattice and colored in maroon-to-blue for high-to-low meta-gene expression values. These mosaic images visualize the transcriptome patterns of each individual sample and therefore can be understood as their molecular portraits exhibiting clusters of coexpressed genes in the samples studied [17]. Mean portraits over replicates were calculated by averaging the meta-gene landscapes of replicated samples while difference portraits between different cell types were obtained by subtracting the respective metagene values to highlight differentially expressed genes. Clusters of coexpressed genes were identified by selecting so-called 'spot-areas' in the SOM portraits using overexpression criteria as described previously [17]. For functional interpretation of the expression-modules, we applied gene set enrichment analysis using the gene set Z-score (GSZ), [17]. Enrichment of functional gene sets in the spot cluster was calculated by applying Fisher's exact test. We considered gene sets related to biological processes (BP) of the gene ontology (GO) classification, standard literature sets [17,18], and literature sets curated by our group.

Downstream analysis methods were described previously [17,19] and are implemented in the R-package 'oposSOM' used for analysis [20].

Pathway activity was analyzed based on pathway topologies and gene expression data using the pathway signal flow method as implemented in oposSOM [21].

2.8. Quantitative Real-Time PCR Analysis of Viral and Cellular RNA

For determination of the mRNA expression level of selected cellular genes, 1.2 µg of total RNA were reverse transcribed with Oligo(dT)18 primer and AMV reverse transcriptase (Promega, Mannheim, Germany) at 42 °C for 1 h. This was followed by an incubation step at 70 °C for 10 min. The carousel-based LightCycler 2.0 (Roche, Mannheim, Germany) was used for quantitative real-time PCR (qRT-PCR) experiments. These experiments included a 1:5 dilution of the respective cDNA samples together with 1 µg BSA and the GoTaq® qPCR master mix (Promega). [22]. Two different approaches for relative expression analysis were pursued. For direct comparison of one sample type after mock- and RV-infection, comparative delta delta Ct (ΔΔCt) was used. For comparison of gene expression levels among different cell types within a large data set, a modified version of the comparative delta delta Ct (ΔΔCt) method was used. The normalized relative quantity (NRQ) values were derived from qbase+ software (Biogazelle, Zulte, Belgium) which are based on the mean expression values of all samples and replicates within a given data set [23].

2.9. Immunofluorescence

For assessment of viral proteins, immunofluorescence was carried out as described [13]. Briefly, cells were fixed with 2% (w/v) paraformaldehyde in PBS and permeabilized with 0.1 Triton X-100 followed by incubation with mAb anti-E1 from Viral Antigens (Viral Antigens Incorporation, Memphis, TN, USA) at a 1:200 dilution as primary antibody.

2.10. Statistical Analysis

All statistical calculations were done with Graph Pad Prism software (GraphPad Software, Inc., La Jolla, CA, USA). Asterisks (* $p < 0.05$, ** $p < 0.01$, *** $p < 0.001$, **** $p < 0.0001$) highlight the level of significance in diagrams which include data as means ± standard deviation (SD). For comparison of normalized mRNA expression levels in RV-infected samples with the corresponding mock controls, a paired Student's t test (consistent ratios of paired values) was applied. Statistical analysis for different samples was based on one-way ANOVA followed by Bonferroni's multiple comparison test.

3. RESULTS

3.1. In the Presence of RV, iPSCs Maintain Pluripotent Properties and Lineage Identity after Initiation of Differentiation

Specification to one of the three germ lineages is the first critical step in directing differentiation to downstream cellular phenotypes. Therefore, directed as well as undirected differentiation was induced in RV-infected iPSCs which were subcultured for two to five passages. Passaging of infected iPSCs results in a homogenous level of infection within iPSC cultures without affecting the protein expression level of the pluripotency marker OCT4 [13]. During passaging of RV-infected iPSCs, replication occurred at a rather constant rate as assessed by viral titer and E1 protein expression rate [13]. Furthermore, passaging allows for adaptation of RV to iPSCs and excludes any possible effects of the differentiation process itself on the otherwise acute infection with RV (Figure 1A). Undirected differentiation is spontaneous and was thus induced to assess whether RV, without a specific differentiation stimulus, directs a gene expression profile different from the mock-infected population. Directed differentiation of RV-infected iPSCs into ecto-, endo-, or mesodermal cells (thereafter referred to as RV-infected) was initiated with the STEMdiff trilineage differentiation kit as an endpoint differentiation approach to determine which of the early cell fate decision pathways could be affected by RV.

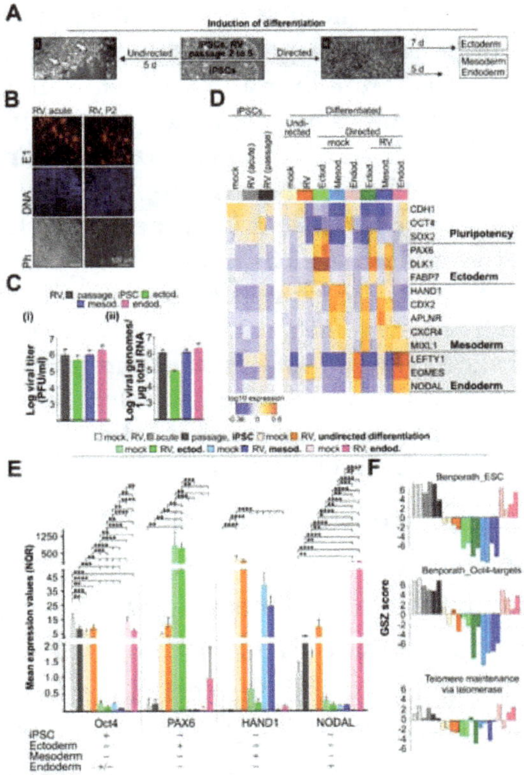

Figure 1. The identity of iPSCs and derived lineages was maintained in the presence of rubella virus (RV). (A) Overview of the methods applied to initiate differentiation in mock- and RV-infected iPSCs. (A) [i] Undirected differentiation in the respective induction medium started at the rim region of iPSC colonies (indicated by white arrows) and extended to their center over time of incubation. (A) [ii] Additionally, directed differentiation into the primary germ layers ectoderm, mesoderm, and endoderm was induced with the STEMdiff differentiation kit. (B) Immunofluorescence analysis with anti-E1 antibody (shown in red) was performed to monitor distribution of RV-positive cells within iPSC colonies. Nuclei are shown in blue. Ph, phase contrast (C) To assess RV replication in iPSCs and derived lineages, (i) virus progeny, and (ii) the amount of genomic viral RNAs was determined by standard plaque assay (n = 11 for passaged iPSCs, otherwise n = 3) and TaqMan-based reverse transcription-quantitative PCR (passaged iPSCs and ectodermal cells n = 3, mesodermal cells n = 7, endodermal cells n = 4), respectively. (D) Expression heatmap of selected marker genes of pluripotency and lineage identity (based on microarray whole transcriptome data) in mock- and RV-infected iPSCs and iPSC-derived lineages. Shading indicates overlap in the expression of some of the

marker genes between iPSCs and iPSC-derived lineage cells, respectively. (E) The expression of indicated target genes in mock- and Wb-12-infected iPSCs and derived cells after initiation of undirected and directed differentiation was determined by real-time quantitative PCR (RT-qPCR) and analyzed by qbase+ software. For normalization, chromosome 1 open reading frame 43 (C1orf43) and hypoxanthine guanine phosphoribosyl transferase 1 (HPRT1) were used. Relative gene expression was calculated as normalized relative quantity (NRQ) and given as means ± SD (n = 3 to 5). As a reference, lineage-specific expression levels were assigned based on literature data and transcriptomics data by Stemcell Technologies for the Trilineage differentiation kit. (F) Analysis of stemness-related expression signatures was based on the GO gene set telomere maintenance. An embryonal stem cell signature and a gene set collecting OCT4 targets were both taken from [25].

RV establishes a noncytopathic infection of iPSCs with a homogenous distribution of infected cells within the respective colony (Figure 1B), [13]. Differentiation into all three embryonic germ layers supported RV replication at a comparable rate (Figure 1C [i,ii]), [13]. As a next step, we generated an expression heatmap of selected marker genes (based on microarray whole transcriptome data) for assessment of pluripotency and lineage identity (Figure 1D). In agreement with the maintenance of OCT4 (octamer-binding transcription factor 4, also known as POU5F1) expression in RV-infected iPSCs [13], high expression of pluripotency markers CDH1 and OCT4 was noted. Their expression was maintained to some degree in endodermal cells, which is in agreement with the conditions of the STEMdiff trilineage differentiation kit. The same applies to the expression of the pluripotency marker SOX2 (SRY (Sex Determining Region Y)-Box 2) in ectodermal cells. The expression profile of lineage-specific markers confirmed ectodermal (PAX6 (Paired Box 6), DLK1 (Delta-Like 1 Homolog), and FABP7 (Fatty Acid Binding Protein 7)), mesodermal (HAND1 (Heart and Neural Crest-Derived Transcript 1), CDX2 (Caudal Type Homeobox 2), APLNR (Apelin Receptor)) and endodermal (LEFTY1 (Left-Right Determination Factor 1), EOMES (Eomesodermin), NODAL (Nodal Growth Differentiation Factor)) identity after initiation of directed differentiation in mock- and RV-infected iPSCs (Figure 1D). Additionally, some overlap between lineages, especially between mesoderm and endoderm, was noted for

RV-infected samples (Figure 1D). Transcriptomic data was confirmed by RT-qPCR of pluripotency marker OCT4 and of selected markers for ectoderm (PAX6), mesoderm (HAND1), and endoderm (NODAL) lineages followed by relative quantification by qbase+ method (Figure 1E). Figure 1D and E indicate that during undirected differentiation, especially mesodermal markers were expressed, which occurred at a comparable level between mock- and RV-infected cells. Among the lineage-specific markers, the expression level of HAND1 was significantly downregulated in mesodermal cells after RV infection. Additionally, RV infection did not alter stemness-related expression signatures as indicated by the transcriptomic activity of the GO gene set telomere maintenance (Figure 1F). Telomere maintenance is active in stem cells, but gets deactivated in differentiated somatic cells [24]. In summary, comparable to the mock-control, RV maintained the pluripotent properties of iPSCs and enabled initiation of differentiation into embryonic germ layer cells as indicated by expression of essential germ layer markers.

3.2. High-Resolution Transcriptomic Maps Reveal Modules of Coregulated Genes Promoted by RV Infection during Endodermal Differentiation

As we found out that RV infection did not affect unspecific differentiation of iPSCs and enabled their lineage-specific differentiation, we wanted to focus on the effect of RV on lineage identity. The self-organizing map (SOM) transcriptome data portrayal provides a high-resolution visualization of the transcriptome landscape of each cell system studied in terms of a quadratic mosaic image and decomposes into clusters of coregulated genes. They are represented as colored spot-like areas where red and blue colors code activated and deactivated gene clusters, respectively. These transcriptomic portraits were then used to evaluate the mutual relatedness between the cell systems by means of a phylogenetic similarity tree (Figure 2A). The tree structure results from the fact that common and different spot patterns in the portraits reflect mutual similarities and differences of the activated cellular programs which enable judging the effect of RV-infection on the different lineages (see the portraits in Figure 2A).

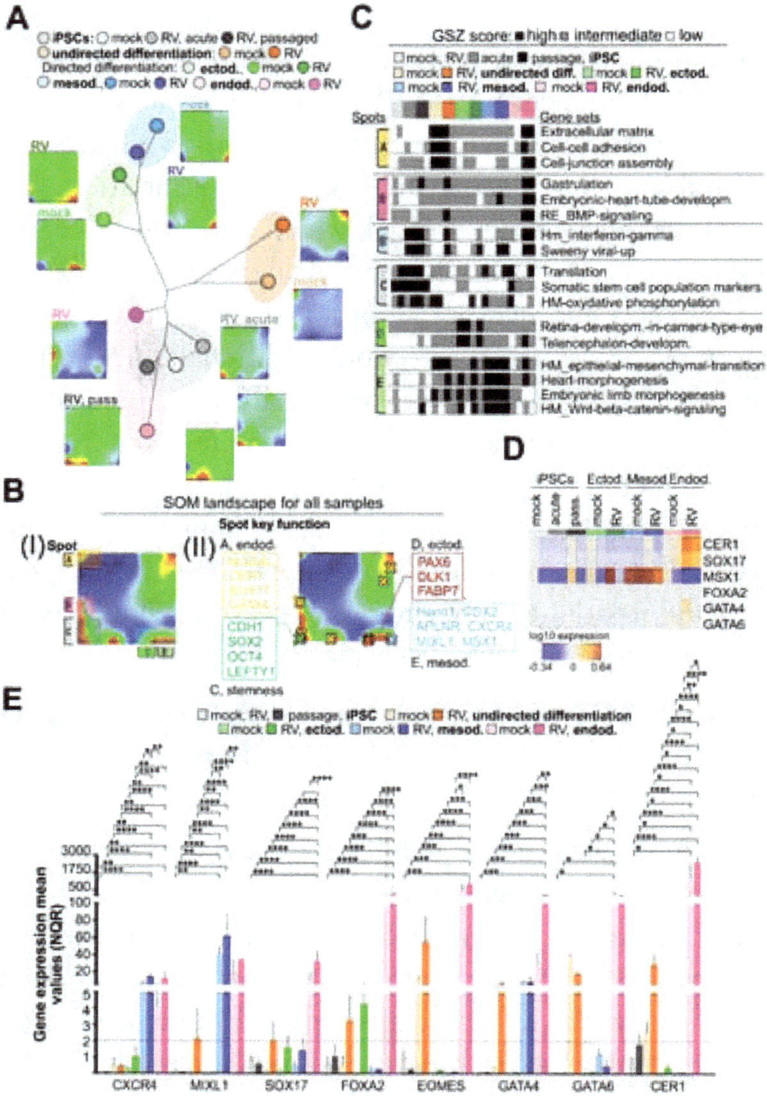

Figure 2. RV supports expression of markers for definite endoderm. (A) Similarity tree of the gene expression portraits of the cell systems studied. Both, mock- and RV-infected mesodermal and ectodermal cells share relatively high mutual similarity of their transcriptomes. In contrast, mock- and RV-infected endodermal cells form a separate branch that is closer to the mock- and RV-infected iPSCs. (B) Spot summary map (I) provides an overview of activated cellular programs and their functional context, which is depicted in more detail in (II). (C) Barplot representation of the expression profiles of gene sets related to different functions. Their genes were enriched in the spots that were identified

in the SOM portraits. They are thus indicated accordingly. (D) Selected marker genes for definite endoderm are illustrated in the heatmap of the transcriptome of mock- and RV-infected iPSCs and iPSC-derived lineages. (E) The expression of marker genes for definite endoderm was determined by RT-qPCR in mock- and Wb-12-infected iPSCs and iPSC-derived lineages by qbase+ software. For normalization, C1orf43 and HPRT1 were used. Relative gene expression was calculated as normalized relative quantity (NRQ) and given as means ± SD (n = 3 to 5). See also Figures S1 and S2.

For an overview, we generated a spot-summary map in Figure 2B which shows the activated spots observed in any of the samples together with their functional context as extracted by means of gene set enrichment analysis of the genes in each of the spot-clusters of coexpressed genes. In total, we identified five relevant spots labeled with capital letters A–E. Each of the spots is characterized by a specific expression profile (Supplement Figures S1 and S2) which, in turn, shows close similarities with the expression profiles of distinct gene sets (shown as 'barcode' plots in Figure 2C).

Mapping of the position of genes into the SOM enables us to deduce their expression profile in the cell systems studied according to the gene's location in or near the spots and the respective spot profiles (Figure 2B). Importantly, key genes of ectoderm and mesoderm development are found in spots D and E, respectively, and are confirmed to be upregulated in the ectoderm and mesoderm cells. On the other hand, genes related to gastrulation and heart-tube development were enriched in spot B and found to be activated in RV-infected endoderm cells. This occurred together with genes that are activated upon interferon response (see next subsection). Stemness key genes locate in spot C together with part of the developmental genes of the endoderm.

Difference portraits clearly indicate that spot B associates with RV infection (Figure 2B), which contains the genes NODAL, CER1, SOX17, and GATA4, reflecting their activation upon RV infection in endodermal cells. In addition, we are able to show that mesoderm and ectoderm cells share similar expression patterns characterized by upregulated spots D and E and downregulated stemness genes (spot C), which, in contrast, are upregulated in endoderm cells. This is in agreement with the expression level of pluripotency genes in the endodermal lineage which was higher than in the remaining two embryonal germ layers (Figure 1D,E). As a consequence, mesoderm and ectoderm cells occupy neighboring positions at one end of the similarity tree, while endoderm cells are found at the opposite end. These findings highlight two important aspects: (I) there is a close similarity of transcriptional patterns among ecto-, mesoderm, and iPSC samples without and with RV infection; and (II) the endodermal lineage is an exception, where RV infection induces a notable shift away from the corresponding mock sample.

To further elucidate this specific effect of RV, we analyzed the transcriptomic data for markers for definitive endoderm (CXCR4 (C-X-C Motif Chemokine Receptor 4), MIXL1, SOX17, FOXA2 (Forkhead Box A2), EOMES, GATA6, CER1 (Cerberus, DAN family BMP Antagonist), and LEFTY, [26]. The expression heatmap shown in Figure 2D indicates an increased expression level of CER1 and SOX17 after RV infection. Figure 2E shows qPCR analysis followed by relative quantification by qbase+ method and highlights that the definite endoderm markers CXCR4, MIXL1, SOX17, GATA4/6, and CER1 were significantly higher expressed in RV-infected endodermal cells as compared to their mock-infected counterparts. While RV infection did not induce these markers above a cut-off of a two-fold increase during mesodermal and ectodermal differentiation, EOMES, GATA4, and CER1 were specifically induced by RV during undirected differentiation.

In conclusion, similarity analysis supports the hypothesis that ectodermal and mesodermal lineage identity was maintained after infection with RV, while endodermal cells derived from RV-infected iPSCs were enriched in markers for definitive endoderm.

3.3. RV Infection Activates IFN Type III Response Pathways on iPSCs and Derived Lineages

Difference portraits (Figure 3A) indicated that RV-infection specifically upregulated genes in spot B, which were associated with "IFN response" characteristics. This is supported by the profiles of gene expression signatures of viral infections such as by influenza virus and pneumonia that is accompanied by interferon activation. These genes were consistently upregulated in RV-infected samples with the largest observed effect in endodermal cells, [27,28,29]. Spot B highlighted in the SOM landscapes contained genes involved in IFN and viral response mechanisms (Figure 3A, a list of genes is given in Figure 3B). This is further emphasized by the expression heatmap shown in Figure 3C. Whereas genes involved in IFN-sensing, including the type III IFN receptor IFNLR1 (IFN lambda receptor 1), were not altered in their expression level, the IFN-signaling components STAT1 (signal transducer and activator of transcription 1) and IRF9 (interferon regulatory factor 9) appeared to be slightly upregulated at their mRNA expression level after infection with RV, especially in endodermal cells. The highest level of upregulation was found for IFN-stimulated genes (ISGs), notably for MX1 (MX dynamin like GTPase 1), IFITM2 (interferon induced transmembrane protein 2), and ISG15 (interferon-stimulated gene 15). Mapping of these genes into SOM space further underlines these findings: The IFN-signaling genes and the ISGs accumulate in and around spots B and D, respectively, while IFN-sensing genes are located outside these spot regions (Figure 3B). The increase in the expression level of selected marker genes of the IFN pathway in the presence of RV was confirmed by RT-qPCR (Figure 3D). Compared to RV-infected iPSCs and ecto- and mesodermal cells, the expression level of IRF9 and STAT1 and selected ISGs (IFITM1/2, IFIT1, and ISG15) was significantly higher in RV-infected endodermal cells. The highest increase in mRNA expression after RV infection was noted for the ISGs IFIT1 and ISG15. Therefore, we determined whether this gene expression pattern was indeed associated with IFN generation during RV infection through quantification of type I (α and β), type II (γ), and type III (λ1 and λ2/3) IFNs by the LEGENDplex assay from the supernatants of RV-infected cells (Figure 3E). In iPSCs as well as iPSC-derived lineage cells, RV infection induced secretion of type III IFNs, namely IFN λ2/3 (Figure 3E). As a positive control, the synthetic dsRNA analog poly I:C was used, which was either transfected into iPSCs or added directly to the supernatant. Either application of poly I:C did not lead to secretion of any type I, II, or III interferons (Figure 3E). The activation of type III IFNs by RV was also confirmed at the mRNA level by RT-qPCR (Figure 3F). Compared to the mock-infected control, RV induced a significant increase in the mRNA expression of IFN λ2/3 in endodermal cells (Figure 3F). Thereafter, we addressed the discrepancy between IFN λ2/3 protein (Figure 3E) and mRNA level (Figure 3F). Gene set analysis revealed that mRNAs associated with the KH type-splicing regulatory protein (KHSRP) were specifically enriched in endodermal cells (Figure 3G). KHSRP is involved in post-transcriptional regulation of mRNA expression, including IFN λ3 [30]. This could explain the discrepancy between IFN λ2/3 protein and mRNA expression level.

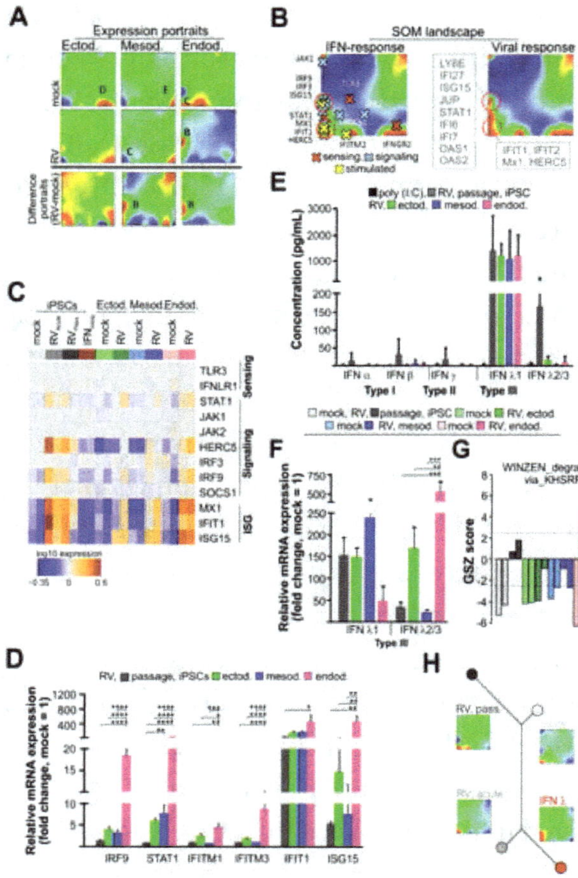

Figure 3. RV infection activates an IFN response in iPSCs and derived lineages. (A) Expression portraits of the embryonal germ layers before and after RV infection and the respective difference portraits reveal characteristic spot patterns, where spots C, D, and E are specific for

endoderm, ectoderm, and mesoderm cells, respectively. Spot B appears after RV infection mainly in endodermal cells. (B) IFN response genes with signaling and stimulated functions in the IFN response pathways accumulate in spots B and C and were upregulated predominantly in iPS and endodermal cells after RV infection as also indicated by the IFN and viral response gene signature profiles. (C) Expression heatmap of selected marker genes involved in IFN-sensing and -signaling in mock- and RV-infected iPSCs and iPSC-derived lineages. Interferon-stimulated genes (ISGs) that were identified by the SOM analysis shown in (A) are included. (D) The mRNA expression level of the IFN-signaling components IRF9 and STAT1 and indicated ISGs was verified by RT-qPCR analysis. Data are given as means ± SD (n = 3 and n = 5 for RV-infected mesoderm, IRF9 and STAT1). (E) The IFN profile for RV-infected iPSCs and derived lineages was determined by the LEGENDplex IFN panel for undiluted supernatants collected after five (iPSCs and mesodermal cells) and seven days (endodermal and ectodermal) of cultivation. (F) The mRNA expression level of type III IFNs was verified by qPCR analysis and given as means ± SD (n = 3). (G) Mean expression of a gene set (gene set Z-score, GSZ) that is controlled by KSRP, which appears to keep inflammatory gene expression within defined limits [31]. (H) Similarity tree of the gene expression portraits of mock- and RV-infected iPSCs in comparison to iPSCs after cultivation in the presence of exogenous type III IFNs. (D,F) For normalization of qRT-PCR data in the 2-ΔΔCt method, the HPRT1 gene was used.

To address the influence of the type III IFNs secreted during RV to the cell culture supernatant on the gene expression landscape of iPSCs, type III IFNs were added exogenously for two weeks of cultivation during daily medium change of iPSCs. The zoom-in similarity tree shown in Figure 3H highlights the relatedness between the expression portraits of mock- and RV-infected iPSCs as well as iPSCs after application of type III IFNs. The gene expression profile of passaged RV-infected cells shifted away from iPSCs after exogenous IFN type III application, but closer to the mock-infected cells, suggesting an adaptation of RV to iPSCs.

In conclusion, in iPSCs and iPSC-derived embryonic lineages, RV infection induced a type III IFN response together with activation of ISGs, notably MX1 and ISG15. This activation appeared to be specifically profound in endodermal cells.

3.4. RV Infection Is Associated with Chromatin Remodeling

Alterations of gene expression patterns during development are governed by epigenetic mechanisms in cooperation with regulation via transcription factor networks [32,33]. Particularly, we found that gene signatures of epigenetic impact, such as targets of the polycomb repressive complex 2 (PRC2), of H3K27me3, and of bivalently (H3K4me3 and H3K27me3) marked gene promoters, have an almost antagonistic expression profile as compared to the stemness signatures (Figure 4A, in comparison to Figure 1F).

Next, we focused on chromatin remodeling which is essential for lineage segregation [9,34]. The analysis of transcription factor networks that act in regions of euchromatin with transcriptionally active genes and regulatory elements revealed that genes in repressed and bivalent states of endoderm progenitors become more rigorously deactivated in RV-infected endoderm cells than genes from these states of mesoderm progenitors in RV-infected mesoderm cells. This suggests that RV infection specifically dedifferentiates endoderm cells by suppressing developmental suppressors. We then studied enzymes affecting methylation of DNA and of arginine and lysine side chains of histones such as H3K4, H3K9, H3K27, and H3K36, with potential impact on chromatin structure. The profiles of the DNA methylation maintenance methyltransferase DNMT1, of PRMT6, a methyltransferase of histone arginine side chains, and of JMJD1c, a H3K9 demethylase, and KDM5b demethylating H3K4me3 correlate with the 'stemness' spot cluster C upregulated in iPS and endodermal cells.

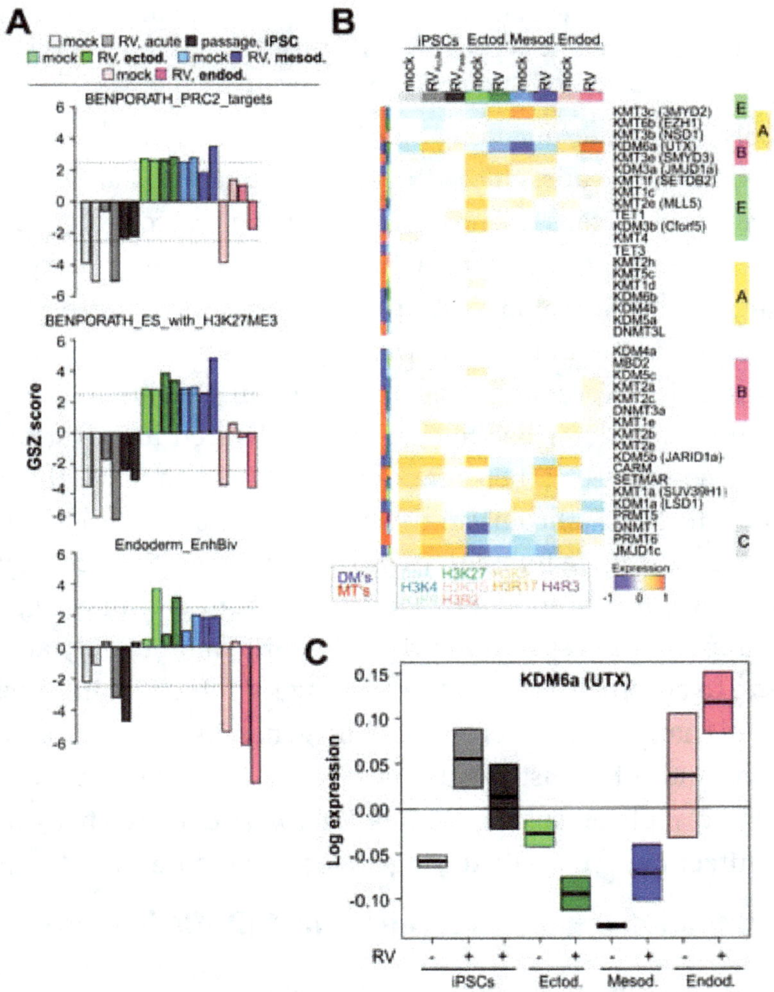

Figure 4. RV infection is accompanied by an altered expression of components of the SWI/SNF and NURF chromatin remodeling complexes. (A) Mean GSZ expression signatures of stemness-related transcriptional programs which act via epigenetic programming (PRC2 targets, repressive and bivalent chromatin marks). (B) Expression heatmap of chromatin modifying enzymes in the cell systems studied. The gene expression data of methyltransferases (MTs) and demethylases (DMs) of DNA cytosines, histone lysine and arginine side chains were assigned to the expression spot-cluster A–E according to their expression profiles. This suggests their involvement in the regulation of chromatin structure as writers and erasers of methylation marks at histone lysine and arginine side chains and affecting DNA methylation. Notably, a very strong variability was observed for KDM6a (alias UTX), a constituent of the SWI/SNF ATP-

dependent chromatin remodeling machinery. (C) Expression of KDM6a directly relates to stemness programs (Figure 1F) and inversely relates to programs repressing stemness functions (part A of the Figure).

Notably, the gene encoding the H3K4 demethylase KDM6a (alias UTX) was markedly upregulated in RV-infected iPSCs and, especially, endoderm-derived cells (Figure 4B). KDM6a is a constituent of the SWI/SNF ATP-dependent chromatin remodeling machinery. Figure 4C highlights that in comparison to iPSCs and ecto- and mesodermal cells, the upregulation of KDM6a was highest in endodermal cells. The alterations of its expression suggest its role in chromatin remodeling after RV infection described above. This motivated us to estimate the expression patterns of other genes encoding components of the SWI/SNF and of the NURF chromatin remodeling complexes [9] by mapping them into the SOM. We found that, indeed, Smarcc2, Smarcd3, and Btpf were all upregulated in endoderm-derived cells after RV infection, which further supports the assumption that SWI/SNF and possibly also NURF contribute to chromatin remodeling during RV infection. In conclusion, expression changes of different sets of genes involved in epigenetic regulation and of constituents of the ATP-dependent chromatin remodeling complexes such as KDM6a-UTX were detected in association with RV-infection, especially during endodermal differentiation.

3.5. RV Infection Impairs Aggregation of iPSCs into Embryoid Bodies

The progression of embryogenesis does not only involve the activation of developmental pathways, but also requires cell–cell interactions based on adhesive forces [35]. The relevance of these observations for RV-infected iPSCs was emphasized by gene ontology analysis regarding focal adhesion and regulation of cell adhesion (Figure 5A). Transcriptomic analysis revealed that THY-1 (also known as cluster of differentiation (CD) 90) was among the targets affected by RV (Figure 5B). The relevance of THY-1 (CD90) for cellular adhesion capacity was highlighted for CD90 negative carcinoma, which compared to their CD90 positive counterparts lack the ability to form spheres [36]. Accordingly, we have addressed whether RV alters the spontaneous aggregation capacity of iPSCs into 3D aggregates called embryoid bodies (EBs). EB formation relies on cell–cell adhesive interactions [35]. Compared to the mock control, EBs generated from RV-infected iPSCs were reduced in diameter and of irregular shape (Figure 5C). Furthermore, during cultivation they lost stability and small-sized debris was generated (Figure 5C). In contrast to RV-infected iPSCs, the mock-infected controls generated viable EBs as indicated by staining with calcein performed after two weeks of cultivation (Figure 5D). In conclusion, RV infection impaired the adhesion capacity of iPSCs as shown by their reduced ability to assemble into EBs. This suggests an impaired cell–cell interaction capacity during lineage segregation.

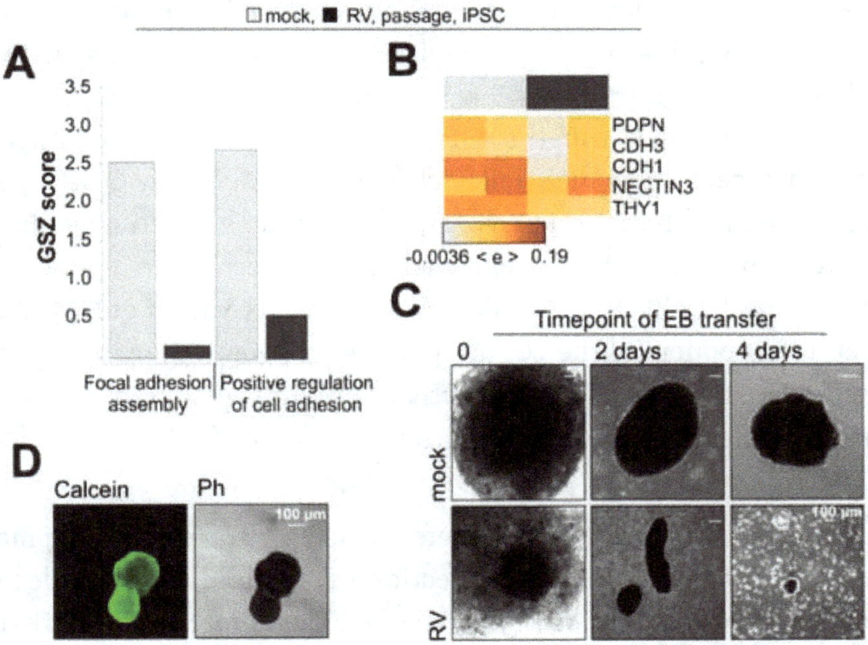

Figure 5. RV impairs the cellular adhesion capacity of iPSCs. (A) Gene signatures related to focal adhesion in mock- and RV-infected iPSCs. (B) Expression heatmap of selected marker genes involved in cellular adhesion. (C) Assessment of 3D stability through embryoid body (EB) formation. Shown are images before and after cultivation in suspension. (D) To verify viability, EBs were stained with calcein after cultivation for two weeks.

3.6. RV Infection Specifically Affects Developmental Pathways during Endodermal Differentiation

For assessment of the effect of RV on global cellular signaling networks, we focused on two important signaling pathways, namely transforming growth factor β (TGF-β) and Wnt/β-catenin (Wnt), (Figure 6A,B, respectively). The TGF-β signaling pathway is involved in cell growth and differentiation during embryogenesis [37]. The Wnt signaling pathway regulates the interaction between cellular pathways involved in primary germ layer formation and is required for mesodermal differentiation from pluripotent stem cells [38]. As expected, the highest TGF-β signaling pathway activity was observed in endodermal cells, while the Wnt signaling pathway was most active in mesodermal cells. Thus, these two lineages were depicted in Figure 6A,B, respectively, to highlight the effect of RV on their activity in comparison to the respective controls. Within the TGF-β signaling pathway, mock-infected endodermal cells show high cell cycle activity induced by CDKN2B and its downstream interaction partners, which became deactivated during RV infection (Figure 6A). Additionally, RV infection in endodermal cells was specifically accompanied by a strong activation of NODAL, an essential component of the TGF-β signaling pathway (Figure 6A). During RV infection, the transcriptional activity of Wnt signaling pathway was reduced in mesodermal cells (Figure 6B), whereas for ectodermal and endodermal cells, almost no alteration in its activity was detected.

As congenital rubella leads to defects in heart and eye development, we analyzed the impact of RV infection on the underlying molecular pathways. For members of the gene annotation embryonic heart tube development and the gene set heart morphogenesis only a slight effect of RV infection was noted, suggesting the involvement of other factors. In mesodermal cells, expression of HAND1, which is involved in embryonic heart tube development, was reduced after RV infection as compared to the mock controls (Figure 6C and Figure 1E). The ectoderm gives rise to components of the eye. At the molecular level, RV infection of ectodermal cells impaired the gene set eye development (Figure 6C), . Specifically, SIX3 and SIX6, as key transcription factors for mammalian eye development [39,40], were reduced in their expression level (Figure 6C). Among others, SIX3, together with RAX, initiates transcription of genes required for lens placode formation [39].

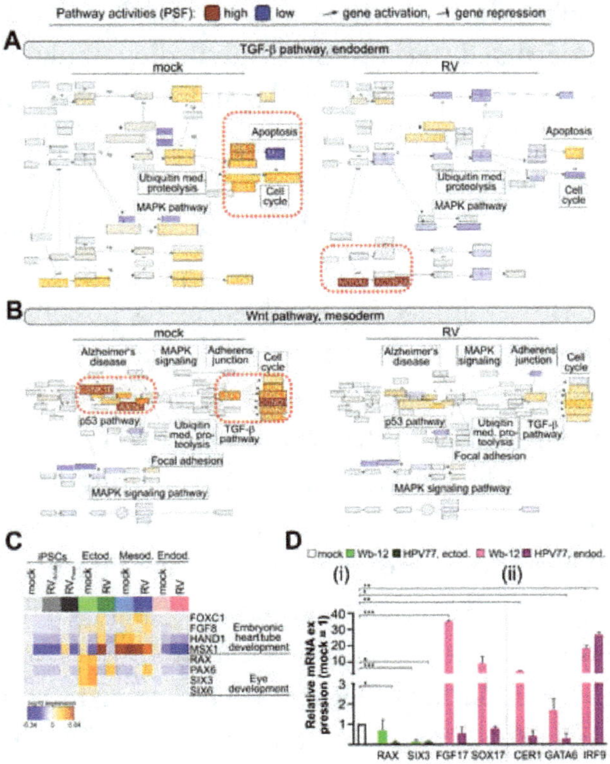

Figure 6. RV exerts lineage-specific effects on developmental signaling pathways and alters expression of transcription and growth factors. Pathway signal flow (PSF) activity plot of (A) the TGF-β signaling pathway in endodermal cells (highlighted are genes with higher (NODAL, ACVR2A, Myc) and lower (RBL1 and E2F4) activity after RV infection) and (B) of the Wnt signaling pathway in mesodermal cells (highlighted are genes with lower (the CSNK1E/AXIN1E and the LEF1/CCND1 axis) activity after RV infection). The calculation of the activity of the nodes was based on the PSF algorithm using the respective gene expression values and the wirings between the nodes [21]. (C) Selected genes within pathways that were specifically affected by RV infection are illustrated in the heatmap of the transcriptome of mock- and RV-infected iPSCs and iPSC-derived lineages. (D) For qRT-PCR expression analysis, the 2-ΔΔCt method based on normalization to HPRT1 gene was used. Values are given as means ± SD (n = 4 for Wb-12-infected ectoderm, n = 2 for HPV77-infected ectoderm, otherwise n = 3).

To further assess developmental pathways with relevance for the teratogenic outcome of RV infection, we determined the mRNA expression of RAX and SIX3 (as important factors for eye development) besides FGF17 (Fibroblast Growth Factor 17) and SOX17 (as contributing factors for endodermal differentiation and cardiovascular development) by RT-qPCR in ectodermal and endodermal cells, respectively (Figure 6D). Here, the vaccine strain HPV77 was used in addition to Wb-12 strain. Attenuated vaccine strains such as HPV77 are not teratogenic as revealed after immunization of unknowingly pregnant women [4]. Thus, any alteration at the molecular level that is present during wild-type Wb-12, but not HPV77 infection, emphasizes its possible contribution to congenital rubella. In comparison to the mock control, a similar reduction in the expression of RAX and SIX3 was detected in ectodermal cells after infection with both RV strains. However, a different picture emerged for FGF17 and SOX17. Figure 6D [i] highlights an increase in the expression of FGF17 and SOX17, which was significant for FGF17 compared to the mock control, but only for endodermal cells derived from Wb-12-infected iPSCs, not for endodermal cells derived from HPV77-infected iPSCs. Moreover, the increase in the expression of the definitive endoderm markers CER and GATA6 after infection with Wb-12 as shown in Figure 2E was not detected after infection with HPV77 (Figure 6D [ii]). However, both RV strains induced a significant increase of the IFN-signaling component IRF9 (Figure 6D [ii]). The endoderm plays an essential role in the crosstalk between the lineages and contributes to the epithelial lining of many organs, including the vascular network. Accordingly, the gene set vasculogenesis, but not angiogenesis, was affected by RV infection. This emphasizes our notion on the correlation between the impact of RV infection on endodermal cells and congenital rubella.

In summary, specific signatures including the TGF-β signaling pathway were affected by RV infection, but in a lineage-specific manner. In ectodermal cells, RV infection significantly reduced expression of SIX3 as key transcription factors for eye field development. Only for the clinical isolate Wb-12, but not for the vaccine strain HPV77, was an impact on the growth factor FGF17 and the endodermal transcription factor SOX17 noted. Figure 7 summarizes the findings of this study in correlation to the main CRS symptoms. The noncytolytic course of infection of RV during directed differentiation is in agreement with its persistence in multiple organs and tissues during congenital rubella. We have not identified any indication at the molecular level that could contribute to the defects in ear development during congenital rubella. However, sensorineural deafness is often a late-onset symptom and could be associated with pathological alterations in the brain of the infected infants [3].

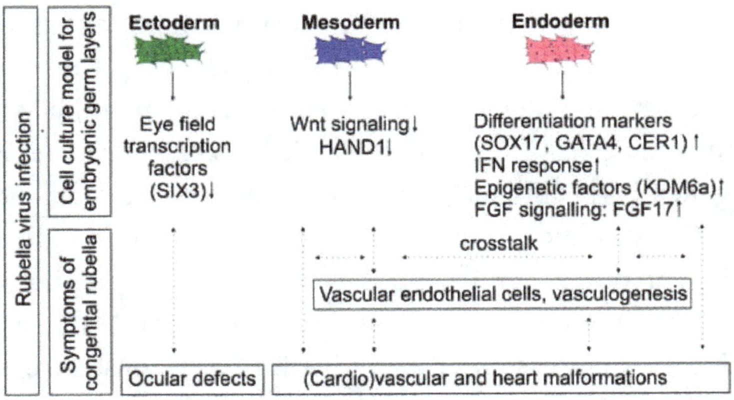

Figure 7. Graphical summary of the identified molecular alterations induced by RV during directed differentiation of iPSCs into the three embryonic germ layer cells. The data was set in a possible relation to the prevailing symptoms of congenital rubella embryopathy. Especially endodermal cells were characterized by profound alterations in their gene expression landscape, including the expression of markers for definitive endoderm and epigenetic factors. This could impair the crosstalk between endodermal and mesodermal cells during differentiation.

4. DISCUSSION

Knowledge on developmental signaling networks is an essential prerequisite to understand congenital abnormalities, either caused by pathogenic, hereditary or environmental risk factors. Models for developmental toxicity testing range from iPSCs to iPSC-derived EBs and three-dimensional organoids. They have different properties regarding high-throughput screening capacity and relevance for in vivo developmental processes [41]. Their proper assessment requires compounds or pathogens with well-known symptoms arising from embryotoxic or teratogenic alterations during embryonal development. Here, we used RV to correlate clinical observations for congenital rubella syndrome with its impact on the differentiation capacity of iPSCs. Although iPSC-based cell culture models reflect only transient stages during human embryogenesis, they allow us to recapitulate essential developmental pathways that are otherwise inaccessible [42].

Among human pathogens, RV is rather exceptional in its ability to replicate noncytopathically in iPSCs, which in general represent a rather restrictive environment to most viral infections [43]. The protection of human development from a pathogenic insult involves several mechanisms, including transcriptional silencing of viruses in pluripotent stem cells [44] and an intrinsic high expression level of IFN-induced genes [14]. This includes interferon-induced transmembrane protein 1 (IFITM1) and its capacity to restrict the potentially harmful reactivation of human endogenous retroviruses [43]. Otherwise, the antiviral innate immune response in iPSCs is rather refractive [15]. The constitutive overexpression of an active IRF7 as a master regulator of the type I IFN system revealed the harmful effects an activated type I IFN response would have on the expression of pluripotency and lineage specific genes, especially of endodermal cells [45]. In contrast to the engineered type I IFN response in iPSCs through overexpression of IRF7, no morphological changes were noted after infection of iPSCs with RV [13]. However, in agreement with the study on the effect of type I IFNS on differentiation capacity of iPSCs [45], the impact of RV on directed differentiation was most profound during endodermal differentiation. The differences in the signaling cascades of type I and III IFNs [46] might explain the milder effects noted after RV-associated type III IFN activation as compared to the severe effects of an engineered type I IFN response [45]. Our data complements a recent study on the impact of Influenza A virus (IAV) on the pluripotency and proteome of hiPSCs [47]. Whereas, in contrast to RV, IAV reduces the pluripotency of iPSCs, both virus infections induce ISG15 and IFN λ1 [47], highlighting this observation as an innate immune mechanism that is already developed in iPSCs. Further studies need to address whether the impact of RV infection on endodermal differentiation is correlated with the activation of the type III IFN signaling pathway and how this affects the course of infection of RV in iPSCs. In ectodermal cells, RV infection was associated with the downregulation of SIX3, an essential transcription factor for early eye development [48]. Together with SIX6, SIX3 suppresses Wnt signaling, which could contribute to the slight activation of this essential developmental signaling pathway in ectodermal cells derived from RV-infected iPSCs [40]. Their functional importance during retinal development and eye field specification was recently shown by the use of iPSC-derived retinal organoids [39]. Our study complements a previous study on the gene expression profile of fetal (HUVEC originating from umbilical cord veins) and adult (HSaVEC derived from the saphenous vein) endothelial cells which revealed a specific enrichment of 18 downregulated genes within the GO terms “sensory organ development”, “eye development”, and “ear development” [49].

Among the embryonic germ layers, especially differentiation to definite endoderm appeared to be affected by RV infection. In addition to its role in formation of organs of the digestive tract, the interaction of endodermal cells with precardiac mesoderm drives specification and differentiation of cardiac myocytes and cells of endocardial endothelium [50]. This is supported by studies on the contribution of signals from endodermal cells and the interactive crosstalk between the endoderm and mesoderm to differentiation of ESCs to a cardiomyogenic lineage [9]. RV infection does not only target the endoderm, but also signals that facilitate this interactive crosstalk. This includes Cerberus as a bone morphogenetic protein (BMP) antagonist [51]. The secretion of Cerberus from endodermal cells initiates differentiation of the neighboring tissue, namely the overlying cardiac mesoderm [51,52]. Furthermore, the analysis of endoderm-depleted frog and avian embryos revealed that the endoderm contributes to vasculogenesis and vascular tube formation [53]. Thus, as summarized in Figure 7, the molecular events identified in

RV-infected endodermal cells could contribute to cardiovascular defects during congenital rubella [2].

Besides the mere expression level of essential components of developmental pathways, post-translational histone modifications are involved in the regulation of gene expression during development. The balance between H3K4me as an active and H3K27 as an inactive state histone modification directs the switch between active and inactive pathways during differentiation [54]. The activity of the KDM6A (UTX) demethylase was especially upregulated in endodermal cells during RV infection. KDM6A demethylase activity was reported to counteract DNA damage response and cell death induction in differentiating ESCs [55], which could also apply to RV-infected endodermal cells.

The presence of RV infection led to an increase in the expression of certain transcription factors that are concentrated in definitive endoderm, such as GATA4, EOMES, and SOX17 [56]. The transcription factor EOMES plays a role in cardiac development and endoderm specification, depending on the environment and dosage [57]. While SOX17-null mice showed a decrease in the activity of many genes related to heart development [58], the excessive production of SOX17 during hematopoiesis hindered the survival of early hematopoietic precursors by triggering apoptosis [59]. This suggests that the proper growth of an embryo, particularly in terms of the formation of the heart, relies on the precise expression of several components [60], which seems to be influenced by a respiratory virus infection.

It is crucial to characterize teratogens like RV on iPSC-based models in order to highlight their potential for evaluating substances that are harmful to embryos, and to find important factors that can enhance their ability to forecast toxicity. Congenital cardiac abnormalities, although not only attributed to infections like RV, are the prevailing type of human developmental problems observed in newborns. iPSC-based models provide unique insights into human development and processes that may disrupt its normal progression, hence expanding our diagnostic and therapy options for congenital abnormalities. Additional research is required to establish a connection between the observed alterations in gene expression and the resulting effects on the pathways that regulate embryonic development.

CONCISE SUMMARY

To investigate congenital viral infections in humans, it is necessary to have appropriate ex vivo platforms that can replicate the species-specific events that occur during embryonic development. An exemplary instance of such illnesses is the rubella virus (RV), which mostly results in malformations in the development of the ear, heart, and eye. We exposed human induced pluripotent stem cells (iPSCs) to a substance known to cause birth defects (teratogenic RV). Afterward, we induced the iPSCs to differentiate into cells representing the three embryonic lineages (ecto-, meso-, and endoderm). This cell culture model was used to simulate the blastocyst and gastrulation stages of development. RV presence led to the expression of differentiation markers specific to each lineage, showing the preservation of lineage identity. Nevertheless, when examining the transcriptome expression fingerprints of all samples, it was found that the endodermal cells infected with the fake virus and the RV virus were less similar to each other compared to their ectodermal and mesodermal counterparts. RV infection led to a rise of markers specifically associated with definitive endoderm. Significant changes in the epigenetic profile, including as modifications in the expression of chromatin remodeling complex components and an increase in type III interferons, were observed, particularly following the conversion of RV-infected iPSCs into endodermal cells. Furthermore, the dysregulated transcripts include the eye field transcription factors RAX and SIX3, as well as components of the gene set vasculogenesis. While the morphology of iPSCs remained intact, their ability to form embryoid bodies as three-dimensional cell aggregates and their cellular adhesion potential were compromised after RV infection. The link between the molecular changes caused by RV during the generation of iPSCs and the clinical manifestations of congenital rubella syndrome implies ways by which the virus hampers human development.

REFERENCES

1. Dudgeon, J.A. Congenital rubella. J. Pediatr. 1975, 87, 1078–1086.
2. Duszak, R.S. Congenital rubella syndrome—Major review. Optometry 2009, 80, 36–43.
3. Lazar, M.; Perelygina, L.; Martines, R.; Greer, P.; Paddock, C.D.; Peltecu, G.; Lupulescu, E.; Icenogle, J.; Zaki, S.R. Immunolocalization and distribution of rubella antigen in fatal congenital rubella syndrome. EBioMedicine 2016, 3, 86–92.
4. Freij, B.J.; South, M.A.; Sever, J.L. Maternal rubella and the congenital rubella syndrome. Clin. Perinatol. 1988, 15, 247–257.
5. Bouthry, E.; Picone, O.; Hamdi, G.; Grangeot-Keros, L.; Ayoubi, J.M.; Vauloup-Fellous, C. Rubella and pregnancy: Diagnosis, management and outcomes. Prenat. Diagn. 2014, 34, 1246–1253.
6. Enders, G.; Nickerl-Pacher, U.; Miller, E.; Cradock-Watson, J.E. Outcome of confirmed periconceptional maternal rubella. Lancet 1988, 1, 1445–1447.
7. Rossant, J.; Tam, P.P.L. Exploring early human embryo development. Science 2018, 360, 1075–1076.
8. Kugler, J.; Huhse, B.; Tralau, T.; Luch, A. Embryonic stem cells and the next generation of developmental toxicity testing. Expert Opin. Drug Metab. Toxicol. 2017, 13, 833–841.
9. Van Vliet, P.; Wu, S.M.; Zaffran, S.; Puceat, M. Early cardiac development: A view from stem cells to embryos. Cardiovasc. Res. 2012, 96, 352–362.
10. Simunovic, M.; Brivanlou, A.H. Embryoids, organoids and gastruloids: New approaches to understanding embryogenesis. Development 2017, 144, 976–985.
11. Meganathan, K.; Jagtap, S.; Wagh, V.; Winkler, J.; Gaspar, J.A.; Hildebrand, D.; Trusch, M.; Lehmann, K.; Hescheler, J.; Schluter, H.; et al. Identification of thalidomide-specific transcriptomics and

proteomics signatures during differentiation of human embryonic stem cells. PLoS ONE 2012, 7, e44228.

12. Luz, A.L.; Tokar, E.J. Pluripotent stem cells in developmental toxicity testing: A review of methodological advances. Toxicol. Sci. 2018, 165, 31–39.

13. Hubner, D.; Jahn, K.; Pinkert, S.; Bohnke, J.; Jung, M.; Fechner, H.; Rujescu, D.; Liebert, U.G.; Claus, C. Infection of ipsc lines with miscarriage-associated coxsackievirus and measles virus and teratogenic rubella virus as a model for viral impairment of early human embryogenesis. ACS Infect. Dis. 2017, 3, 886–897.

14. Wu, X.; Dao Thi, V.L.; Huang, Y.; Billerbeck, E.; Saha, D.; Hoffmann, H.H.; Wang, Y.; Silva, L.A.V.; Sarbanes, S.; Sun, T.; et al. Intrinsic immunity shapes viral resistance of stem cells. Cell 2018, 172, 423–438.

15. Hong, X.X.; Carmichael, G.G. Innate immunity in pluripotent human cells: Attenuated response to interferon-beta. J. Biol. Chem. 2013, 288, 16196–16205.

16. Dziedzicka, D.; Markouli, C.; Barbe, L.; Spits, C.; Sermon, K.; Geens, M. A high proliferation rate is critical for reproducible and standardized embryoid body formation from laminin-521-based human pluripotent stem cell cultures. Stem Cell Rev. 2016, 12, 721–730.

17. Wirth, H.; Loffler, M.; von Bergen, M.; Binder, H. Expression cartography of human tissues using self organizing maps. BMC Bioinform. 2011, 12, 306.

18. Subramanian, A.; Tamayo, P.; Mootha, V.K.; Mukherjee, S.; Ebert, B.L.; Gillette, M.A.; Paulovich, A.; Pomeroy, S.L.; Golub, T.R.; Lander, E.S.; et al. Gene set enrichment analysis: A knowledge-based approach for interpreting genome-wide expression profiles. Proc. Natl. Acad. Sci. USA 2005, 102, 15545–15550.

19. Wirth, H.; von Bergen, M.; Murugaiyan, J.; Rosler, U.; Stokowy, T.; Binder, H. Maldi-typing of infectious algae of the genus prototheca using som portraits. J. Microbiol. Methods 2012, 88, 83–97.

20. Loffler-Wirth, H.; Kalcher, M.; Binder, H. Opossom: R-package for high-dimensional portraying of genome-wide expression landscapes on bioconductor. Bioinformatics 2015, 31, 3225–3227.

21. Nersisyan, L.; Löffler-Wirth, H.; Arakelyan, A.; Binder, H. Gene set- and pathway- centered knowledge discovery assigns transcriptional activation patterns in brain, blood, and colon cancer: A bioinformatics perspective. Int. J. Knowl. Discov. Bioinform. 2014, 4, 46–69.

22. Claus, C.; Bergs, S.; Emmrich, N.C.; Hubschen, J.M.; Mankertz, A.; Liebert, U.G. A sensitive one-step taqman amplification approach for detection of rubella virus clade i and ii genotypes in clinical samples. Arch. Virol. 2017, 162, 477–486.

23. Hellemans, J.; Mortier, G.; De Paepe, A.; Speleman, F.; Vandesompele, J. Qbase relative quantification framework and software for management and automated analysis of real-time quantitative pcr data. Genome Biol. 2007, 8, R19.

24. Wang, H.; Zhang, K.; Liu, Y.; Fu, Y.; Gao, S.; Gong, P.; Wang, H.; Zhou, Z.; Zeng, M.; Wu, Z.; et al. Telomere heterogeneity linked to metabolism and pluripotency state revealed by simultaneous analysis of telomere length and rna-seq in the same human embryonic stem cell. BMC Biol. 2017, 15, 114.

25. Ben-Porath, I.; Thomson, M.W.; Carey, V.J.; Ge, R.; Bell, G.W.; Regev, A.; Weinberg, R.A. An embryonic stem cell-like gene expression signature in poorly differentiated aggressive human tumors. Nat. Genet. 2008, 40, 499–507.

26. Chu, L.F.; Leng, N.; Zhang, J.; Hou, Z.; Mamott, D.; Vereide, D.T.; Choi, J.; Kendziorski, C.; Stewart, R.; Thomson, J.A. Single-cell rna-seq reveals novel regulators of human embryonic stem cell differentiation to definitive endoderm. Genome Biol. 2016, 17, 173.

27. Hopp, L.; Loeffler-Wirth, H.; Nersisyan, L.; Arakelyan, A.; Binder, H. Footprints of sepsis framed within community acquired pneumonia in the blood transcriptome. Front Immunol 2018, 9, 1620.

28. Andres-Terre, M.; McGuire, H.M.; Pouliot, Y.; Bongen, E.; Sweeney, T.E.; Tato, C.M.; Khatri, P. Integrated, multi-cohort analysis identifies conserved transcriptional signatures across multiple respiratory viruses. Immunity 2015, 43, 1199–1211.

29. Sweeney, T.E.; Wong, H.R.; Khatri, P. Robust classification of bacterial and viral infections via integrated host gene expression diagnostics. Sci. Transl. Med. 2016, 8, 346ra91.

30. Schmidtke, L.; Schrick, K.; Saurin, S.; Kafer, R.; Gather, F.; Weinmann-Menke, J.; Kleinert, H.; Pautz, A. The kh-type splicing regulatory protein (ksrp) regulates type iii interferon expression post-transcriptionally. Biochem. J. 2019, 476, 333–352.

31. Winzen, R.; Thakur, B.K.; Dittrich-Breiholz, O.; Shah, M.; Redich, N.; Dhamija, S.; Kracht, M.; Holtmann, H. Functional analysis of ksrp interaction with the au-rich element of interleukin-8 and identification of inflammatory mrna targets. Mol. Cell. Biol. 2007, 27, 8388–8400.

32. Dambacher, S.; Hahn, M.; Schotta, G. Epigenetic regulation of development by histone lysine methylation. Heredity 2010, 105, 24–37.

33. Thalheim, T.; Hopp, L.; Binder, H.; Aust, G.; Galle, J. On the cooperation between epigenetics and transcription factor networks in the specification of tissue stem cells. Epigenomes 2018, 2, 20.

34. Grandy, R.A.; Whitfield, T.W.; Wu, H.; Fitzgerald, M.P.; VanOudenhove, J.J.; Zaidi, S.K.; Montecino, M.A.; Lian, J.B.; van Wijnen, A.J.; Stein, J.L.; et al. Genome-wide studies reveal that h3k4me3 modification in bivalent genes is dynamically regulated during the pluripotent cell cycle and stabilized upon differentiation. Mol. Cell. Biol. 2016, 36, 615–627.

35. Bratt-Leal, A.M.; Carpenedo, R.L.; McDevitt, T.C. Engineering the embryoid body microenvironment to direct embryonic stem cell differentiation. Biotechnol. Prog. 2009, 25, 43–51.

36. Zhang, K.T.; Che, S.Y.; Su, Z.; Zheng, S.Y.; Zhang, H.Y.; Yang, S.L.; Li, W.D.; Liu, J.P. Cd90 promotes cell migration, viability and sphere-forming ability of hepatocellular carcinoma cells. Int. J. Mol. Med. 2018, 41, 946–954.

37. Liu, C.; Peng, G.; Jing, N. Tgf-beta signaling pathway in early mouse development and embryonic stem cells. Acta Biochim. Biophys. Sin. 2018, 50, 68–73.

38. Lindsley, R.C.; Gill, J.G.; Kyba, M.; Murphy, T.L.; Murphy, K.M. Canonical wnt signaling is required for development of embryonic stem cell-derived mesoderm. Development 2006, 133, 3787–3796.

39. Weed, L.S.; Mills, J.A. Strategies for retinal cell generation from human pluripotent stem cells. Stem Cell Investig. 2017, 4, 65.

40. Diacou, R.; Zhao, Y.; Zheng, D.; Cvekl, A.; Liu, W. Six3 and six6 are jointly required for the maintenance of multipotent retinal progenitors through both positive and negative regulation. Cell Rep. 2018, 25, 2510–2523.

41. Worley, K.E.; Rico-Varela, J.; Ho, D.; Wan, L.Q. Teratogen screening with human pluripotent stem cells. Integr. Biol. 2018, 10, 491–501.

42. Rathjen, J. The states of pluripotency: Pluripotent lineage development in the embryo and in the dish. ISRN Stem Cells 2014, 2014, 19.

43. Fu, Y.; Zhou, Z.; Wang, H.; Gong, P.; Guo, R.; Wang, J.; Lu, X.; Qi, F.; Liu, L. Ifitm1 suppresses expression of human endogenous retroviruses in human embryonic stem cells. FEBS Open Bio 2017, 7, 1102–1110.

44. Wolf, D.; Goff, S.P. Embryonic stem cells use zfp809 to silence retroviral dnas. Nature 2009, 458, 1201–1204.

45. Eggenberger, J.; Blanco-Melo, D.; Panis, M.; Brennand, K.J.; tenOever, B.R. Type i interferon response impairs differentiation

potential of pluripotent stem cells. Proc. Natl. Acad. Sci. USA 2019, 116, 1384–1393.

46. Pervolaraki, K.; Talemi, S.R.; Albrecht, D.; Bormann, F.; Bamford, C.; Mendoza, J.L.; Garcia, K.C.; McLauchlan, J.; Hofer, T.; Stanifer, M.L.; et al. Differential induction of interferon stimulated genes between type i and type iii interferons is independent of interferon receptor abundance. PLoS Pathog. 2018, 14, e1007420.
47. Zahedi-Amiri, A.; Sequiera, G.L.; Dhingra, S.; Coombs, K.M. Influenza a virus-triggered autophagy decreases the pluripotency of human-induced pluripotent stem cells. Cell Death Dis. 2019, 10, 337.
48. Heavner, W.; Pevny, L. Eye development and retinogenesis. Cold Spring Harb. Perspect. Biol. 2012, 4, a008391.
49. Geyer, H.; Bauer, M.; Neumann, J.; Ludde, A.; Rennert, P.; Friedrich, N.; Claus, C.; Perelygina, L.; Mankertz, A. Gene expression profiling of rubella virus infected primary endothelial cells of fetal and adult origin. Virol. J. 2016, 13, 21.
50. Lough, J.; Sugi, Y. Endoderm and heart development. Dev. Dyn. 2000, 217, 327–342.
51. Mulloy, B.; Rider, C.C. The bone morphogenetic proteins and their antagonists. Vitam. Horm. 2015, 99, 63–90.
52. Foley, A.C.; Korol, O.; Timmer, A.M.; Mercola, M. Multiple functions of cerberus cooperate to induce heart downstream of nodal. Dev. Biol. 2007, 303, 57–65.
53. Vokes, S.A.; Krieg, P.A. Endoderm is required for vascular endothelial tube formation, but not for angioblast specification. Development 2002, 129, 775–785.
54. Mikkelsen, T.S.; Ku, M.; Jaffe, D.B.; Issac, B.; Lieberman, E.; Giannoukos, G.; Alvarez, P.; Brockman, W.; Kim, T.K.; Koche, R.P.; et al. Genome-wide maps of chromatin state in pluripotent and lineage-committed cells. Nature 2007, 448, 553–560.
55. Hofstetter, C.; Kampka, J.M.; Huppertz, S.; Weber, H.; Schlosser, A.; Muller, A.M.; Becker, M. Inhibition of kdm6 activity during murine

esc differentiation induces DNA damage. J. Cell Sci. 2016, 129, 788–803.

56. Fisher, J.B.; Pulakanti, K.; Rao, S.; Duncan, S.A. Gata6 is essential for endoderm formation from human pluripotent stem cells. Biol Open 2017, 6, 1084–1095.

57. Pfeiffer, M.J.; Quaranta, R.; Piccini, I.; Fell, J.; Rao, J.; Ropke, A.; Seebohm, G.; Greber, B. Cardiogenic programming of human pluripotent stem cells by dose-controlled activation of eomes. Nat. Commun. 2018, 9, 440.

58. Pfister, S.; Jones, V.J.; Power, M.; Truisi, G.L.; Khoo, P.L.; Steiner, K.A.; Kanai-Azuma, M.; Kanai, Y.; Tam, P.P.; Loebel, D.A. Sox17-dependent gene expression and early heart and gut development in sox17-deficient mouse embryos. Int. J. Dev. Biol. 2011, 55, 45–58.

59. Serrano, A.G.; Gandillet, A.; Pearson, S.; Lacaud, G.; Kouskoff, V. Contrasting effects of sox17- and sox18-sustained expression at the onset of blood specification. Blood 2010, 115, 3895–3898.

60. George, R.M.; Firulli, A.B. Hand factors in cardiac development. Anat. Rec. 2019, 302, 101–107.

20 AN ONCOLYTIC VIRUS THAT CARRIES THE GENE FOR A POTENT PRO-INFLAMMATORY CYTOKINE, INTERLEUKIN 12, FOR USE IN CANCER IMMUNOTHERAPY

1. INTRODUCTION: AN OVERVIEW

Interleukin 12 (IL-12) is a potent controller of both the natural and acquired immune responses against tumors. As a cytokine composed of two different subunits, it has various effects that help fight against tumors. These effects include promoting the growth and cytotoxic activity of natural killer (NK) cells and T cells (both CD4+ and CD8+), causing CD4+ T cells to differentiate into Th1 phenotype, increasing the production of IFN-γ from NK and T cells, and preventing the formation of blood vessels that supply tumors. Although preclinical research showed promising results [4], the initial phases of IL-12 clinical trials fell short of expectations. The phase II clinical trial of intravenous IL-12 treatment was halted by the FDA when severe adverse events were observed in 15 out of 17 patients, including two occurrences of death [11,12]. While cutaneous T cell lymphoma variations, AIDS-related Kaposi sarcoma, and non-Hodgkin's lymphoma have shown some success with IL-12 based therapy [13,14,15,16], the bulk of oncology clinical trials have found that the severity of adverse effects outweighs the benefit of these treatments [17]. Researchers are currently investigating various methods to specifically target the expression of IL-12 to the tumor microenvironment, with the goal of improving effectiveness and ensuring safety. Recent investigations indicate that the systemic toxicity of IL-12 is restricted when it is produced by oncolytic viruses (OVs) within the tumors [18,19,20] and in the brains of non-human primates [21].

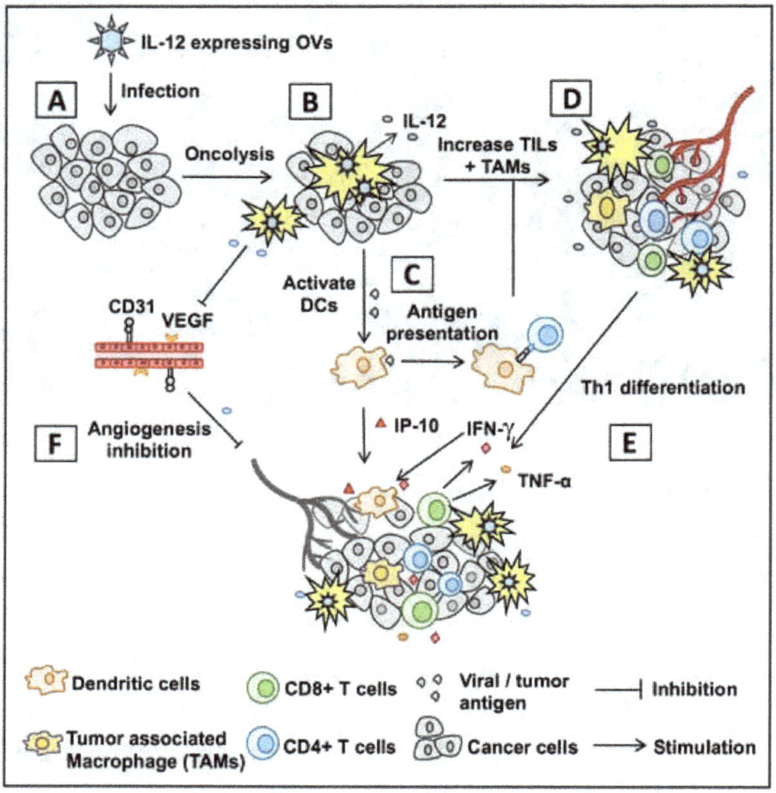

Figure 1. Graphic presentation of mechanism of action of oncolytic virus encoding IL-12. (A) Infection of tumor with oncolytic virus encoding IL-12 (OV-IL12). (B) OV-IL12 replicates in and kills cancer cells (i.e., oncolysis) and releases IL-12 in the tumor microenvironment. (C) Neoantigens from lysed cancer cells activate and recruit dendritic cells (DCs) into the tumor microenvironment. DCs process neoantigens, travel to nearest lymphoid organs, and present the antigen to T cells (CD4+ and CD8+ T cells). (D,E) T cells migrate to the site of infection (referred as tumor-infiltrating T cells or TILs), differentiate into Th1 cells, produce anti-tumor cytokines and kill cancer cells. (F) IL-12-induced production of IFN-γ and interferon inducible protein 10 (IP-10) produces anti-angiogenetic effect through reduction of tumoral vascular endothelial growth factor (VEGF) and CD31+ tumor endothelial cells.

Oncolytic viruses (OVs) are a specific group of anti-cancer drugs that have unique ways of working. Firstly, they are able to selectively replicate in and destroy cancer cells without causing harm to healthy cells or tissues. This process is known as oncolysis. Secondly, OVs also have the ability to expose viral and tumor antigens, which triggers a series of natural and adaptive immune responses against the tumor. This effect is similar to that of a vaccine and occurs within the tumor itself. The effects of the OV-induced vaccination can be augmented by including the viral expression of anti-tumor cytokines, such as IL-12 [18,19,20], as demonstrated in Figure 1. Cancer immunotherapy utilizing oncolytic viruses (OVs) is a developing and extensively studied therapeutic strategy for cancer treatment [27,28]. Oncolytic herpes simplex virus (OHSV) is the most advanced among all oncolytic viruses (OVs) and has been licensed by the FDA for treating metastatic melanoma [29].

In order to harness the therapeutic benefits of IL-12, many oncolytic herpes simplex viruses (OHSVs) that carry the genetic code for IL-12 have been created by genetic manipulation and evaluated in different types of cancer (refer to Table 1). Furthermore, apart from OHSVs, various other oncolytic viruses (OVs) including adenoviruses, measles virus, maraba virus, Newcastle disease virus, Semliki forest virus, vesicular stomatitis virus, and Sindbis virus are being manipulated to produce IL-12 (Table 2). Our literature review has identified that there are now 25 distinct variations of OV-IL12s that are being actively researched or have been recently investigated. Please refer to Table 1 and Table 2 for further details. Although there is a growing interest in IL-12 expressing viruses, there is currently a lack of reviews that examine their potential effectiveness and safety in translating to human subjects. This review contains up-to-date information on the topic and offers a fundamental understanding of OV-IL12 as a viable treatment method in cancer immunotherapy. This could potentially encourage further study in the future. In this review, we will explicitly address the safety, tumor-specificity, and anti-tumor/anti-vascular effects of OHSV-IL12 as either a standalone treatment or in conjunction with other therapies. We will examine not only OHSV-IL12 viruses but also additional oncolytic viruses that express IL-12, and explore their use in cancer treatment.

Table 1. List of OHSV-IL12s and their efficacy in pre-clinical cancer models.

Virus	Genomic Modification	Cancer Model	RoA	Dose (pfu)	Efficacy	Ref.
G47Δ-mIL12	ΔICP6, ΔΔICP34.5, ΔICP47, ◦LacZ, ◦mIL-12	Intracranial 005 GSC (Glioblastoma)	I.T.	5×10^5	Inhibited intracranial tumor growth and extended survival. Promoted IL-12 expression, stimulated IFN-γ production, upregulated IP-10, and inhibited VEGF. Polarized T_H1 response and inhibited T-regs function.	[19,30]
T-mfIL12	ΔICP6, ΔΔICP34.5, ΔICP47, ◦mIL-12	Intracerebral Neuro2a (neuroblastma)	I.V.	5×10^6	Prolonged survival (Mock vs. T-mfIL12, $p < 0.05$), although not statistically significant versus T-01 treatments.	[31]
NV1042	ΔICP0, ΔICP4, ΔICP34.5, ΔUL56, ΔICP47, Us11Δ, Us10Δ, UL56 (duplicated), ◦mIL-12	Subcutaneous SCC VII (Squamous Cell Carcinoma)	I.T.	1×10^7	Reduced tumor volume and improved survival (3 doses of 2×10^7 pfu). 57% of mice from NV1042 group rejected subsequent SCC re-challenge in the contralateral flank compared with 14% in NV1023 or NV1034 group.	[32]
			I.V.	5×10^7	NV1042 treatment resulted in 100% survival, in contrast to 70% of NV1023 and 0% of PBS.	[33]

Virus	Genomic Modification	Cancer Model	RoA	Dose (pfu)	Efficacy	Ref.
M002	ΔΔICP34.5, °mIL-12	Intracranial X21415 (Pediatric embryonal tumor); D456 (pediatric glioblastoma); GBM-12 and UAB106 (adult glioblastoma)	I.T.	1×10^7	M002 significantly prolonged survival in mice bearing all 4 types of tumor compared to saline. No difference in survival was observed compared with G207, excluding X21415 with high levels of nectin-1	[34]
		Intracranial SCK (brain metastasized breast cancer)	I.T.	1.5×10^7	Single injection of M002 extended the survival of treated animals more effectively than a non-cytokine control virus.	[35]
		Xenograft SK-N-AS and SK-N-BE (2) (human neuroblastoma); subcutaneous Neuro-2a (murine neuroblastoma)	I.T.	1×10^7	Significant decrease in tumor growth were observed in both SK-N-AS and SK-N-BE (2) cell lines. Extended median survival compared to the parent R3659.	[36]
		HuH6 (human hepatoblastoma; G401 (human malignant rhabdoid kidney tumor); SK-NEP-1 (renal Ewing sarcoma)	I.T.	1×10^7	M002 significantly reduced tumor volume and increased survival over those treated with vehicle alone in all three different xenograft models.	[37]

Virus	Genomic Modification	Cancer Model	RoA	Dose (pfu)	Efficacy	Ref.
R-115	Virulent with retargeted HER-2, °mIL-12	pLV-HER2-nectin-puro	I.P.	1×10^8 to 2×10^9	Induced greater local and systemicanti-tumorimmunity and durable response than unarmed R-LM113 in both early and late schedule. All mice that survived from primary tumor challenge were protected from the distant challenge tumor and subsequent re-challenge. Increased number of CD8+ and CD141+ cells, PD-L1+ tumor cells, and Treg with a decrease in the number of CD11b+ cells. Enhanced Th1 polarization and increased expression of IFN-γ, IL-2, Granzyme B, T-bet and TNFα and tumor infiltrating lymphocytes	[38]
		Orthotopic mHGG^{pdgf}-hHER2 (glioblastoma)	I.T.	Low dose: 2×10^6 High dose: 1×10^8	27% of mice treated with R-115 ($n=6$, 4 low-dose arm and 2 high-dose arm) alive 100 days after the virus treatment versus all mice died within 48 days. Increased infiltrating CD4+ and CD8+, and expression of IFN-γ	[39]

Virus	Genomic Modification	Cancer Model	RoA	Dose (pfu)	Efficacy	Ref.
vHSV-IL-12	ΔICP6, ΔΔICP34.5, ∘mIL-12	Subcutaneous Neuro2a (neuroblastoma)	I.T.	1×10^4	Significantly reduced tumor growth versus vHSV-null and other cytokine armed groups.	[40]
T2850	ΔIR 15,091bp, ∘mIL-12	Subcutaneous A20 (Murine B Lymphoma), MC38 (colon adenocarcinoma), MFC (Murine Forestomach Carcinoma)	I.T.	1×10^7	Reduced tumor volume compared to IL-12 unarmed parental group. IFN-γ level was markedly increased in the tumor bed and sera of mice infected with both T2850 and T3855 by day 4.	[41]
T3855	ΔIR 15,091bp, ∘mIL-12, ∘mPD-1	Subcutaneous B16 (melanoma)		5×10^6, 1×10^7, 3×10^7		
T3011	ΔIR15,091bp, ∘hIL-12, ∘hPD-1	Subcutaneous B16 (melanoma)	I.T.	5×10^6, 1×10^7 or 3×10^7	Reduced tumor volume as compared with control group.	[41]

Δ—deletion, ∘ insertion, RoA—Route of administration, I.T—intratumorally, IP—intraperitoneally, pfu—plaque forming unit, ref.—reference, VEGF—vascular endothelial growth factor.

Table 2. List of IL-12 expressing oncolytic viruses (other than OHSVs) and their efficacy in pre-clinical cancer models.

Virus	Strain	Cancer Model	RoA	Dose	Efficacy	Ref
Adenovirus	Ad5-yCD/ mutTKSR39 rep-mIL12	Subcutaneous TRAMP (C2 prostate adenocarcinoma)	I.T.	5×10^8 pfu	Improved local and metastatic tumor control. Increased NK and CTL cytolytic activities. Significantly increased survival, levels of IL-12 and IFN-γ in serum and tumor.	[42,43]
	Ad-TD-IL-12, Ad-TD-nsIL-12	Subcutaneous HPD1NR (pancreatic cancer)	I.T.	1×10^9 pfu	100% tumor eradication and survival of both IL-12 modified Adenovirus treated animals. Ad-TD-IL-12, but not Ad-TD-nsIL-12 resulted in a significant increase in $CD3^+CD4^-CD8^+$ populations in the spleen. Level of splenic IFN-γ, IP-10 and lymph node IFN-γ were lower in Ad-TD-nsIL-12 compared to Ad-TD-IL-12 treated hamster.	[44]
	Ad-ΔB7/IL12/ GMCSF	Subcutaneous B16-F10 (melanoma)	I.T.	5×10^7 pfu	Primary tumor growth was better controlled in Ad-ΔB7/IL12/GMCSF and Ad-ΔB7/IL12 compared to Ad-ΔB7GMCSF or PBS. Increased tumor infiltrating $CD86^+$ APCs and enhanced CD4+ and CD8+ T cell-mediated Th1 antitumor immune response.Reduced VEGF expression in the tumor treated with oncolytic Ad co-expressing IL-12 and GM-CSF or IL-12 alone. IFN-γ, TNF-α and IL-6 were higher in Ad-ΔB7/IL12/GMCSF and Ad-ΔB7/IL12 compared to Ad-ΔB7GMCSF or PBS.	[45]

Virus	Strain	Cancer Model	RoA	Dose	Efficacy	Ref
	RdB/IL-12/IL-18	Subcutaneous B16-F10 (melanoma)	I.T.	1×10^8 pfu	95% and 99% tumor growth inhibition was observed in treatment with RdB/IL-12 and RdB/IL-12/IL-18, respectively. Increased Th1/Th2 cytokine ratio and increased tumor infiltration of $CD4^+$ T, $CD8^+$ T and NK cells. Promoted differentiation of T cells expressing IL-12Rβ2 or IL-18Rα.	[46]
	RdB/IL12/DCN	Orthotopic 4T1 (Triple negative breast cancer)	I.T.	2×10^{10} VP	Both of the IL-12-expressing oncolytic Ads showed similar tumor growth inhibition up to day 9 after initial treatment. RdB/IL12/DCN increased upregulation of IFN-γ, TNF-α, infiltrating cytotoxic lymphocytes, downregulation of TGF-β expression and T-regs compared to RdB/IL12 and RdB/DCN.	[47]
	YKL-IL12/B7	Subcutaneous B16-F10 (melanoma)	I.T.	5×10^8 pfu	Tumor growth was suppressed in both YKL-IL12 and YKL-IL12/B7 treated mice vs PBS. YKL-IL12- or YKL-IL12/B7-treated mice produced a significantly greater level of IFN-γ, infiltrating APCs, CD4+, CD8+ compared with PBS.	[48]
	Ad-ΔB7/IL-12/4-1BBL	Subcutaneous B16-F10 (melanoma)	I.T.	5×10^9 VP	100% of mice in the Ad-ΔB7/IL-12/4-1BBL group survived >30 days after initial viral injection compared with 20% of that in virus expressing either IL-12 or 4-1BBL. Mice treated with Ad-ΔB7/IL-12 or Ad-ΔB7/IL-12/4-1BBL had greater amount of tumor infiltrating CD4+ and CD8+ compared to Ad-ΔB7/4-1BBL and Ad-ΔB7.	[49]

Virus	Strain	Cancer Model	RoA	Dose	Efficacy	Ref
Measles virus	MeVac FmIL-12	Subcutaneous MC38ce (colon carcinoma)	I.T.	5×10^5–1×10^6 ciu	Tumor remissions in 90% of animals. Driven polarization of Th1-associated immune response and increased tumor infiltrating $CD8^+$ T cells. Increased IFN-γ and TNF-α, and polarization of Th1-associated immune response. Co-expression of IL-12 and IL-15 showed synergistic effect.	[20,50]
Maraba Virus	MG1-IL12-ICV	CT26 and B16F10 peritoneal carcinomatosis	I.P	Seeding dose 5×10^5, then 1×10^4 on day 3	Reduced tumor burden and improved mouse survival. Activated and matured DCs to secrete IP-10, and activated and recruited NK cells. Increased production of IFN-γ	[51]
Newcastle disease virus	rClone30–IL-12	Orthotopic H22 (hepatocarcinoma)	I.T.	1×10^7 pfu	Reduced tumor volume and improved percentage of survival. Increased IFN-γ and IP-10. Co-expression of IL-12 and IL-2 showed synergistic effect.	[52]
Semliki Forest virus	rSFV/IL12	Subcutaneous B16 (melanoma)	I.T.	10^7 IU	Single injection with SFV-IL12 resulted in significant tumor regression.2 days after injection, IFN-γ production increased with inhibition of tumor vascularization. Splenic IP-10 and MIG expression was increased.	[53]
	SFV/IL12	Subcutaneous P815 (mastocytoma)	I.T.	10^6 IU	Significantly delayed P815 tumor growth.40–53% of mice exhibited complete tumor regressions.Induced high levels of IFN-γ production in draining lymph nodes.	[54]

Virus	Strain	Cancer Model	RoA	Dose	Efficacy	Ref
	SFV/IL12	Subcutaneous MC38 (colon adenocarcinoma)	I.T.	10^8 particles	Reduced tumor volume and improved percentage of survival. Increased tumor-specific CD8+ T lymphocytes.Enhanced the expression of CD11c, CD8α, CD40, and CD86 of tumor-infiltrating M-MDSCs in the presence of an intact endogenous IFN-I system.	[55]
	SFV-VLP-	Syngeneic RG2 (rat glioma)	I.T.	5×10^7 (low-dose) or 5×10^8 (high-dose)	Reduction in tumor volume (70%—low dose; 87%—high dose)	[56]
Vesicular stomatitis virus (VSV)	VSV-IL12	Orthotopic SCC VII (squamous cell Carcinoma)	I.T.	MOI 0.01	Significant reduction in tumor volume, and prolonged survival.	[57]
Sindbis virus	Sin/IL12	Orthotopic ES-2 cells (ovarian clear cell Carcinoma)	I.P	10^7 pfu	Reduced tumor growth and improved survival. Activated and matured DCs, activated and recruited NK cells. Increased production of IFN-γ.	[58]
Δ—deletion, RoA—Route of administration, I.T—intratumorally, IP—intraperitoneally, pfu—plaque forming unit, ref. —reference, MOI—multiplicity of infection, VP—viral particle, IU—infectious units.						

2. GENETICALLY ENGINEERED IL-12 EXPRESSING OHSVS AND THEIR THERAPEUTIC EFFICACY

2.1. IL-12 Insertion Does Not Compromise Safety and Tumor Specificity of Genetically Modified OHSV-IL12 Viruses

Herpes simplex virus type 1 (HSV-1) has a large 152 kb genome, and therefore, deletion or mutation of various genes (Figure 2) does not fundamentally alter its functional properties (such as infectivity, viral replication, etc.), but rather confers tumor cell-specific replication, with no or reduced toxicity (e.g., neuropathogenicity) [59]. For example, G47Δ-IL12 (an IL-12 expressing OHSV) has three genomic modifications that endow its tumor-specificity and safety: ICP6 inactivation, γ-(ICP)34.5 and ICP47 deletion, and IL-12 insertion [30,60]. ICP6, encodes for viral ribonucleotide reductase, controls nucleotide metabolism and helps HSV to replicate in healthy or non-dividing cells that are inherently lacking sufficient nucleotide pools [59,61] (Figure 3A). ICP6 inactivation by fusion of LacZ does not hamper DNA synthesis in cancer cells (Figure 3B) [59,62] but the lack of ribonucleotide reductase results in no nucleotide metabolism and no viral DNA replication in healthy cells (Figure 3C) [59,61,62,63]. Therefore, mutation in the ICP6 gene makes viral infection and replication tumor-specific and thereby increases safety. Similar to ICP6 inactivation, deletion of γ-34.5 also increases safety and cancer selectivity [22,59,64]. Healthy cells have various anti-viral defense mechanisms. For example, protein kinase R (PKR) phosphorylates eukaryotic translation initiation factor eIF2α, which shuts down synthesis of foreign proteins or viral antigens (Figure 4A) [65,66]. HSV with intact γ-34.5 overturns anti-viral defense in healthy cells through γ-34.5-mediated dephosphorylation of eIF2α and helps in viral protein synthesis and viral replication in healthy cells (Figure 4B) [67,68]. Therefore, γ-34.5 deletion results in no eIF2α dephosphorylation, and thereby, no protein synthesis and viral replication (Figure 4C) [67,68]. However, in cancer cells, γ-34.5-deleted HSV can freely replicate, since cancer cells usually have defects in anti-viral pathways such as PKR-eIF2α pathway (Figure 4D). ICP47 downregulates MHC class I presentation through inhibition of transporter-associated protein

(TAP) channel [69,70] and prevents detection of OHSV-infected cancer cells by virus-specific CD8+ cells [71]. Thus, deletion of ICP47 enhances MHC class I expression and immune response against virus-infected tumor cells and the host's anti-tumor immunity (innate and adaptive) [59,60,72]. ICP47 deletion also complements the loss of γ-34.5 through immediate early (IE) expression of unique short sequence US11 under the control of ICP47-IE promoter [59,73]. US11 is a true late gene that binds with PKR, preventing it from phosphorylating eIF2a [59,74,75]. IE expression of US11 keeps eIF2α dephosphorylated and helps viral protein translation and synthesis [59,74,75]. Finally, in order to improve efficacy, the master anti-tumor cytokine IL-12 has been inserted into the ICP6 region of OHSV G47Δ to create G47Δ-IL12 [30]. Viral expression of IL-12 does not compromise its tumor specificity and safety, but rather significantly improves its anti-tumor properties [19,26,30,76]. T-mfIL12 (another IL-12 expressing OHSV) has the same backbone as G47Δ but with mIL-12 inserted in the ICP6 deletion region [77]. In vivo safety data showed that intravenous application of T-mfIL12 (5 × 106 pfu) in mice bearing subcutaneous, intracerebral or intravenously disseminated tumors is as safe as non-IL12 expressing OHSV [31].

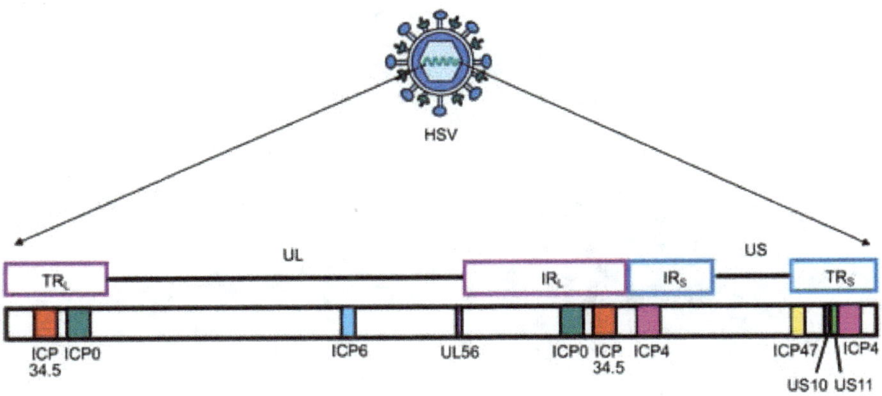

Figure 2. Schematic presentation of herpes simplex virus (HSV) genome with unique long (UL) and short (US) sequences. TRL or S—terminal repeat long or short; IRL or S—internal repeat long or short. Only genes that are modified and/or deleted during construction of OHSV-IL12 are presented. ICP, infected cell protein.

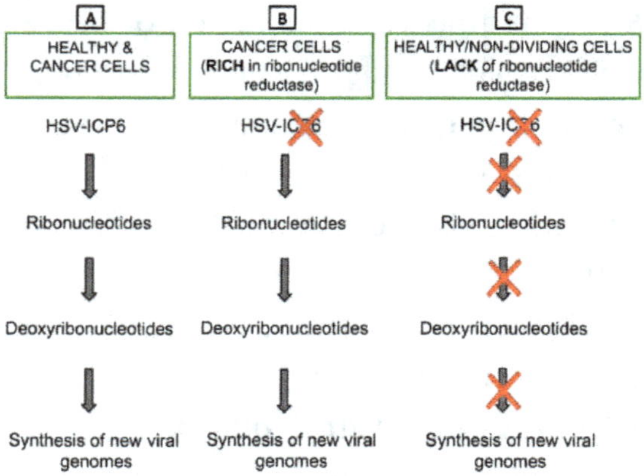

Figure 3. ICP6 inactivation drives tumor-specific replication of OHSV-IL12. (A) ICP6 encodes for large subunit of ribonucleotide reductase, which controls nucleotide metabolism and helps HSV to replicate in normal or healthy host cells that are inherently lacking or have insufficient nucleotide pools. (B) Cancer cells are rich in ribonucleotide reductase, thus HSV with an inactivated ICP6 does not hamper DNA synthesis in cancer cells. (C) Healthy or non-dividing cells lack ribonucleotide reductase. Thus, infection of healthy cells with an ICP6-inactivated HSV leads to no nucleotide metabolism and no viral DNA replication.

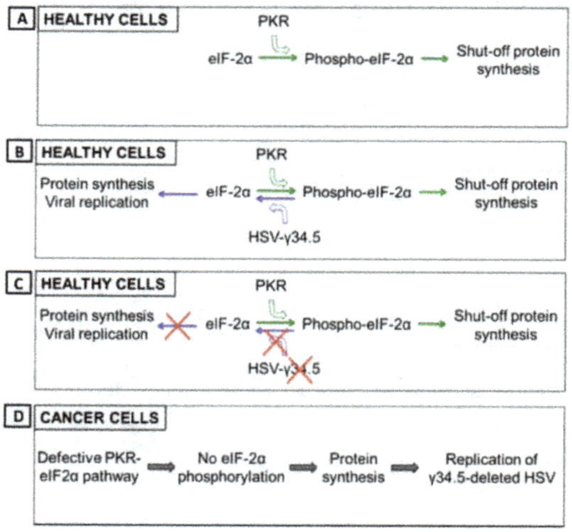

Figure 4. γ-34.5 deletion enhances safety and tumor-specificity of OHSV-IL12. (A) Healthy cells have inherent anti-viral defense mechanisms, such as protein kinase R (PKR). PKR phosphorylates translation initiation factor eIF2α, which shuts down synthesis of foreign proteins or viral antigens. (B) OHSV with an intact γ-34.5 overturns anti-viral defense in healthy cells through γ-34.5-mediated dephosphorylation of eIF2α and helps in viral protein synthesis/viral replication in healthy cells,

leading to development of disease. (C) γ-34.5 deletion results in no eIF2α dephosphorylation in normal or healthy cells, and thereby, no protein synthesis and viral replication. (D) Cancer cells usually have defective PKR-eIF2α pathway, thus no inhibition of foreign protein synthesis. Therefore, γ-34.5-deleted OHSV can freely replicate in cancer cells.

Similar to OHSV G47Δ-IL12, a HSV-1/2 recombinant vaccine strain encoding IL-12 is constructed (designated NV1042), which has also multiple deletions/mutations: (i) deletion of one copy of ICP0, ICP4, ICP34.5, and one copy of UL56 at the UL/S junction, (ii) insertion of Esherichia coli LacZ gene under the control of the α47 promoter at the α47 locus, (iii) deletion of ICP47, and (iv) insertion of mIL-12 under the control of a hybrid a4-TK (thymidine kinase) promoter [32,59,78,79]. ICP0 is an important immediate early (IE) protein in switching viral lytic and latent phases that affects defense mechanisms of the host by blocking nuclear factor kappa B (NF-κB)-mediated transcription of immunomodulatory cytokines, inhibiting interferon regulatory factor 3 (IRF3) translocation to the nucleus, inhibiting gamma-interferon inducible protein 16 (IFI16), and degrading mature dendritic cell (DC) markers (CD83) [24,80]. After translocating to the host's nucleus, ICP0 modulates different overlapping cellular pathways to regulate intrinsic and innate antiviral defense mechanism of host cells, allowing the virus to replicate and persist [80,81]. ICP4 blocks apoptosis and positively regulates many other genes in the HSV-1 genome necessary for viral growth [82]. Function of UL56 has not been fully studied but is thought to be involved in neuro-invasiveness of HSV-1 [78]. Therefore, removal of ICP0, ICP4, ICP34.5 and UL56 attenuates virulence and ensures selective viral replication in cancer. In vivo experiment shows no toxicity after intravenous administration of NV1042 (5 × 107 pfu), as demonstrated by lack of cytopathic effects in vital organs (such as lung, brain, spleen, liver, and pancreas) during three months follow up [33]. However, its safety and tumor-selective replication is still a major concern especially for the treatment of tumors located in the central nervous system, since it has 1 intact copy of γ-34.5 (responsible for neuropathogenicity) and intact ribonucleotide reductase ICP6.

The OHSV M002 and M032 have deletion of both copies of γ-34.5, with murine and human IL-12 cDNA (p35 and p40 subunits, connected by an IRES), respectively, inserted into each of the γ-34.5 deleted regions [83,84,85,86]. M002 has been reported to be safe with no significant toxicity seen after intracerebral inoculation into mice or HSV-sensitive primate Aotus nancymae, despite long-term persistence of viral DNA [87]. M032, with demonstrated safety in non-human primates [21], is now in clinical trial in patients with recurrent glioblastoma (GBM) (see clinical section) [88].

Introducing multiple mutations or deletions in the OHSV genome to confer safety and cancer selectivity may lead to over-attenuation or undermine replication efficiency in cancer cells as opposed to its wild-type or lowly mutated/deleted HSV counterparts [38]. To address this issue, a recent next-generation retargeted IL-12-expressing OHSV known as R-115 has been developed. This OHSV contains no major mutation or deletion and expresses mouse IL-12 under a CMV promoter [38,89]. IL-12-armed R-115 is a derivative of R-LM113 [90]. R-LM113 is a recombinant human epidermal growth factor receptor 2 (HER2) retargeted OHSV with no IL-12 expression, and is successfully engineered by deleting amino acid residues 6 to 38 and by moving the site of single-chain antibody insertion in front of the nectin 1 interacting surface (i.e., at residue 39) [90]. Because of retargeting, it enters and spreads from cancer cell to cell solely via HER2 receptors, and has lost the ability to enter cells through natural glycoprotein D (gD) receptors, herpes virus entry mediator (HVEM) and nectin 1 [90]. Safety profile of R-115 is evaluated in immunocompetent (wt-C57BL/6) model and HER2-transgenic/tolerant counterparts. Mice receiving R-LM113 or R-115 resist very high intraperitoneal OHSV dose of 2x109 PFU, which is a lethal dose for wild-type HSV that kills 83% animals [38]. In addition, 4 consecutive intratumoral injections of R-115 at 3–4 days interval shows no viral DNA in vital organs (blood, brain, heart, kidney, liver, brain and spleen) [38]. This indicates that IL12-armed R-115 is safe in mice. However, HER2 specificity makes R-115 applicable only in HER2-expressing tumors, such as mammary tumors, and merely suitable for the treatment of other lethal cancers, such as glioblastoma or other non-HER2-expressing tumors [90]. Recently, another IL-12-expressing OHSV MH1006 (ICP47 deletion and human IL-12 insertion) has been developed and found to be safe in an immunocompetent subcutaneous model of neuroblastoma [91]. However, MH1006 has an intact neurovirulence gene γ-34.5 and an intact ribonucleotide reductase ICP6, thus raising safety concerns in the brain [91]. In Table 1, we have listed all IL-12-expressing OHSVs with their genomic modifications and preclinical applications.

2.2. IL-12 Expressing OHSVs Produce Superior Anti-Tumor Immunity/Efficacy than Non-IL12 OHSVs

In an orthotopic, immunocompetent intracranial glioblastoma (GBM) stem-like cell model (005 GSC model), OHSV G47Δ-IL12 therapy provides superior results compared to OHSV G47Δ treatment alone. For example, G47Δ-IL12 causes significant extension of survival of mice compared with either G47Δ-empty (i.e., an OHSV with no IL-12 transgene expression; $p < 0.005$) or mock treatment ($p < 0.001$, >50% increase in median survival), with 10% of mice surviving long term [30]. Anti-tumor efficacy of G47Δ-IL12 is associated with a significant reduction of tumor cells (i.e., GFP+ 005 GBM stem-like cells or GSCs) and a robust immune alteration in the tumor, following single intra-tumoral virus injection [19]. The immune alteration mainly includes, but is not limited to, increased tumor infiltration of CD3+ and CD8+ T cells, reduction of immunosuppressive FoxP3+ regulatory T cells, and an increased ratio of CD8+ T cells/regulatory T cells (CD4+FoxP3+) (a hall mark of clinical efficacy) (Figure 5) [19,30]. The role of T cells in therapeutic efficacy is further investigated in athymic nude mice (i.e., T cell-deficient mice) bearing orthotopic brain tumors [30]. In the absence of T cells, G47Δ-IL12 treatment is unable to significantly enhance survival over G47Δ-empty treatment, indicating a critical role of T cells in the IL12-mediated anti-tumor activity [30]. While replicating in vivo in the tumor, G47Δ-IL12 treatment causes increased local production of IL-12 in the tumor, which is accompanied by a marked release of downstream Th1 mediator interferon gamma (IFN-γ) in the tumors and, to a lesser extent, in the blood [30]. IL-12/IFN-γ promotes differentiation of T cells towards Th1 phenotype [92], which further produces IFN-γ and anti-tumor immune effects, as opposed to Th2 type T cells [9,93].

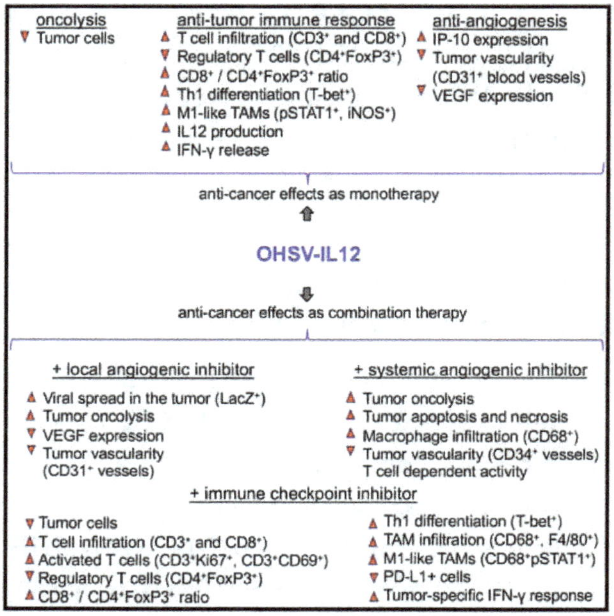

Figure 5. Anti-tumor effects of OHSV-IL12 treatment as mono- and combination-therapies. OHSV-IL12 treatment as monotherapy leads to three distinct anti-cancer effects: 1. Oncolysis, leading to reduction of cancer cells; 2. Induction of anti-tumor immunity, which is characterized by increased intratumoral infiltration of T cells, reduction of regulatory T cells, increased T effector/regulatory T cell ratio, enhanced Th1 differentiation, increased polarization of macrophages toward anti-tumoral M1-phenotype, and increased production of IL-12 and IFN-γ; and 3. Inhibition of tumor angiogenesis as demonstrated by reduced CD31+ blood vessels and increased expression of vascular endothelial growth factor (VEGF) and interferon inducible protein 10 (IP-10). Because of these three aforementioned unique anti-cancer potentials, OHSV-IL12 was tested in combination with local or systemic antiangiogenic inhibitors and immune checkpoint blockade. The combination of OHSV-IL12 + local angiogenic inhibitor produces anti-tumor effects by increasing intratumoral virus spread (as determined by X-gal staining for viral LacZ expression) and oncolysis, and by reducing CD31+ tumor vascularity and VEGF expression. The anti-tumor effects of the OHSV-IL12 + systemic angiogenic inhibitor are characterized by increased lysis of cancer cells and macrophage (CD68+) infiltration into tumors, increased apoptosis and necrosis in

the tumor microenvironment, reduced tumor vascularity, and T cell dependent anti-tumor activity. OHSV-IL12 + immune checkpoint inhibitor produces robust and multifaceted anti-cancer activities, which include: oncolysis, increased infiltration of T cells and activated T cells into tumors, reduction of immunosuppressive regulatory T cells, increased effector (CD8+)/regulatory T cell (CD4+FoxP3+) ratio, Th1 differentiation, tumor-associated macrophage (TAM) infiltration and macrophage polarization towards M1-type, reduction of immune checkpoint PD-L1 positive cells, and induction of tumor-specific IFN-γ response. Upward and downward triangles indicate 'increase' and 'decrease', respectively.

Similarly, T-bet+ Th1 type cells are increased in the tumor following intra-tumoral G47Δ-IL12 treatment [19,26], though it has not been determined whether this increase is directly associated with OHSV-IL12-mediated IFN-γ production in the tumor. G47Δ-IL12 treatment promotes polarization of macrophages from pro-tumoral M2 towards anti-tumoral M1 (e.g., increased expression of iNOS+ and pSTAT1+ cells) without affecting total tumor-associated macrophage (TAM) population (Figure 5) [19], possibly because of IL-12 induced M1-polarizing IFN-γ expression [30]. G47Δ-IL12-mediated anti-cancer immune responses, i.e., in situ vaccine effect opens the door to combination treatment strategies involving other cancer immunotherapy drugs. In orthotopic malignant peripheral nerve sheath tumor (MPNST) models, a single intra-tumoral injection of G47Δ in sciatic nerve tumors, derived from human MPNST stem-like cells in athymic mice or mouse MPNST cells in immunocompetent mice, significantly inhibits tumor growth and prolongs survival, as compared to mock treatment [94]. Local IL-12 expression (i.e., G47Δ-IL12) further significantly improves the efficacy of G47Δ in an immunocompetent orthotopic MPNST model, indicating that IL-12 expression induces anti-MPNST immune responses and improves overall efficacy [94]. These studies support the application of G47Δ-IL12 in combination immunotherapies for MPNST tumors.

Similar to anti-tumor efficacy with G47Δ-IL12, NV1042 (i.e., another OHSV with IL-12 expression) treatment results in a striking reduction in squamous cell carcinoma (SCC) tumor volume compared with the tumors treated with NV1023 (i.e., OHSV lacking IL-12 expression) and NV1034 (i.e., OHSV lacking IL-12, but with GM-CSF expression) [32]. Fifty-seven percent of mice treated with NV1042 reject subsequent SCC re-challenge in the contralateral flank, indicating strong global anti-cancer immune response, as opposed to 14% mice treated with NV1023 or NV1034 [32]. Besides local application, NV1042 was intravenously administered for the treatment of spontaneous primary and metastatic prostate cancer in the transgenic TRAMP mice. Systemic IL12-expressing NV1042 was significantly more efficacious than non-IL12 expressing OHSV NV1023 in reducing the frequency of prostate cancer development and lung metastases [95]. NV1042 DNA was detected in primary and metastatic tumors at 2 weeks after the final systemic virus injection but not in liver or blood [95]. Similarly, anti-cancer efficacy of intravenously delivered NV1042 was also observed in disseminated pulmonary SCC. Compared to PBS and parental NV1023, the group treated with IL-12 expressing NV1042 completely showed no sign of pulmonary nodules at day 12. In a low tumor burden model, NV1042 treatment resulted in 100% survival, in contrast to 70% in NV1023-treated group and 0% in PBS-treated group [33]. Depletion of CD4+ and CD8+ T cells reduces anti-cancer efficacy of IL-12 expressing NV1042, which is similar to anti-cancer effects of non-IL12-expressing NV1023, indicating IL-12 expression plays an important role in enhancing oncolytic efficacy through immune modulation [33].

M002 treatment resulted in prolonged survival in both pediatric and adult intracranial patient-derived tumor xenograft models [34]. The better survival benefit is associated with OHSV receptor nectin-1 expression in tumor cells, which is usually higher in pediatric brain tumors than in adult GBMs [34]. In an immunocompetent breast cancer metastasis model, IL-12-armed M002 treatment significantly improved survival of mice over its parental unarmed OHSV R3659 (no IL-12 expression) [35], indicating IL-12 played a critical role for anti-tumor efficacy. In a syngeneic neuroblastoma model, single intracranial injection of M002 produced a minimal survival benefit over untreated mice [85], indicating the need for IL-12 in immunocompetent models. Similarly, mice bearing intracranial neuroblastoma treated intramuscularly (IM) with M002-infected irradiated neuroblastoma cells did not show any survival advantage over mice treated with non-infected irradiated tumor cells. However, a prime-boost vaccine strategy, such as IM injection of M002-infected irradiated tumor cells seven days prior to tumor implantation and seven days post-tumor implantation, produced sustained anti-tumor T-cell responses and significant survival advantage, as opposed to irradiated control tumor cells [85]. Because an important control group is missing in this experimental setup (i.e., unarmed OHSV-infected irradiated tumor cells), it is not clear whether this anti-cancer vaccine effect in neuroblastoma was due to OHSV, local IL-12 expression, or both. In syngeneic sarcoma models, M002 did not produce any survival benefit compared to its parental virus R3659 (no IL-12 expression) [96], despite M002 inducing a significant anti-tumor immune effect over R3659 treatment, such as an increased percentage of intra-tumoral CD8+ T cells and activated monocytes, a decreased percentage of myeloid-derived suppressor cells (MDSCs), and increased CD8:MDSC and CD8:T regulatory cell ratios [96]. In recently performed pilot experiments in an ovarian cancer metastatic model, systemic intraperitoneal application of M002 resulted in a robust tumor-antigen specific CD8+ T cell response in the peritoneal cavity and the omentum [97], which are the primary sites of ovarian cancer metastasis [98]. Because of the tumor-specific immunity, M002 treatment was more successful in controlling ovarian cancer metastasis and produced a significantly longer overall survival than mock treatment [97]. Whether the anti-tumor efficacy is minimal or better, local IL-12 expression (M002) creates a more favorable immune-active tumor microenvironment than unarmed OHSV, which makes tumors more responsive to other forms of immunotherapies, such as immune checkpoint blockades.

IL-12-armed R115 was superior in inducing local and systemic anti-tumor immunity and durable response over unarmed R-LM113 in both early and late schedules [38]. All mice that survived the primary tumor were protected from the distant tumor challenge and subsequent re-challenge. Treatment with R115 drove Th1 polarization, increased immunomodulatory cytokines such as IFN-γ, IL-2, Granzyme B, T-bet and TNF-α, and tumor infiltrating lymphocytes [38]. Tumor microenvironment of R115 group showed an increase in number of CD8+ and CD141+ cells, PD-L1+ tumor cells, and FoxP3+ T regulatory cells with a decrease in the number of CD11b+ cells [38]. In another study, a single R-115 injection in established tumors resulted in complete tumor eradication in about 30% of animals [39]. The treatment also induced a significant improvement in the overall median survival time of mice and a resistance to recurrence from the same neoplasia [39]. Interestingly, treatment with R-115 increases the number of CD4+ and CD8+ T cells infiltrating into the tumor microenvironment, while the vast majority of CD4+ and CD8+ T cells in the R-LM113 treatment group accumulated at the edge of the tumors, indicating the effects of IL-12 [39].

2.3. Anti-Tumor Anti-Vascular Effects of OHSV-IL12

IL-12 does not only enhances the anti-tumor immune effects of virotherapy, but it also suppresses the development of new blood vessels, a process termed angiogenesis [99], making it an anti-angiogenic cytokine. IL-12 elicits its anti-angiogenic effects through release of IFN-γ, which activates IFN-inducible protein 10 [IP-10 or CXC chemokine ligand (CXCL) 10], a chemokine that mediates chemotaxis of lymphocytes and angiostatic effects [10,17,100]. It has been demonstrated that IL-12-armed OHSV produces significantly higher level of anti-angiogenic effects (i.e., reduction of CD31+ blood vessels in the tumors) through IFN-γ/IP-10 pathway in an immunocompetent brain tumor model (Figure 5) [30], compared to non-IL12 OHSV. IL12-armed OHSV treatment also causes a reduction in vascular endothelial growth factor (VEGF) expression, another likely contributor in tumor angiogenesis (Figure 5) [30].

In a model of prostate cancer in transgenic TRAMP mice, treatment with IL-12 armed NV1042 significantly reduces expression of CD31+ vascularity compared to either NV1023 or mock treated tumors [95]. Anti-angiogenic property of NV1042 is confirmed by another study in a SCC model. Intratumoral delivery of NV1042 results in release of a high level of IL-12, as well as other secondary angiogenic mediators such as IFN-γ, monokine induced by gamma interferon (MIG), and IP-10. In contrast, IL-12 unarmed parental NV1023 treatment shows no increase in IL-12 expression and lower level of secondary angiogenic mediators [101]. These studies indicate that IL-12 gene transfer could significantly enhance unique anti-tumor and anti-angiogenic effects of virotherapy. These anti-angiogenic features allow OHSV-IL12 to be tested with other local or systemic angiogenesis inhibitors for an improved therapeutic outcome.

2.4. Inhibition of Tumor Angiogenesis Enhances Anti-Tumor Potential of OHSV-IL12 Treatment

Angiogenesis is one of the hallmarks of cancer. It plays a key role in cancer progression [102,103,104,105,106,107,108,109,110] and anti-angiogenic therapy has been an interesting target to control tumor growth [108,111,112]. Efforts to disrupt the vascular supply and starve the tumor from nutrients and oxygen have resulted in 11 anti-VEGF drugs approved for certain advanced cancers, either alone or in combination with chemotherapy or other targeted therapies. Unfortunately, this success has had only limited impact on overall survival of cancer patients, and rarely resulted in durable responses. Bevacizumab (Avastin), an FDA approved anti-angiogenic drug (anti-VEGF), did not show significant improvement in overall survival [112,113,114]. Therefore, other anti-angiogenic agents and combinatorial strategies are being tested to target complex tumor microenvironment.

Because OHSV G47Δ-IL12 does not only induce anti-tumor immunity but also produce anti-angiogenic activities [30], it is hypothesized that anti-tumor effects of G47Δ-IL12 treatment would synergize with anti-vascular drugs. Axitinib (AG-013736) is an FDA approved, orally administered potent small molecule tyrosine kinase inhibitor (TKI), which inhibits VEGF receptor (VEGFR) 1-3, platelet-derived growth factor receptor beta (PDGFR-β) and receptor tyrosine kinase c-KIT (CD117) [115], and shows promising anti-vascular and anti-tumor activity in a variety of advanced stage cancers, including GBM [116,117,118]. In addition to anti-vascular effects, it also induces anti-tumor immune effects [119,120]. Therefore, anti-vascular/immune axitinib was tested in combination with anti-angiogenic/immune stimulatory G47Δ-IL12 in highly angiogenic patient and mouse GSC-derived GBM models [76]. This combination significantly extends survival in both models and involves multifaceted anti-tumor activities including: direct oncolysis of tumor cells, extensive tumor apoptosis and necrosis, increased macrophage infiltration to the tumor, greatly reduced tumor vascularity (i.e., CD34+ blood vessels) and inhibition of angiogenic PDGFR/ERK pathway in patient GSC-derived GBM model, and T cell dependent activity in mouse GSC-derived GBM model (Figure 5) [76]. Since the anti-tumor efficacy of the dual combination therapy (G47Δ-IL12+axitinib) was T cell dependent, it is hypothesized that ICI (i.e., anti-PD-1 or anti-CTLA4) will improve the therapeutic outcome of G47Δ-IL12+axitinib dual combination. Interestingly, ICI did not improve anti-tumor effects of axitinib or G47Δ-IL12+axitinib combination therapy. This is in sharp contrast with the findings from other investigators, since they observed synergistic anti-tumor immune effects following axitinib + ICI combination therapy in preclinical syngeneic tumor models [121]. The underlying mechanism(s) behind the failure of ICI combination therapy in orthotopic brain tumor model is not clear and warrants further investigation. It is speculated that reduced vascular permeability after axitinib therapy may inhibit extravasation of T cells into the tumor [122]. Indeed, axitinib treatment significantly reduces T cell (CD3+ and CD8+) infiltration into brain tumors [76]. Because both axitinib and OHSV are already in clinical trials for brain tumors as monotherapy with limited efficacy, dual combination therapy (OHSV-IL12+axitinib) that shows anti-tumor efficacy in both immune deficient and immune competent orthotopic brain tumor models has translational relevance [76]. Because systemic anti-angiogenic therapy (i.e., axitinib) is often associated with renal toxicities [123,124], G47Δ-IL12 was tested in combination with a local OHSV expression of angiostatin (OHSV G47Δ-angio), an anti-angiogenic polypeptide, in hypervascular human GBM models [125]. The combination of two OHSVs (G47Δ-IL12+G47Δ-angio) significantly prolongs survival compare to each armed OHSV alone, which is associated with increased viral spread and reduced expression of VEGF and CD31+ blood vessels in the tumor (Figure 5) [125]. This study supports further engineering of OHSV to express both IL-12 and angiostatin locally in the tumor. Use of one virus rather than two is practical in the context of future FDA approval. Similar to G47Δ-IL12 and anti-angiogenesis studies, the combination of another IL-12 expressing OHSV NV1042 with the anti-cancer chemotherapy drug vinblastine results in significant reduction of tumor burden in athymic mice bearing subcutaneous CWR22 prostate tumors, which is most likely associated with diminishing the number of CD31+ endothelial cells [126].

2.5. Immune Checkpoint Inhibition Enhances OHSV-IL12 Treatment-Induced Anti-Tumor Immunity

Though anti-tumor effects of OHSV-IL12 therapy is multifaceted, virotherapy alone does not improve significant survival in cancer [19,30,76,127]. For example, G47Δ-IL12 monotherapy shows limited efficacy in preclinical immune competent models of prostate and malignant peripheral nerve sheath tumors [94]. Since OHSV-IL12 treatment changes immune phenotypes of the tumor microenvironment, it is tested in combination with other forms of immunotherapies (e.g., ICIs) to obtain a better therapeutic response [19,76]. ICIs, such as cytotoxic T lymphocyte antigen 4 (CTLA-4) and programmed death 1 (PD-1) suppress T cell-mediated anti-tumor immune responses (Figure 6), leading to tumor progression [128]. ICI antibodies (i.e., anti-CTLA-4 or anti-PD-1) are effective in unleashing tumor-induced immunosuppression and activating effector immune cells (Figure 6) [129]. Since OHSV-IL12 induces robust anti-tumor immunity [19], it is hypothesized that OHSV-IL12 will synergize with ICI antibodies and will improve the anti-cancer efficacy of G47Δ-IL12. Indeed, in 005 GSC-derived orthotopic brain tumor models, dual combination modestly extends survival compared to ICI antibody alone or G47Δ-IL12 therapy alone [19]. The modest anti-tumor efficacy of the dual combination is not due to the inability of ICI antibodies to cross the blood-brain barrier, since ICI antibodies were detected in the tumor [19]. Because CTLA-4 and PD-1 regulate anti-tumor immunity via distinct and non-redundant immune evasion mechanisms [130,131,132], it is hypothesized that combinatorial blockade of two immune inhibitory pathways will produce enhanced anti-tumor immune effects and will synergize with the anti-tumor efficacy of G47Δ-IL12 (i.e., triple combination therapy: G47Δ-IL12+anti-PD-1+anti-CTLA-4). Indeed, triple combination therapy leads to a significant percentage (89%) of long-term survivors (i.e., survived six months post-tumor implantation) [19]. These survivors remain protected following lethal tumor re-challenge in the contralateral hemisphere, surviving another three months until the experiment was terminated, which indicates development of long-term memory protection [19]. These unprecedented findings were reproduced in a second aggressive immune competent CT-2A GBM model [19]. The survival efficacy of the triple combination therapies is associated with a significant but complex immune alteration in the tumor microenvironment, as opposed to mock, single or dual combination therapies, which includes: i) increase tumor infiltration of T cells (CD3+ and CD8+); ii) increase number of proliferating T cells (CD3+Ki67+); iii) increase activated T cells (CD8+CD69+); iv) reduce regulatory T cells (FoxP3+); v) increase T effector (CD8+)/regulatory T cell (CD4+FoxP3+) ratio; vi) increase TAMs (CD68+, F4/80+); vii) skew TAMs towards M1-phenotypes (iNOS+, pSTAT1+, CD68+pSTAT1+); viii) increase Th1 differentiation (T-bet+); ix) reduce PD-L1+ cells; x) increase tumor-cell specific IFN-γ response; and (xi) reduce tumor cells (Figure 5) [19,26]. Depletion and inhibition of immune cell subtypes (i.e., CD4+ cell depletion by anti-CD4, CD8+ cell depletion by anti-CD8, peripheral macrophage depletion by liposomal clodronate, or CSF-1R inhibition by brain penetrant drug BLZ945 to target TAMs) confirms the necessity of CD4+ cells, CD8+ cells, and macrophages in the therapeutic efficacy, with CD4+ cells playing the critical role [19]. It remains to be determined how immune cells, especially CD4+ cells and M1-like macrophage polarization contribute to therapeutic efficacy. IL-12 appears to be critical for this exceptional anti-tumor efficacy, since another triple combination therapy involving the base G47Δ (without IL-12 expression) plus two systemic ICI antibodies results in only 13% long-term survivors [25], as opposed to 89% from G47Δ-IL12 + anti-PD-1 + anti-CTLA-4 combination [19].

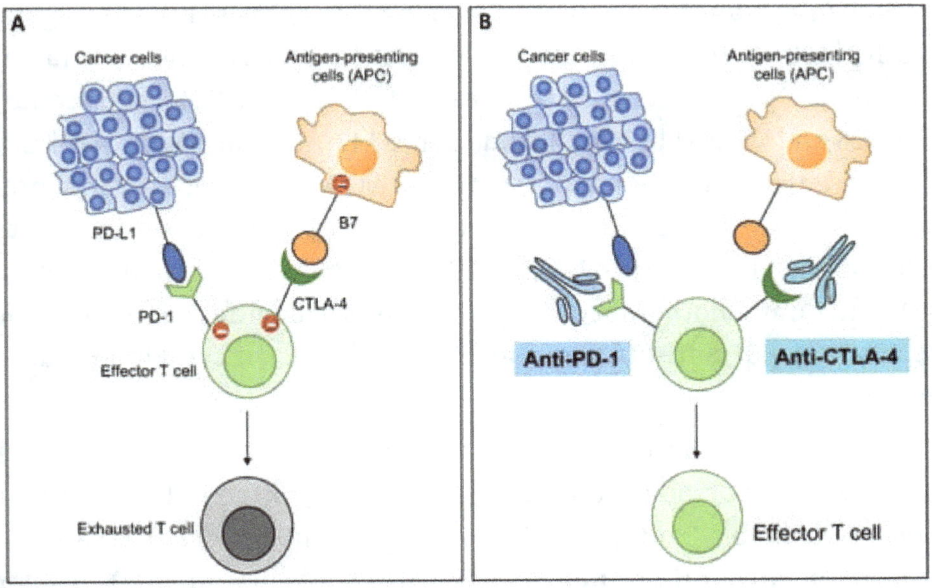

Figure 6. Graphic presentation of mechanism of action of immune checkpoint inhibitors (ICIs). (A) T cells express immune checkpoints programmed death 1 (PD-1) and cytotoxic T lymphocyte antigen 4 (CTLA-4), which interact with their corresponding ligands, i.e., programmed death ligand 1 (PD-L1) on cancer cells and B7 molecules on antigen presenting cells (APC), respectively. PD-1:PD-L1 and CTLA-4:B7 interactions send negative signals to immune cells, leading to exhaustion of T cells and no effector activity. (B) Antibodies to PD-1 (anti-PD-1) and CTLA-4 (anti-CTLA4) block those interactions and unleash anti-tumor immunity by enhancing activity of effector T cells.

3. ANTI-CANCER POTENTIAL OF OTHER OVS ENCODING IL-12

As IL-12 is a master anti-tumor cytokine and has distinct multifaceted anti-cancer properties [1,2,3,6,8,10], IL-12 expression by OVs is not limited to OHSVs. Several other OVs encoding IL-12 have been developed (Table 2), including adenoviruses [18,42,46,47,48,49,133,134,135,136,137 ,138], measles virus [20,50], maraba virus [51], Newcastle disease virus [52], Semliki forest virus [53,54,55,56,139,140,141], vesicular stomatitis virus [57,142], and Sindbis virus [58]. In these above-mentioned studies, IL-12 is expressed either alone [18,20,51,57,137,138,139] or co-expressed alongside with GM-CSF [134,135,136], pericellular matrix proteoglycan decorin [47], tumor necrosis factor-related apoptosis-inducing ligand (TRAIL) [133], IL-2 proinflammatory cytokine [52], IFN-γ inducing factor IL-18 [46], T cell co-stimulatory ligand 4-1BBL [49], CD8+ co-receptor for CD28 and CTLA-4 [48], or suicide genes [42]. The anti-tumor efficacy of these engineered OVs are tested in various mouse or hamster preclinical cancer models and produce superior anti-tumor immunity either alone [18,20,42,46,47,48,51,52,57,133,135,137] or in combination with other immunotherapeutic agents such as dendritic cell (DC) vaccine [49,134,136], ICI anti-PD-1 and anti-PD-L1 [139], or cytokine-induced killer cells [138].

The tumor-targeting oncolytic adenovirus (Ad-TD) delivers either wild-type IL-12 or non-secreting IL-12 (nsIL-12) directly to pancreatic tumor cells (Table 2) [44]. In a Syrian hamster tumor model, treatment with both Ad-TD-IL12 and Ad-TD-nsIL-12 results in 100% tumor eradication and animal survival, and treatment with Ad-TD-IL12 in particular produces a significant increase in populations of CD3+CD4-CD8+ in the spleen [44]. In addition, treatment with Ad-TD-nsIL-12 results in lower levels of lymph node IFN-γ, and splenic IFN-γ and IP-10 production [44]. Another oncolytic adenovirus expressing IL-12 (RdB/IL-12) inhibits tumor growth in murine melanoma lines by 95%, while adenovirus expressing both IL-12 and IL-18 (RdB/IL-12/IL-18) inhibits growth by 99% [46]. RdB/IL-12/IL-18 also increases the cytokine ratio of Th1/Th2, increases tumor infiltration of CD4+ T, CD8+ T and NK cells, and promotes differentiation of T cells expressing IL-12Rβ2 or IL-18Rα [46]. In another engineered oncolytic adenovirus (Ad-ΔB7/IL-12/4-1BBL), co-expression of both IL-12 and the cytokine 4-1BB ligand (4-1BBL) produces significantly higher survival in mice, with 100% of mice surviving more than 30 days after viral injection [49]. This is considerable when in comparison to the 20% survival rate of mice treated with virus expressing only IL-12 or 4-1BBL. In this study, mice treated with either Ad-ΔB7/IL-12 or Ad-ΔB7/IL-12/4-1BBL have a higher amount of tumor infiltrating CD4+ and CD8+ cells in comparison to treatment with Ad-ΔB7/4-1BBL or Ad-ΔB7 [49].

An oncolytic measles virus encoding an IL-12 fusion protein (MeVac FmIL-12) as a single agent produces potent anti-tumor immune effects in an immunocompetent colon cancer model with 90% complete remission [20]. This robust anti-cancer efficacy is dependent on T cells, particularly CD8+ cells, and is associated with activation of early NK cells and effector T cells, and upregulation of effector anti-tumor cytokines IFN-γ and TNF-α [20]. The findings are similar to what we observed in GBM with OHSV G47Δ-IL12 virus [19,30]. Although it was not examined whether the long-term survivors in MeVac FmIL-12 group developed any memory protection, the potent anti-tumor immune efficacy indicates that MeVac FmIL-12 is also an attractive candidate in cancer therapy [20]. Similar to the MeVac FmIL-12 virus, IL12-expressing oncolytic maraba virus (designated MG1-IL12-ICV) treatment also leads to complete eradication of peritoneal carcinomatosis, which is associated with significant recruitment of NK cells in the tumor microenvironment [51]. The lentogenic Newcastle disease virus Clone30 strain was generated to express IL-12 (designated rClone30-IL-12) that displays improved survival and reduced tumor volume [52]. In this case, co-expression of two cytokines (IL-12 and IL-2) produced synergistic effects on treatment, with the rClone30-IL-12-IL-2 virus inducing the greatest release of IFN-γ and IP-10 [52].

The Semliki Forest virus (SFV) has a broad host range and is suicidal, causing apoptosis in infected cells, which makes it a promising viral vector [53]. In a murine tumor model using B16 cells, a single intratumoral injection of SFV expressing IL-12 (SFV-IL12) results in significant tumor regression seven days after injection. This follows a distinct inhibition of tumor vascularization and an increase in IFN-γ production at two days post-injection [53]. Furthermore, treatment with SFV-IL12 results in an increase in expression of splenic IP-10 and monokine induced by interferon-γ (MIG) [53]. Similar results are obtained on the murine colon adenocarcinoma cell line, MC38, with SFV-IL12 treatment resulting in increased tumor-specific CD8+ T lymphocytes, reduced tumor volume, and improved survival [55]. Of particular notice, treatment of MC38 cells with SFV-IL12 results in tumor-infiltrating monocytic myeloid-derived suppressor cells (M-MDSCs) displaying increased expression of CD11c, CD8α, CD40, and CD86 in the presence of an intact, endogenous host type-I interferon (IFN-I) system [55].

Vesicular stomatitis virus (VSV) is another example of a viral vector [57]. Oncolytic VSV carrying IL-12 (rVSV-IL12) succeeds in effectively reducing tumor volume and prolonging survival in both human and murine SCC tumors, with 40% of mice treated with rVSV-IL12 surviving past 100 days post-injection, in comparison to mice treated with the fusogenic OV, rVSV-F, surviving only to 60 days post-injection ($p < 0.0001$) [57]. Similarly, treatment with the Sindbis viral vector carrying IL-12 (Sin/IL12) increases production of IFN-γ, and results in reduced tumor growth and improved survival in ovarian clear cell carcinoma [58]. Treatment with Sin/IL12 also modulates the regulatory functions of NK cells, increasing activation and recruitment of the cells [58]. These and afore-mentioned studies clearly suggest that IL-12 is a useful anti-cancer agent for oncolytic immunovirotherapy, and boosts anti-cancer immune properties of OVs.

4. CLINICAL PERSPECTIVES

Regardless of cancer types, IL-12 expressing OVs have been tested and found effective against various cancers (such as glioma, neuroblastoma, squamous cell carcinoma, metastatic breast cancer, hepatoblastoma, sarcoma, kidney cancer, lymphoma, prostate cancer, pancreatic cancer, colon cancer, ovarian cancer, melanoma, etc.) (Table 1 and Table 2). This clearly suggests that IL-12 expressing OVs are an attractive therapeutic candidate for clinical translation against any forms of cancer. In general, when translating results from bench to bed side, safety remains the most concerning aspect, along with dose, route of administration, viral pharmacokinetics and resistance mechanism of host cells [143]. Recent FDA approval of OHSV T-VEC in 2015 has heightened the field of oncolytic virus-based immunotherapy. The FDA approved OHSV expresses GM-CSF instead of IL-12. The safety and efficacy of T-VEC in immune-privileged organs, such as brain, has not been extensively elucidated. Moreover, T-VEC has not demonstrated durable responses in a majority of advanced melanoma patients [29], especially those with visceral metastases [144,145], raising questions about its possible long-term efficacy in patients with difficult-to-treat metastatic cancers.

Safety and anti-tumor efficacy of OHSV-IL12 as monotherapy or combination therapy has been demonstrated in preclinical tumor models [19,30,76]. Currently, G47Δ expressing human IL-12 is under development for clinical use. Safety of M002 has been established in the brain after intracerebral administration to non-human primates [21]. M032 is now in a phase I clinical trial (NCT02062827) in patients with recurrent/progressive GBM, anaplastic astrocytoma, or GBM [88]. Similar to OHSV-IL12, an adenovirus expressing human IL-12 is also under clinical trial investigation as monotherapy in prostate cancer (NCT02555397, NCT00406939), pancreatic cancer (NCT03281382), breast cancer (NCT00849459) and melanoma (NCT01397708). In addition, Ad-RTS-hIL-12 (another adenovirus encoding for IL-12) is under clinical trial evaluation in combination with veledimex (an oral activator ligand to promote release of IL-12 by an OV) in pediatric brain tumor (NCT03330197) and adult glioblastoma or malignant glioma (NCT02026271). Triple combination of Ad-RTS-hIL-12, veledimex and nivolumab (anti-PD-1) are also in active status in glioblastoma patients (NCT03636477). In our opinion, since OV-IL12 treatment induces prominent anti-tumor immunity including increased expression of PD-L1 in the tumor microenvironment, OV-IL12 therapy may improve the response rate to anti-PD-L1 treatment, especially in cancer patients who inherently lack PD-L1 expression and/or previously unresponsive to anti-PD-L1 treatment. Thus, combination studies involving OV-IL12 and ICI warrant clinical investigation and could be an attractive treatment strategy for cancer patients.

5. CONCISE SUMMARY

The ability to manipulate and comprehend the genetic makeup of pathogens has rendered oncolytic viruses (OVs) a compelling option for cancer treatment. Out of all oncolytic viruses (OVs), OHSV (Oncolytic Herpes Simplex Virus) has received approval from the FDA for cancer treatment and is the most advanced in terms of clinical development [29,144,145,146]. Furthermore, the presence of antiviral medications, such as acyclovir, enhances the safety of OHSV as a potential treatment for cancer compared to other oncolytic viruses [59,147]. Nevertheless, it is well acknowledged that the production of IL-12 by OHSV and its use as a standalone treatment does not yield the expected therapeutic results, as evidenced by various preclinical cancer models [30,84,86,148,149]. In order to get a more effective therapeutic outcome, it is necessary to use a synergistic or additive combination approach with other anti-cancer medications [19,25,76,148,150,151]. Like OHSV-IL12, other oncolytic viruses (OVs) that encode IL-12 likewise necessitate a combination immunotherapeutic strategy to augment their anti-tumor effectiveness [49,134,136,138,139].

While ICI-based cancer immunotherapy is advancing quickly in the field of oncology, the use of combination therapy with one or more ICIs often leads to substantial toxicity in people [152,153,154,155,156]. In addition, ICI immunotherapy has shown limited efficacy in treating aggressive cancer types such as GBM (CheckMate-143) [157,158] or triple-negative breast cancer (KEYNOTE-119) [159]. On the other hand, when the virus is produced locally in the tumor, OV expressing either a single or multiple immune stimulator does not result in toxicity [21,52,160,161,162,163]. This is true even in the immuno-privileged brain [21]. Instead, it stimulates a strong immune response against the tumor. Therefore, by creating viral vectors that are well-designed and more powerful, and that express multiple immune stimulators along with the master anti-tumor cytokine IL-12, it is possible to generate stronger local anti-tumor immune responses. This approach can potentially reduce the reliance on multiple systemic immune checkpoint inhibitors (ICI) or other systemic drugs, and ultimately overcome the current limitations of systemic cancer immunotherapy.

In addition to the development of new viral vectors, another challenge that must be resolved in order to fully harness the capabilities of oncolytic viruses (OVs) is the restricted dissemination of the virus within tumors. This limitation is caused by the presence of genes that are resistant to viral infection or the increased expression of proteins that counteract viral activity after OV treatment. The restricted multiplication and spread of the virus within the tumor may lead to a decrease in tumor destruction and a limited effect of the vaccine within the tumor itself. Discovering and suppressing anti-viral factors will enable the development of novel immunotherapeutic approaches using oncolytic viruses (OVs) to boost viral replication and dissemination within tumors, while also stimulating strong anti-tumor immune responses [164]. The key to achieving clinical success lies in developing an improved justification for combining OV-based treatments while maintaining safety.Oncolytic viruses (OVs) are viruses that have been either genetically manipulated or occur naturally. They have the ability to selectively reproduce in and destroy cancer cells, while leaving healthy cells unharmed. Additionally, they stimulate the immune system to respond against tumors. Tumor immunity produced by oncolytic viruses (OVs) can be augmented by viral production of anti-tumor cytokines, such as interleukin 12 (IL-12). IL-12 is a powerful substance that fights against cancer by encouraging the development of T-helper 1 (Th1) cells, aiding in the destruction of cancer cells by T-cells, and preventing the growth of blood vessels in tumors. Although systemic IL-12 therapy has shown promise in preclinical models, it is also linked to substantial harm in humans. In order to harness the therapeutic benefits of IL-12 in cancer treatment using oncolytic viruses (OVs), researchers have developed 25 different OVs that express IL-12 (referred to as OV-IL12s). These OVs have been genetically modified to produce IL-12 locally and have undergone preclinical testing in different cancer models. Out of all the OV-IL12s, the oncolytic herpes simplex virus expressing IL-12 (OHSV-IL12) is the most advanced in clinical development. Local expression of IL-12 within the tumors prevents harmful effects throughout the body, while also promoting a strong immune response against the tumor. Additionally, it enhances the effectiveness of medications that inhibit blood vessel growth or modulate the immune system, without causing any safety concerns. Although there is a growing interest in OV-IL12s, there are currently no reviews available that explore their potential effectiveness and safety in human subjects. This paper will examine the safety, tumor-specificity, and anti-tumor immune/anti-angiogenic effects of OHSV-IL12 as both mono- and combination-therapies. We will examine not only OHSV-IL12 viruses, but also additional oncolytic viruses that express IL-12, and explore their use in cancer treatment.

REFERENCES

1. Berraondo, P.; Etxeberria, I.; Ponz-Sarvise, M.; Melero, I. Revisiting Interleukin-12 as a Cancer Immunotherapy Agent. Clin. Cancer Res. 2018, 24, 2716–2718.
2. Lu, X. Impact of IL-12 in Cancer. Curr. Cancer Drug Targets 2017, 17, 682–697.
3. Zeh, H.J., 3rd; Hurd, S.; Storkus, W.J.; Lotze, M.T. Interleukin-12 promotes the proliferation and cytolytic maturation of immune effectors: Implications for the immunotherapy of cancer. J. Immunother. Emphasis Tumor Immunol. 1993, 14, 155–161.
4. Lasek, W.; Zagozdzon, R.; Jakobisiak, M. Interleukin 12: Still a promising candidate for tumor immunotherapy? Cancer Immunol. Immunother. 2014, 63, 419–435.
5. Lehmann, D.; Spanholtz, J.; Sturtzel, C.; Tordoir, M.; Schlechta, B.; Groenewegen, D.; Hofer, E. IL-12 directs further maturation of ex vivo differentiated NK cells with improved therapeutic potential. PLoS ONE 2014, 9, e87131.
6. Trinchieri, G.; Wysocka, M.; D'Andrea, A.; Rengaraju, M.; Aste-Amezaga, M.; Kubin, M.; Valiante, N.M.; Chehimi, J. Natural killer cell stimulatory factor (NKSF) or interleukin-12 is a key regulator of immune response and inflammation. Prog. Growth Factor Res. 1992, 4, 355–368.
7. Trinchieri, G. Interleukin-12 and the regulation of innate resistance and adaptive immunity. Nat. Rev. Immunol. 2003, 3, 133–146.
8. Otani, T.; Nakamura, S.; Toki, M.; Motoda, R.; Kurimoto, M.; Orita, K. Identification of IFN-gamma-producing cells in IL-12/IL-18-treated mice. Cell Immunol. 1999, 198, 111–119.

9. Tugues, S.; Burkhard, S.H.; Ohs, I.; Vrohlings, M.; Nussbaum, K.; Vom Berg, J.; Kulig, P.; Becher, B. New insights into IL-12-mediated tumor suppression. Cell Death Differ. 2015, 22, 237–246.

10. Angiolillo, A.L.; Sgadari, C.; Tosato, G. A role for the interferon-inducible protein 10 in inhibition of angiogenesis by interleukin-12. Ann. N. Y. Acad. Sci. 1996, 795, 158–167.

11. Lamont, A.G.; Adorini, L. IL-12: A key cytokine in immune regulation. Immunol. Today 1996, 17, 214–217.

12. Jenks, S. After initial setback, IL-12 regaining popularity. J. Natl. Cancer Inst. 1996, 88, 576–577.

13. Duvic, M.; Sherman, M.L.; Wood, G.S.; Kuzel, T.M.; Olsen, E.; Foss, F.; Laliberte, R.J.; Ryan, J.L.; Zonno, K.; Rook, A.H. A phase II open-label study of recombinant human interleukin-12 in patients with stage IA, IB, or IIA mycosis fungoides. J. Am. Acad. Dermatol. 2006, 55, 807–813.

14. Rook, A.H.; Wood, G.S.; Yoo, E.K.; Elenitsas, R.; Kao, D.M.; Sherman, M.L.; Witmer, W.K.; Rockwell, K.A.; Shane, R.B.; Lessin, S.R.; et al. Interleukin-12 therapy of cutaneous T-cell lymphoma induces lesion regression and cytotoxic T-cell responses. Blood 1999, 94, 902–908.

15. Little, R.F.; Pluda, J.M.; Wyvill, K.M.; Rodriguez-Chavez, I.R.; Tosato, G.; Catanzaro, A.T.; Steinberg, S.M.; Yarchoan, R. Activity of subcutaneous interleukin-12 in AIDS-related Kaposi sarcoma. Blood 2006, 107, 4650–4657.

16. Younes, A.; Pro, B.; Robertson, M.J.; Flinn, I.W.; Romaguera, J.E.; Hagemeister, F.; Dang, N.H.; Fiumara, P.; Loyer, E.M.; Cabanillas, F.F.; et al. Phase II clinical trial of interleukin-12 in patients with relapsed and refractory non-Hodgkin's lymphoma and Hodgkin's disease. Clin. Cancer Res. 2004, 10, 5432–5438.

17. Del Vecchio, M.; Bajetta, E.; Canova, S.; Lotze, M.T.; Wesa, A.; Parmiani, G.; Anichini, A. Interleukin-12: Biological properties and clinical application. Clin. Cancer Res. 2007, 13, 4677–4685.

18. Poutou, J.; Bunuales, M.; Gonzalez-Aparicio, M.; Garcia-Aragoncillo, E.; Quetglas, J.I.; Casado, R.; Bravo-Perez, C.; Alzuguren, P.; Hernandez-Alcoceba, R. Safety and antitumor effect of oncolytic and helper-dependent adenoviruses expressing interleukin-12 variants in a hamster pancreatic cancer model. Gene Ther. 2015, 22, 696–706.
19. Saha, D.; Martuza, R.L.; Rabkin, S.D. Macrophage Polarization Contributes to Glioblastoma Eradication by Combination Immunovirotherapy and Immune Checkpoint Blockade. Cancer Cell 2017, 32, 253–267.e255.
20. Veinalde, R.; Grossardt, C.; Hartmann, L.; Bourgeois-Daigneault, M.C.; Bell, J.C.; Jager, D.; von Kalle, C.; Ungerechts, G.; Engeland, C.E. Oncolytic measles virus encoding interleukin-12 mediates potent antitumor effects through T cell activation. Oncoimmunology 2017, 6, e1285992.
21. Roth, J.C.; Cassady, K.A.; Cody, J.J.; Parker, J.N.; Price, K.H.; Coleman, J.M.; Peggins, J.O.; Noker, P.E.; Powers, N.W.; Grimes, S.D.; et al. Evaluation of the safety and biodistribution of M032, an attenuated herpes simplex virus type 1 expressing hIL-12, after intracerebral administration to aotus nonhuman primates. Hum. Gene Ther. Clin Dev. 2014, 25, 16–27.
22. Saha, D.; Ahmed, S.S.; Rabkin, S.D. Exploring the Antitumor Effect of Virus in Malignant Glioma. Drugs Future 2015, 40, 739–749.
23. Saha, D.; Martuza, R.L.; Curry, W.T. Viral oncolysis of glioblastoma. In Neurotropic viral infections, 2nd ed.; Reiss, C.S., Ed.; Springer: New York, NY, USA, 2016; Volume 2, pp. 481–517.
24. Saha, D.; Wakimoto, H.; Rabkin, S.D. Oncolytic herpes simplex virus interactions with the host immune system. Curr. Opin. Virol. 2016, 21, 26–34.
25. Saha, D.; Martuza, R.L.; Rabkin, S.D. Oncolytic herpes simplex virus immunovirotherapy in combination with immune checkpoint blockade to treat glioblastoma. Immunotherapy 2018, 10, 779–786.

26. Saha, D.; Martuza, R.L.; Rabkin, S.D. Curing glioblastoma: Oncolytic HSV-IL12 and checkpoint blockade. Oncoscience 2017, 4, 67–69.

27. Kaufman, H.L.; Bommareddy, P.K. Two roads for oncolytic immunotherapy development. J. Immunother. Cancer 2019, 7, 26.

28. Raja, J.; Ludwig, J.M.; Gettinger, S.N.; Schalper, K.A.; Kim, H.S. Oncolytic virus immunotherapy: Future prospects for oncology. J. Immunother. Cancer 2018, 6, 140.

29. Andtbacka, R.H.; Kaufman, H.L.; Collichio, F.; Amatruda, T.; Senzer, N.; Chesney, J.; Delman, K.A.; Spitler, L.E.; Puzanov, I.; Agarwala, S.S.; et al. Talimogene Laherparepvec Improves Durable Response Rate in Patients With Advanced Melanoma. J. Clin. Oncol. 2015, 33, 2780–2788.

30. Cheema, T.A.; Wakimoto, H.; Fecci, P.E.; Ning, J.; Kuroda, T.; Jeyaretna, D.S.; Martuza, R.L.; Rabkin, S.D. Multifaceted oncolytic virus therapy for glioblastoma in an immunocompetent cancer stem cell model. Proc. Natl. Acad. Sci. USA 2013, 110, 12006–12011.

31. Guan, Y.; Ino, Y.; Fukuhara, H.; Todo, T. Antitumor Efficacy of Intravenous Administration of Oncolytic Herpes Simplex Virus Expressing Interleukin 12. Mol. Ther. 2006, 13, S108.

32. Wong, R.J.; Patel, S.G.; Kim, S.; DeMatteo, R.P.; Malhotra, S.; Bennett, J.J.; St-Louis, M.; Shah, J.P.; Johnson, P.A.; Fong, Y. Cytokine gene transfer enhances herpes oncolytic therapy in murine squamous cell carcinoma. Hum. Gene. Ther. 2001, 12, 253–265.

33. Wong, R.J.; Chan, M.K.; Yu, Z.; Kim, T.H.; Bhargava, A.; Stiles, B.M.; Horsburgh, B.C.; Shah, J.P.; Ghossein, R.A.; Singh, B.; et al. Effective intravenous therapy of murine pulmonary metastases with an oncolytic herpes virus expressing interleukin 12. Clin. Cancer Res. 2004, 10, 251–259.

34. Friedman, G.K.; Bernstock, J.D.; Chen, D.; Nan, L.; Moore, B.P.; Kelly, V.M.; Youngblood, S.L.; Langford, C.P.; Han, X.; Ring, E.K.; et al. Enhanced Sensitivity of Patient-Derived Pediatric High-Grade Brain Tumor Xenografts to Oncolytic HSV-1 Virotherapy Correlates with Nectin-1 Expression. Sci. Rep. 2018, 8, 13930.

35. Cody, J.J.; Scaturro, P.; Cantor, A.B.; Yancey Gillespie, G.; Parker, J.N.; Markert, J.M. Preclinical evaluation of oncolytic deltagamma(1)34.5 herpes simplex virus expressing interleukin-12 for therapy of breast cancer brain metastases. Int. J. Breast Cancer 2012, 2012, 628697.

36. Gillory, L.A.; Megison, M.L.; Stewart, J.E.; Mroczek-Musulman, E.; Nabers, H.C.; Waters, A.M.; Kelly, V.; Coleman, J.M.; Markert, J.M.; Gillespie, G.Y.; et al. Preclinical evaluation of engineered oncolytic herpes simplex virus for the treatment of neuroblastoma. PLoS ONE 2013, 8, e77753.

37. Megison, M.L.; Gillory, L.A.; Stewart, J.E.; Nabers, H.C.; Mroczek-Musulman, E.; Waters, A.M.; Coleman, J.M.; Kelly, V.; Markert, J.M.; Gillespie, G.Y.; et al. Preclinical evaluation of engineered oncolytic herpes simplex virus for the treatment of pediatric solid tumors. PLoS ONE 2014, 9, e86843.

38. Leoni, V.; Vannini, A.; Gatta, V.; Rambaldi, J.; Sanapo, M.; Barboni, C.; Zaghini, A.; Nanni, P.; Lollini, P.L.; Casiraghi, C.; et al. A fully-virulent retargeted oncolytic HSV armed with IL-12 elicits local immunity and vaccine therapy towards distant tumors. PLoS Pathog 2018, 14, e1007209.

39. Alessandrini, F.; Menotti, L.; Avitabile, E.; Appolloni, I.; Ceresa, D.; Marubbi, D.; Campadelli-Fiume, G.; Malatesta, P. Eradication of glioblastoma by immuno-virotherapy with a retargeted oncolytic HSV in a preclinical model. Oncogene 2019, 38, 4467–4479.

40. Ino, Y.; Saeki, Y.; Fukuhara, H.; Todo, T. Triple combination of oncolytic herpes simplex virus-1 vectors armed with interleukin-12, interleukin-18, or soluble B7-1 results in enhanced antitumor efficacy. Clin. Cancer Res. 2006, 12, 643–652.

41. Yan, R.; Zhou, X.; Chen, X.; Liu, X.; Ma, J.; Wang, L.; Liu, Z.; Zhan, B.; Chen, H.; Wang, J.; et al. Enhancement of Oncolytic Activity of oHSV Expressing IL-12 and Anti PD-1 Antibody by Concurrent Administration of Exosomes Carrying CTLA-4 miRNA. Immunotherapy 2019, 5, 1–10.

42. Freytag, S.O.; Barton, K.N.; Zhang, Y. Efficacy of oncolytic adenovirus expressing suicide genes and interleukin-12 in preclinical model of prostate cancer. Gene Ther. 2013, 20, 1131–1139.

43. Freytag, S.O.; Zhang, Y.; Siddiqui, F. Preclinical toxicology of oncolytic adenovirus-mediated cytotoxic and interleukin-12 gene therapy for prostate cancer. Mol. Ther. Oncolytics. 2015, 2.

44. Wang, P.; Li, X.; Wang, J.; Gao, D.; Li, Y.; Li, H.; Chu, Y.; Zhang, Z.; Liu, H.; Jiang, G.; et al. Re-designing Interleukin-12 to enhance its safety and potential as an anti-tumor immunotherapeutic agent. Nat. Commun. 2017, 8, 1395.

45. Choi, K.J.; Zhang, S.N.; Choi, I.K.; Kim, J.S.; Yun, C.O. Strengthening of antitumor immune memory and prevention of thymic atrophy mediated by adenovirus expressing IL-12 and GM-CSF. Gene Ther. 2012, 19, 711–723.

46. Choi, I.K.; Lee, J.S.; Zhang, S.N.; Park, J.; Sonn, C.H.; Lee, K.M.; Yun, C.O. Oncolytic adenovirus co-expressing IL-12 and IL-18 improves tumor-specific immunity via differentiation of T cells expressing IL-12Rbeta2 or IL-18Ralpha. Gene Ther. 2011, 18, 898–909.

47. Oh, E.; Choi, I.K.; Hong, J.; Yun, C.O. Oncolytic adenovirus coexpressing interleukin-12 and decorin overcomes Treg-mediated immunosuppression inducing potent antitumor effects in a weakly immunogenic tumor model. Oncotarget 2017, 8, 4730–4746.

48. Lee, Y.S.; Kim, J.H.; Choi, K.J.; Choi, I.K.; Kim, H.; Cho, S.; Cho, B.C.; Yun, C.O. Enhanced antitumor effect of oncolytic adenovirus expressing interleukin-12 and B7-1 in an immunocompetent murine model. Clin. Cancer Res. 2006, 12, 5859–5868.

49. Huang, J.H.; Zhang, S.N.; Choi, K.J.; Choi, I.K.; Kim, J.H.; Lee, M.G.; Kim, H.; Yun, C.O. Therapeutic and tumor-specific immunity induced by combination of dendritic cells and oncolytic adenovirus expressing IL-12 and 4-1BBL. Mol. Ther. 2010, 18, 264–274.

50. Backhaus, P.S.; Veinalde, R.; Hartmann, L.; Dunder, J.E.; Jeworowski, L.M.; Albert, J.; Hoyler, B.; Poth, T.; Jager, D.; Ungerechts, G.; et al. Immunological Effects and Viral Gene Expression Determine the

Efficacy of Oncolytic Measles Vaccines Encoding IL-12 or IL-15 Agonists. Viruses 2019, 11, 914.

51. Alkayyal, A.A.; Tai, L.H.; Kennedy, M.A.; de Souza, C.T.; Zhang, J.; Lefebvre, C.; Sahi, S.; Ananth, A.A.; Mahmoud, A.B.; Makrigiannis, A.P.; et al. NK-Cell Recruitment Is Necessary for Eradication of Peritoneal Carcinomatosis with an IL12-Expressing Maraba Virus Cellular Vaccine. Cancer Immunol. Res. 2017, 5, 211–221.

52. Ren, G.; Tian, G.; Liu, Y.; He, J.; Gao, X.; Yu, Y.; Liu, X.; Zhang, X.; Sun, T.; Liu, S.; et al. Recombinant Newcastle Disease Virus Encoding IL-12 and/or IL-2 as Potential Candidate for Hepatoma Carcinoma Therapy. Technol. Cancer Res. Treat. 2016, 15, NP83-94.

53. Asselin-Paturel, C.; Lassau, N.; Guinebretiere, J.M.; Zhang, J.; Gay, F.; Bex, F.; Hallez, S.; Leclere, J.; Peronneau, P.; Mami-Chouaib, F.; et al. Transfer of the murine interleukin-12 gene in vivo by a Semliki Forest virus vector induces B16 tumor regression through inhibition of tumor blood vessel formation monitored by Doppler ultrasonography. Gene Ther. 1999, 6, 606–615.

54. Colmenero, P.; Chen, M.; Castanos-Velez, E.; Liljestrom, P.; Jondal, M. Immunotherapy with recombinant SFV-replicons expressing the P815A tumor antigen or IL-12 induces tumor regression. Int. J. Cancer 2002, 98, 554–560.

55. Melero, I.; Quetglas, J.I.; Reboredo, M.; Dubrot, J.; Rodriguez-Madoz, J.R.; Mancheno, U.; Casales, E.; Riezu-Boj, J.I.; Ruiz-Guillen, M.; Ochoa, M.C.; et al. Strict requirement for vector-induced type I interferon in efficacious antitumor responses to virally encoded IL12. Cancer Res. 2015, 75, 497–507.

56. Roche, F.P.; Sheahan, B.J.; O'Mara, S.M.; Atkins, G.J. Semliki Forest virus-mediated gene therapy of the RG2 rat glioma. Neuropathol. Appl. Neurobiol. 2010, 36, 648–660.

57. Shin, E.J.; Wanna, G.B.; Choi, B.; Aguila, D., 3rd; Ebert, O.; Genden, E.M.; Woo, S.L. Interleukin-12 expression enhances vesicular stomatitis virus oncolytic therapy in murine squamous cell carcinoma. Laryngoscope 2007, 117, 210–214.

58. Granot, T.; Venticinque, L.; Tseng, J.C.; Meruelo, D. Activation of cytotoxic and regulatory functions of NK cells by Sindbis viral vectors. PLoS ONE 2011, 6, e20598.

59. Peters, C.; Rabkin, S.D. Designing Herpes Viruses as Oncolytics. Mol Ther Oncolytics 2015, 2.

60. Todo, T.; Martuza, R.L.; Rabkin, S.D.; Johnson, P.A. Oncolytic herpes simplex virus vector with enhanced MHC class I presentation and tumor cell killing. Proc. Natl. Acad. Sci. USA 2001, 98, 6396–6401.

61. Goldstein, D.J.; Weller, S.K. Factor(s) present in herpes simplex virus type 1-infected cells can compensate for the loss of the large subunit of the viral ribonucleotide reductase: characterization of an ICP6 deletion mutant. Virology 1988, 166, 41–51.

62. Aghi, M.; Visted, T.; Depinho, R.A.; Chiocca, E.A. Oncolytic herpes virus with defective ICP6 specifically replicates in quiescent cells with homozygous genetic mutations in p16. Oncogene 2008, 27, 4249–4254.

63. Cameron, J.M.; McDougall, I.; Marsden, H.S.; Preston, V.G.; Ryan, D.M.; Subak-Sharpe, J.H. Ribonucleotide reductase encoded by herpes simplex virus is a determinant of the pathogenicity of the virus in mice and a valid antiviral target. J. Gen. Virol. 1988, 69, 2607–2612.

64. Markert, J.M.; Razdan, S.N.; Kuo, H.C.; Cantor, A.; Knoll, A.; Karrasch, M.; Nabors, L.B.; Markiewicz, M.; Agee, B.S.; Coleman, J.M.; et al. A phase 1 trial of oncolytic HSV-1, G207, given in combination with radiation for recurrent GBM demonstrates safety and radiographic responses. Mol. Ther. 2014, 22, 1048–1055.

65. Pasieka, T.J.; Baas, T.; Carter, V.S.; Proll, S.C.; Katze, M.G.; Leib, D.A. Functional genomic analysis of herpes simplex virus type 1 counteraction of the host innate response. J. Virol. 2006, 80, 7600–7612.

66. Wylie, K.M.; Schrimpf, J.E.; Morrison, L.A. Increased eIF2alpha phosphorylation attenuates replication of herpes simplex virus 2

vhs mutants in mouse embryonic fibroblasts and correlates with reduced accumulation of the PKR antagonist ICP34.5. J. Virol. 2009, 83, 9151–9162.

67. He, B.; Gross, M.; Roizman, B. The gamma(1)34.5 protein of herpes simplex virus 1 complexes with protein phosphatase 1alpha to dephosphorylate the alpha subunit of the eukaryotic translation initiation factor 2 and preclude the shutoff of protein synthesis by double-stranded RNA-activated protein kinase. Proc. Natl. Acad. Sci. USA 1997, 94, 843–848.

68. Li, Y.; Zhang, C.; Chen, X.; Yu, J.; Wang, Y.; Yang, Y.; Du, M.; Jin, H.; Ma, Y.; He, B.; et al. ICP34.5 protein of herpes simplex virus facilitates the initiation of protein translation by bridging eukaryotic initiation factor 2alpha (eIF2alpha) and protein phosphatase 1. J. Biol. Chem. 2011, 286, 24785–24792.

69. Orr, M.T.; Edelmann, K.H.; Vieira, J.; Corey, L.; Raulet, D.H.; Wilson, C.B. Inhibition of MHC class I is a virulence factor in herpes simplex virus infection of mice. PLoS Pathog 2005, 1, e7.

70. Eggensperger, S.; Tampe, R. The transporter associated with antigen processing: A key player in adaptive immunity. Biol. Chem. 2015, 396, 1059–1072.

71. Goldsmith, K.; Chen, W.; Johnson, D.C.; Hendricks, R.L. Infected cell protein (ICP)47 enhances herpes simplex virus neurovirulence by blocking the CD8+ T cell response. J. Exp. Med. 1998, 187, 341–348.

72. Liu, B.L.; Robinson, M.; Han, Z.Q.; Branston, R.H.; English, C.; Reay, P.; McGrath, Y.; Thomas, S.K.; Thornton, M.; Bullock, P.; et al. ICP34.5 deleted herpes simplex virus with enhanced oncolytic, immune stimulating, and anti-tumour properties. Gene Ther. 2003, 10, 292–303.

73. Wollmann, G.; Ozduman, K.; van den Pol, A.N. Oncolytic virus therapy for glioblastoma multiforme: concepts and candidates. Cancer J. 2012, 18, 69–81.

74. Mulvey, M.; Poppers, J.; Sternberg, D.; Mohr, I. Regulation of eIF2alpha phosphorylation by different functions that act during discrete phases in the herpes simplex virus type 1 life cycle. J. Virol. 2003, 77, 10917–10928.

75. Cassady, K.A.; Gross, M.; Roizman, B. The herpes simplex virus US11 protein effectively compensates for the gamma1(34.5) gene if present before activation of protein kinase R by precluding its phosphorylation and that of the alpha subunit of eukaryotic translation initiation factor 2. J. Virol. 1998, 72, 8620–8626.

76. Saha, D.; Wakimoto, H.; Peters, C.W.; Antoszczyk, S.J.; Rabkin, S.D.; Martuza, R.L. Combinatorial effects of VEGFR kinase inhibitor axitinib and oncolytic virotherapy in mouse and human glioblastoma stem-like cell models. Clin. Cancer Res. 2018.

77. Todo, T. "Armed" oncolytic herpes simplex viruses for brain tumor therapy. Cell Adh. Migr. 2008, 2, 208–213.

78. Kelly, K.J.; Wong, J.; Fong, Y. Herpes simplex virus NV1020 as a novel and promising therapy for hepatic malignancy. Expert Opin. Investig. Drugs 2008, 17, 1105–1113.

79. Bennett, J.J.; Malhotra, S.; Wong, R.J.; Delman, K.; Zager, J.; St-Louis, M.; Johnson, P.; Fong, Y. Interleukin 12 secretion enhances antitumor efficacy of oncolytic herpes simplex viral therapy for colorectal cancer. Ann. Surg. 2001, 233, 819–826.

80. Smith, M.C.; Boutell, C.; Davido, D.J. HSV-1 ICP0: Paving the way for viral replication. Future Virol. 2011, 6, 421–429.

81. Lanfranca, M.P.; Mostafa, H.H.; Davido, D.J. HSV-1 ICP0: An E3 Ubiquitin Ligase That Counteracts Host Intrinsic and Innate Immunity. Cells 2014, 3, 438–454.

82. Jacobs, A.; Breakefield, X.O.; Fraefel, C. HSV-1-based vectors for gene therapy of neurological diseases and brain tumors: Part I. HSV-1 structure, replication and pathogenesis. Neoplasia 1999, 1, 387–401.

83. Parker, J.N.; Pfister, L.A.; Quenelle, D.; Gillespie, G.Y.; Markert, J.M.; Kern, E.R.; Whitley, R.J. Genetically engineered herpes simplex viruses that express IL-12 or GM-CSF as vaccine candidates. Vaccine 2006, 24, 1644–1652.

84. Parker, J.N.; Gillespie, G.Y.; Love, C.E.; Randall, S.; Whitley, R.J.; Markert, J.M. Engineered herpes simplex virus expressing IL-12 in the treatment of experimental murine brain tumors. Proc. Natl. Acad. Sci. USA 2000, 97, 2208–2213.

85. Bauer, D.F.; Pereboeva, L.; Gillespie, G.Y.; Cloud, G.A.; Elzafarany, O.; Langford, C.; Markert, J.M.; Lamb, L.S., Jr. Effect of HSV-IL12 Loaded Tumor Cell-Based Vaccination in a Mouse Model of High-Grade Neuroblastoma. J. Immunol. Res. 2016, 2016, 2568125.

86. Hellums, E.K.; Markert, J.M.; Parker, J.N.; He, B.; Perbal, B.; Roizman, B.; Whitley, R.J.; Langford, C.P.; Bharara, S.; Gillespie, G.Y. Increased efficacy of an interleukin-12-secreting herpes simplex virus in a syngeneic intracranial murine glioma model. Neuro. Oncol. 2005, 7, 213–224.

87. Markert, J.M.; Cody, J.J.; Parker, J.N.; Coleman, J.M.; Price, K.H.; Kern, E.R.; Quenelle, D.C.; Lakeman, A.D.; Schoeb, T.R.; Palmer, C.A.; et al. Preclinical evaluation of a genetically engineered herpes simplex virus expressing interleukin-12. J. Virol. 2012, 86, 5304–5313.

88. Patel, D.M.; Foreman, P.M.; Nabors, L.B.; Riley, K.O.; Gillespie, G.Y.; Markert, J.M. Design of a Phase I Clinical Trial to Evaluate M032, a Genetically Engineered HSV-1 Expressing IL-12, in Patients with Recurrent/Progressive Glioblastoma Multiforme, Anaplastic Astrocytoma, or Gliosarcoma. Hum. Gene Ther. Clin. Dev. 2016, 27, 69–78.

89. Menotti, L.; Avitabile, E.; Gatta, V.; Malatesta, P.; Petrovic, B.; Campadelli-Fiume, G. HSV as A Platform for the Generation of Retargeted, Armed, and Reporter-Expressing Oncolytic Viruses. Viruses 2018, 10, 352.

90. Menotti, L.; Cerretani, A.; Hengel, H.; Campadelli-Fiume, G. Construction of a fully retargeted herpes simplex virus 1

recombinant capable of entering cells solely via human epidermal growth factor receptor 2. J. Virol. 2008, 82, 10153–10161.

91. Liu, X.J.; Wang, X.Y.; Guo, J.X.; Zhu, H.J.; Zhang, C.R.; Ma, Z.H. Oncolytic property of HSV-1 recombinant viruses carrying the human IL-12. Zhonghua Yi Xue Za Zhi 2017, 97, 2135–2140.

92. Athie-Morales, V.; Smits, H.H.; Cantrell, D.A.; Hilkens, C.M. Sustained IL-12 signaling is required for Th1 development. J. Immunol. 2004, 172, 61–69.

93. Grivennikov, S.I.; Greten, F.R.; Karin, M. Immunity, inflammation, and cancer. Cell 2010, 140, 883–899.

94. Antoszczyk, S.; Spyra, M.; Mautner, V.F.; Kurtz, A.; Stemmer-Rachamimov, A.O.; Martuza, R.L.; Rabkin, S.D. Treatment of orthotopic malignant peripheral nerve sheath tumors with oncolytic herpes simplex virus. Neuro. Oncol. 2014, 16, 1057–1066.

95. Varghese, S.; Rabkin, S.D.; Nielsen, G.P.; MacGarvey, U.; Liu, R.; Martuza, R.L. Systemic therapy of spontaneous prostate cancer in transgenic mice with oncolytic herpes simplex viruses. Cancer Res. 2007, 67, 9371–9379.

96. Ring, E.K.; Li, R.; Moore, B.P.; Nan, L.; Kelly, V.M.; Han, X.; Beierle, E.A.; Markert, J.M.; Leavenworth, J.W.; Gillespie, G.Y.; et al. Newly Characterized Murine Undifferentiated Sarcoma Models Sensitive to Virotherapy with Oncolytic HSV-1 M002. Mol. Ther. Oncolytics 2017, 7, 27–36.

97. Thomas, E.D.; Meza-Perez, S.; Bevis, K.S.; Randall, T.D.; Gillespie, G.Y.; Langford, C.; Alvarez, R.D. IL-12 Expressing oncolytic herpes simplex virus promotes anti-tumor activity and immunologic control of metastatic ovarian cancer in mice. J. Ovarian Res. 2016, 9, 70.

98. Krist, L.F.; Kerremans, M.; Broekhuis-Fluitsma, D.M.; Eestermans, I.L.; Meyer, S.; Beelen, R.H. Milky spots in the greater omentum are predominant sites of local tumour cell proliferation and accumulation in the peritoneal cavity. Cancer Immunol. Immunother. 1998, 47, 205–212.

99. Sorensen, E.W.; Gerber, S.A.; Frelinger, J.G.; Lord, E.M. IL-12 suppresses vascular endothelial growth factor receptor 3 expression on tumor vessels by two distinct IFN-gamma-dependent mechanisms. J. Immunol. 2010, 184, 1858–1866.

100. Cicchelero, L.; Denies, S.; Haers, H.; Vanderperren, K.; Stock, E.; Van Brantegem, L.; de Rooster, H.; Sanders, N.N. Intratumoural interleukin 12 gene therapy stimulates the immune system and decreases angiogenesis in dogs with spontaneous cancer. Vet. Comp. Oncol. 2017, 15, 1187–1205.

101. Wong, R.J.; Chan, M.K.; Yu, Z.; Ghossein, R.A.; Ngai, I.; Adusumilli, P.S.; Stiles, B.M.; Shah, J.P.; Singh, B.; Fong, Y. Angiogenesis inhibition by an oncolytic herpes virus expressing interleukin 12. Clin. Cancer Res. 2004, 10, 4509–4516.

102. Das, S.; Marsden, P.A. Angiogenesis in glioblastoma. N Engl. J. Med. 2013, 369, 1561–1563.

103. Gatson, N.N.; Chiocca, E.A.; Kaur, B. Anti-angiogenic gene therapy in the treatment of malignant gliomas. Neurosci. Lett. 2012, 527, 62–70.

104. Hardee, M.E.; Zagzag, D. Mechanisms of glioma-associated neovascularization. Am. J. Pathol. 2012, 181, 1126–1141.

105. Jain, R.K.; di Tomaso, E.; Duda, D.G.; Loeffler, J.S.; Sorensen, A.G.; Batchelor, T.T. Angiogenesis in brain tumours. Nat. Rev. Neurosci. 2007, 8, 610–622.

106. Zirlik, K.; Duyster, J. Anti-Angiogenics: Current Situation and Future Perspectives. Oncol. Res. Treat. 2018, 41, 166–171.

107. Ramjiawan, R.R.; Griffioen, A.W.; Duda, D.G. Anti-angiogenesis for cancer revisited: Is there a role for combinations with immunotherapy? Angiogenesis 2017, 20, 185–204.

108. Rajabi, M.; Mousa, S.A. The Role of Angiogenesis in Cancer Treatment. Biomedicines 2017, 5, 34.

109. Melegh, Z.; Oltean, S. Targeting Angiogenesis in Prostate Cancer. Int. J. Mol. Sci. 2019, 20, 2676.

110. van Moorselaar, R.J.; Voest, E.E. Angiogenesis in prostate cancer: Its role in disease progression and possible therapeutic approaches. Mol. Cell Endocrinol. 2002, 197, 239–250.

111. Batchelor, T.T.; Reardon, D.A.; de Groot, J.F.; Wick, W.; Weller, M. Antiangiogenic therapy for glioblastoma: Current status and future prospects. Clin. Cancer Res. 2014, 20, 5612–5619.

112. Wick, W.; Chinot, O.L.; Bendszus, M.; Mason, W.; Henriksson, R.; Saran, F.; Nishikawa, R.; Revil, C.; Kerloeguen, Y.; Cloughesy, T. Evaluation of pseudoprogression rates and tumor progression patterns in a phase III trial of bevacizumab plus radiotherapy/temozolomide for newly diagnosed glioblastoma. Neuro. Oncol. 2016, 18, 1434–1441.

113. Lombardi, G.; Pambuku, A.; Bellu, L.; Farina, M.; Della Puppa, A.; Denaro, L.; Zagonel, V. Effectiveness of antiangiogenic drugs in glioblastoma patients: A systematic review and meta-analysis of randomized clinical trials. Crit. Rev. Oncol. Hematol. 2017, 111, 94–102.

114. Kelly, W.K.; Halabi, S.; Carducci, M.; George, D.; Mahoney, J.F.; Stadler, W.M.; Morris, M.; Kantoff, P.; Monk, J.P.; Kaplan, E.; et al. Randomized, double-blind, placebo-controlled phase III trial comparing docetaxel and prednisone with or without bevacizumab in men with metastatic castration-resistant prostate cancer: CALGB 90401. J. Clin. Oncol. 2012, 30, 1534–1540.

115. Hu-Lowe, D.D.; Zou, H.Y.; Grazzini, M.L.; Hallin, M.E.; Wickman, G.R.; Amundson, K.; Chen, J.H.; Rewolinski, D.A.; Yamazaki, S.; Wu, E.Y.; et al. Nonclinical antiangiogenesis and antitumor activities of axitinib (AG-013736), an oral, potent, and selective inhibitor of vascular endothelial growth factor receptor tyrosine kinases 1, 2, 3. Clin. Cancer Res. 2008, 14, 7272–7283.

116. Ho, A.L.; Dunn, L.; Sherman, E.J.; Fury, M.G.; Baxi, S.S.; Chandramohan, R.; Dogan, S.; Morris, L.G.; Cullen, G.D.; Haque, S.; et al. A phase II study of axitinib (AG-013736) in patients with incurable adenoid cystic carcinoma. Ann. Oncol. 2016, 27, 1902–1908.

117. McNamara, M.G.; Le, L.W.; Horgan, A.M.; Aspinall, A.; Burak, K.W.; Dhani, N.; Chen, E.; Sinaei, M.; Lo, G.; Kim, T.K.; et al. A phase II trial of second-line axitinib following prior antiangiogenic therapy in advanced hepatocellular carcinoma. Cancer 2015, 121, 1620–1627.

118. Schiller, J.H.; Larson, T.; Ou, S.H.; Limentani, S.; Sandler, A.; Vokes, E.; Kim, S.; Liau, K.; Bycott, P.; Olszanski, A.J.; et al. Efficacy and safety of axitinib in patients with advanced non-small-cell lung cancer: Results from a phase II study. J. Clin. Oncol. 2009, 27, 3836–3841.

119. Du Four, S.; Maenhout, S.K.; Benteyn, D.; De Keersmaecker, B.; Duerinck, J.; Thielemans, K.; Neyns, B.; Aerts, J.L. Disease progression in recurrent glioblastoma patients treated with the VEGFR inhibitor axitinib is associated with increased regulatory T cell numbers and T cell exhaustion. Cancer Immunol. Immunother. 2016, 65, 727–740.

120. Du Four, S.; Maenhout, S.K.; De Pierre, K.; Renmans, D.; Niclou, S.P.; Thielemans, K.; Neyns, B.; Aerts, J.L. Axitinib increases the infiltration of immune cells and reduces the suppressive capacity of monocytic MDSCs in an intracranial mouse melanoma model. Oncoimmunology 2015, 4, e998107.

121. Laubli, H.; Muller, P.; D'Amico, L.; Buchi, M.; Kashyap, A.S.; Zippelius, A. The multi-receptor inhibitor axitinib reverses tumor-induced immunosuppression and potentiates treatment with immune-modulatory antibodies in preclinical murine models. Cancer Immunol. Immunother. 2018, 67, 815–824.

122. Wilmes, L.J.; Pallavicini, M.G.; Fleming, L.M.; Gibbs, J.; Wang, D.; Li, K.L.; Partridge, S.C.; Henry, R.G.; Shalinsky, D.R.; Hu-Lowe, D.; et al. AG-013736, a novel inhibitor of VEGF receptor tyrosine kinases, inhibits breast cancer growth and decreases vascular permeability as detected by dynamic contrast-enhanced magnetic resonance imaging. Magn Reson Imaging 2007, 25, 319–327.

123. Chen, Y.; Rini, B.I.; Motzer, R.J.; Dutcher, J.P.; Rixe, O.; Wilding, G.; Stadler, W.M.; Tarazi, J.; Garrett, M.; Pithavala, Y.K. Effect of

Renal Impairment on the Pharmacokinetics and Safety of Axitinib. Target Oncol. 2016, 11, 229–234.

124. Gross-Goupil, M.; Francois, L.; Quivy, A.; Ravaud, A. Axitinib: A review of its safety and efficacy in the treatment of adults with advanced renal cell carcinoma. Clin. Med. Insights Oncol. 2013, 7, 269–277.

125. Zhang, W.; Fulci, G.; Wakimoto, H.; Cheema, T.A.; Buhrman, J.S.; Jeyaretna, D.S.; Stemmer Rachamimov, A.O.; Rabkin, S.D.; Martuza, R.L. Combination of oncolytic herpes simplex viruses armed with angiostatin and IL-12 enhances antitumor efficacy in human glioblastoma models. Neoplasia 2013, 15, 591–599.

126. Passer, B.J.; Cheema, T.; Wu, S.; Wu, C.L.; Rabkin, S.D.; Martuza, R.L. Combination of vinblastine and oncolytic herpes simplex virus vector expressing IL-12 therapy increases antitumor and antiangiogenic effects in prostate cancer models. Cancer Gene Ther. 2013, 20, 17–24.

127. Chen, C.Y.; Hutzen, B.; Wedekind, M.F.; Cripe, T.P. Oncolytic virus and PD-1/PD-L1 blockade combination therapy. Oncolytic Virother 2018, 7, 65–77.

128. Huang, J.; Liu, F.; Liu, Z.; Tang, H.; Wu, H.; Gong, Q.; Chen, J. Immune Checkpoint in Glioblastoma: Promising and Challenging. Front. Pharmacol. 2017, 8, 242.

129. Sharma, P.; Allison, J.P. Immune checkpoint targeting in cancer therapy: Toward combination strategies with curative potential. Cell 2015, 161, 205–214.

130. Twyman-Saint Victor, C.; Rech, A.J.; Maity, A.; Rengan, R.; Pauken, K.E.; Stelekati, E.; Benci, J.L.; Xu, B.; Dada, H.; Odorizzi, P.M.; et al. Radiation and dual checkpoint blockade activate non-redundant immune mechanisms in cancer. Nature 2015, 520, 373–377.

131. Curran, M.A.; Montalvo, W.; Yagita, H.; Allison, J.P. PD-1 and CTLA-4 combination blockade expands infiltrating T cells and reduces regulatory T and myeloid cells within B16 melanoma tumors. Proc. Natl. Acad. Sci. USA 2010, 107, 4275–4280.

132. Topalian, S.L.; Drake, C.G.; Pardoll, D.M. Immune checkpoint blockade: A common denominator approach to cancer therapy. Cancer Cell 2015, 27, 450–461.

133. El-Shemi, A.G.; Ashshi, A.M.; Na, Y.; Li, Y.; Basalamah, M.; Al-Allaf, F.A.; Oh, E.; Jung, B.K.; Yun, C.O. Combined therapy with oncolytic adenoviruses encoding TRAIL and IL-12 genes markedly suppressed human hepatocellular carcinoma both in vitro and in an orthotopic transplanted mouse model. J. Exp. Clin. Cancer Res. 2016, 35, 74.

134. Zhang, S.N.; Choi, I.K.; Huang, J.H.; Yoo, J.Y.; Choi, K.J.; Yun, C.O. Optimizing DC vaccination by combination with oncolytic adenovirus coexpressing IL-12 and GM-CSF. Mol. Ther. 2011, 19, 1558–1568.

135. Kim, W.; Seong, J.; Oh, H.J.; Koom, W.S.; Choi, K.J.; Yun, C.O. A novel combination treatment of armed oncolytic adenovirus expressing IL-12 and GM-CSF with radiotherapy in murine hepatocarcinoma. J. Radiat. Res. 2011, 52, 646–654.

136. Oh, E.; Oh, J.E.; Hong, J.; Chung, Y.; Lee, Y.; Park, K.D.; Kim, S.; Yun, C.O. Optimized biodegradable polymeric reservoir-mediated local and sustained co-delivery of dendritic cells and oncolytic adenovirus co-expressing IL-12 and GM-CSF for cancer immunotherapy. J. Control. Release 2017, 259, 115–127.

137. Bortolanza, S.; Bunuales, M.; Otano, I.; Gonzalez-Aseguinolaza, G.; Ortiz-de-Solorzano, C.; Perez, D.; Prieto, J.; Hernandez-Alcoceba, R. Treatment of pancreatic cancer with an oncolytic adenovirus expressing interleukin-12 in Syrian hamsters. Mol. Ther. 2009, 17, 614–622.

138. Yang, Z.; Zhang, Q.; Xu, K.; Shan, J.; Shen, J.; Liu, L.; Xu, Y.; Xia, F.; Bie, P.; Zhang, X.; et al. Combined therapy with cytokine-induced killer cells and oncolytic adenovirus expressing IL-12 induce enhanced antitumor activity in liver tumor model. PLoS ONE 2012, 7, e44802.

139. Quetglas, J.I.; Labiano, S.; Aznar, M.A.; Bolanos, E.; Azpilikueta, A.; Rodriguez, I.; Casales, E.; Sanchez-Paulete, A.R.; Segura, V.;

Smerdou, C.; et al. Virotherapy with a Semliki Forest Virus-Based Vector Encoding IL12 Synergizes with PD-1/PD-L1 Blockade. Cancer Immunol. Res. 2015, 3, 449–454.

140. Schirmacher, V.; Forg, P.; Dalemans, W.; Chlichlia, K.; Zeng, Y.; Fournier, P.; von Hoegen, P. Intra-pinna anti-tumor vaccination with self-replicating infectious RNA or with DNA encoding a model tumor antigen and a cytokine. Gene Ther. 2000, 7, 1137–1147.

141. Quetglas, J.I.; Dubrot, J.; Bezunartea, J.; Sanmamed, M.F.; Hervas-Stubbs, S.; Smerdou, C.; Melero, I. Immunotherapeutic synergy between anti-CD137 mAb and intratumoral administration of a cytopathic Semliki Forest virus encoding IL-12. Mol. Ther. 2012, 20, 1664–1675.

142. Klas, S.D.; Robison, C.S.; Whitt, M.A.; Miller, M.A. Adjuvanticity of an IL-12 fusion protein expressed by recombinant deltaG-vesicular stomatitis virus. Cell Immunol. 2002, 218, 59–73.

143. Lawler, S.E.; Speranza, M.C.; Cho, C.F.; Chiocca, E.A. Oncolytic Viruses in Cancer Treatment: A Review. JAMA Oncol. 2017, 3, 841–849.

144. Senzer, N.N.; Kaufman, H.L.; Amatruda, T.; Nemunaitis, M.; Reid, T.; Daniels, G.; Gonzalez, R.; Glaspy, J.; Whitman, E.; Harrington, K.; et al. Phase II clinical trial of a granulocyte-macrophage colony-stimulating factor-encoding, second-generation oncolytic herpesvirus in patients with unresectable metastatic melanoma. J. Clin. Oncol 2009, 27, 5763–5771.

145. Johnson, D.B.; Puzanov, I.; Kelley, M.C. Talimogene laherparepvec (T-VEC) for the treatment of advanced melanoma. Immunotherapy 2015, 7, 611–619.

146. Bommareddy, P.K.; Patel, A.; Hossain, S.; Kaufman, H.L. Talimogene Laherparepvec (T-VEC) and Other Oncolytic Viruses for the Treatment of Melanoma. Am. J. Clin. Dermatol 2017, 18, 1–15.

147. Vere Hodge, R.A.; Field, H.J. Antiviral agents for herpes simplex virus. Adv. Pharmacol. 2013, 67, 1–38.

148. Parker, J.N.; Meleth, S.; Hughes, K.B.; Gillespie, G.Y.; Whitley, R.J.; Markert, J.M. Enhanced inhibition of syngeneic murine tumors by combinatorial therapy with genetically engineered HSV-1 expressing CCL2 and IL-12. Cancer Gene Ther. 2005, 12, 359–368.

149. Ghouse, S.M.; Bommareddy, P.K.; Nguyen, H.M.; Guz-Montgomery, K.; Saha, D. Oncolytic herpes simplex virus encoding IL12 controls triple-negative breast cancer growth and metastasis in CD8-dependent manner. In Proceedings of the International Oncolytic Virus Conference, Rochester, MN, USA, 9–12 October 2019.

150. Esaki, S.; Nigim, F.; Moon, E.; Luk, S.; Kiyokawa, J.; Curry, W., Jr.; Cahill, D.P.; Chi, A.S.; Iafrate, A.J.; Martuza, R.L.; et al. Blockade of transforming growth factor-beta signaling enhances oncolytic herpes simplex virus efficacy in patient-derived recurrent glioblastoma models. Int. J. Cancer 2017, 141, 2348–2358.

151. Ning, J.; Wakimoto, H.; Peters, C.; Martuza, R.L.; Rabkin, S.D. Rad51 Degradation: Role in Oncolytic Virus-Poly(ADP-Ribose) Polymerase Inhibitor Combination Therapy in Glioblastoma. J. Natl. Cancer Inst. 2017, 109, 1–13.

152. Johnson, D.B.; Chandra, S.; Sosman, J.A. Immune Checkpoint Inhibitor Toxicity in 2018. JAMA 2018, 320, 1702–1703.

153. Lyon, A.R.; Yousaf, N.; Battisti, N.M.L.; Moslehi, J.; Larkin, J. Immune checkpoint inhibitors and cardiovascular toxicity. Lancet Oncol. 2018, 19, e447–e458.

154. Pai, C.S.; Simons, D.M.; Lu, X.; Evans, M.; Wei, J.; Wang, Y.H.; Chen, M.; Huang, J.; Park, C.; Chang, A.; et al. Tumor-conditional anti-CTLA4 uncouples antitumor efficacy from immunotherapy-related toxicity. J. Clin. Investig. 2019, 129, 349–363.

155. Palmieri, D.J.; Carlino, M.S. Immune Checkpoint Inhibitor Toxicity. Curr. Oncol Rep. 2018, 20, 72.

156. Rota, E.; Varese, P.; Agosti, S.; Celli, L.; Ghiglione, E.; Pappalardo, I.; Zaccone, G.; Paglia, A.; Morelli, N. Concomitant myasthenia gravis, myositis, myocarditis and polyneuropathy, induced by

immune-checkpoint inhibitors: A life-threatening continuum of neuromuscular and cardiac toxicity. eNeurologicalSci 2019, 14, 4–5.

157. Filley, A.C.; Henriquez, M.; Dey, M. Recurrent glioma clinical trial, CheckMate-143: The game is not over yet. Oncotarget 2017, 8, 91779–91794.

158. Omuro, A.; Vlahovic, G.; Lim, M.; Sahebjam, S.; Baehring, J.; Cloughesy, T.; Voloschin, A.; Ramkissoon, S.H.; Ligon, K.L.; Latek, R.; et al. Nivolumab with or without ipilimumab in patients with recurrent glioblastoma: Results from exploratory phase I cohorts of CheckMate 143. Neuro. Oncol. 2018, 20, 674–686.

159. Emens, L.A. Breast Cancer Immunotherapy: Facts and Hopes. Clin. Cancer Res. 2018, 24, 511–520.

160. Conry, R.M.; Westbrook, B.; McKee, S.; Norwood, T.G. Talimogene laherparepvec: First in class oncolytic virotherapy. Hum. Vaccin. Immunother. 2018, 14, 839–846.

161. Siurala, M.; Havunen, R.; Saha, D.; Lumen, D.; Airaksinen, A.J.; Tahtinen, S.; Cervera-Carrascon, V.; Bramante, S.; Parviainen, S.; Vaha-Koskela, M.; et al. Adenoviral Delivery of Tumor Necrosis Factor-alpha and Interleukin-2 Enables Successful Adoptive Cell Therapy of Immunosuppressive Melanoma. Mol. Ther. 2016, 24, 1435–1443.

162. Tahtinen, S.; Blattner, C.; Vaha-Koskela, M.; Saha, D.; Siurala, M.; Parviainen, S.; Utikal, J.; Kanerva, A.; Umansky, V.; Hemminki, A. T-Cell Therapy Enabling Adenoviruses Coding for IL2 and TNFalpha Induce Systemic Immunomodulation in Mice With Spontaneous Melanoma. J. Immunother. 2016, 39, 343–354.

163. Martinet, O.; Divino, C.M.; Zang, Y.; Gan, Y.; Mandeli, J.; Thung, S.; Pan, P.Y.; Chen, S.H. T cell activation with systemic agonistic antibody versus local 4-1BB ligand gene delivery combined with interleukin-12 eradicate liver metastases of breast cancer. Gene Ther. 2002, 9, 786–792.

164. Kurokawa, C.; Iankov, I.D.; Anderson, S.K.; Aderca, I.; Leontovich, A.A.; Maurer, M.J.; Oberg, A.L.; Schroeder, M.A.; Giannini, C.; Greiner, S.M.; et al. Constitutive Interferon Pathway Activation in Tumors as an Efficacy Determinant Following Oncolytic Virotherapy. J. Natl. Cancer Inst. 2018, 110, 1123–1132.

21

CREATION OF ORGANOIDS DERIVED FROM THE ILEUM AND COLON OF FELINES AND THEIR POTENTIAL APPLICATION IN AIDING FELINE CORONAVIRUS INFECTION

1. INTRODUCTION: AN OVERVIEW

The Coronaviridae family, which is characterized by its diversity, is responsible for causing infections in a broad spectrum of mammals, birds, and humans. Feline coronavirus (FCoV) is a widely distributed member of the Coronaviridae family that is commonly found in domestic and wild cat populations across the globe [1]. FCoVs exist in two distinct biotypes: feline enteric coronavirus (FECV) and feline infectious peritonitis virus (FIPV). Feline enteric coronavirus (FECV) infections typically present as moderate or asymptomatic infections of the digestive system in cats. Infections in domestic cats are frequently long-lasting and include the periodic release of the virus over extended periods. This is a significant factor in the high levels of seropositivity observed in domestic cats. Among cats living alone, 20-60% exhibit indications of virus exposure, while up to 90% of cats in communities with multiple cats test positive for the virus [2,3,4]. Feline infectious peritonitis (FIP) is a devastating clinical disease that arises from mutations in the innocuous feline enteric coronavirus (FECV) [5,6,7,8,9]. Despite limited understanding of the molecular pathophysiology of FIP [10,11,12], new studies have identified promising treatment strategies [13,14,15,16,17,18]. It should be emphasized that there are two serotypes [19,20,21] that encompass both biotypes. Serotype II FCoVs are formed through recombination between a serotype I FCoV and a closely related canine coronavirus (CCoV) [22,23,24,25] and can be readily cultured in a laboratory setting. On the other hand, the serotype I FCoVs, which are more important and commonly found, cannot be grown in cell culture. Historically, serotype II viruses were frequently utilized to obtain understanding of FCoV biology, instead of serotype I FCoVs. In order to clarify the molecular cause of FIP, laboratory strains of serotype I FIPV that had been acclimated to cell culture were acquired over a period of time [26]. Nevertheless, these viruses were shown to be inappropriate for studying the development of FIP since they lost their ability to cause disease after being adapted to cell culture [1,9,26]. Tekes et al. (2008) developed the initial reverse genetic approach for manipulating the complete genome of FCoV. They used a vaccinia viral vector to work with the serotype I FIPV laboratory strain Black [26,27,28]. Nevertheless, animal tests demonstrated that serotype I FIPV Black, similar to other laboratory strains, no longer possessed the potential to produce FIP [26]. In contrast, the laboratory strain 79-1146, which belongs to serotype II of FIPV and is frequently employed, is significantly more pathogenic. As a result, it does not accurately represent the majority of field strains. Given the absence of appropriate in vitro systems for field serotype I FECVs, it is crucial to develop a suitable in vitro system that allows for the cultivation of serotype I FECV. This culture system for serotype I FECV field viruses would not only offer comprehension of the molecular process by which FECVs endure in the gut for an extended duration, but it may also contribute to comprehending how FIPV can develop from FECV during a benign persistent infection. In recent years, organoids have been used as an in vitro system to facilitate the growth of many human viruses that could not be cultivated using conventional cell culture methods [31,32]. Organoids are created by cultivating and differentiating induced pluripotent stem cells (iPSCs) or tissue-derived stem cells into three-dimensional constructs that accurately mimic the cellular makeup and functions of the original organ [33]. Tissue-derived organoids depend on the capacity to separate crypts that contain stem cells. Subsequently, these crypts are cultivated in the presence of differentiation factors (Wnt3a, R-Spondin, Noggin, and EGF), facilitating their growth into three-dimensional mini-gut organoids [33]. As these intricate cultures increasingly mimic the various types of cells seen in their real tissue equivalents, they frequently possess essential elements necessary for the replication and spread of viruses that are absent in conventional cell cultures. In order to assess the suitability of these model systems for facilitating the growth of the relevant serotype I FECV, we created a culture system using cat intestine organoids. Our findings demonstrate that this system is capable of supporting infection with GFP-expressing recombinant serotype I FECV produced by reverse genetics. This model will facilitate the investigation of the molecular mechanism underlying the persistence of serotype I FECV in its natural intestinal environment.

2. MATERIALS AND METHODS

2.1. Viruses and Cell Lines

Serotype I recFECV-GFP and recFECV-S79-GFP were produced in vitro using the reverse genetic system for FCoV field strains described previously [34]. Recombinant virus stocks of recFECV-S79-GFP were titrated by plaque assay on routinely used felis catus whole fetus (FCWF) cells [26,27,28,35,36]. Virus stocks of recFECV-GFP which cannot be cultivated in standard cell culture systems were quantified by comparative Western blot analysis of the FCoV M protein together with recFECV-S79-GFP [34]. The cell line FCWF was provided by the diagnostic laboratory at the Justus Liebig University Giessen and maintained in culture media (DMEM with 1× penicillin/streptomycin (Thermo, Waltham, MA, USA) and 10% FBS (Biochrom, Cambridge, UK)). According to our experience with propagation of FCoV laboratory strains, the FCWF cells were used at a confluency of approximately 90% for the infection with recFECV-S79-GFP.

2.2. Animals

Handling of the animals used for enteric tissue donation was performed according to the guidelines of the Hungarian legislation on animal protection. The protocol was approved by the Pest Megyei Kormany-hivatal, Budapest (assurance numbers PE/EA/2441-6/2016 and TMF/657-12/2016). The animals were euthanized according to the designate protocol. Female specific-pathogen-free (SPF) cats were raised and housed in pathogen free conditions for laboratory use. These were not cats taken from outside veterinary practices. The SPF cats were euthanized at an age of 25 weeks and tissue samples from the gut were collected, 6 donors were used for this study. The cats were not tested prior to isolation specifically for FCoV. Animals had access to feed ad libitum prior to euthanization and tissue collection.

2.3. Chemicals and Solutions

Conditioned media containing Wnt3a, R-Spondin and Noggin was produced from the L-WRN cell line (ATCC CRL-3276) as per manufacturer's instructions. A 293T cell line which produces only R-Spondin was a kind gift from Calvin Kuo (Standford University) and conditioned media was made as previously described [37]. All organoid media were made from a base of advanced DMEM/F12 (Thermo, Waltham, MA, USA) which contained 1% penicillin/streptomycin (Thermo), 2 mM GlutaMAX (Thermo) and 10 mM HEPES (Thermo) and is referred to as Ad DMEM/F12++. All other organoid media components are found in Table 1.

Table 1. Human and mouse media compositions tested for their ability to support feline ileum and colon organoid growth. Final concentrations and manufactures of each component are listed.

Reagent	Company	Final Concentration	Human Media	Mouse Intestine Media	Mouse Colon Media
L-WRN (Wnt3s, R-Spondin, Noggin containing conditioned media)	Made in the lab	50%	X		X
Ad DMEM/F12++	Thermo	50% (Human and mouse colon) 90% for mouse intestine	X	X	X

Reagent	Company	Final Concentration	Human Media	Mouse Intestine Media	Mouse Colon Media
B27	Thermo	1X	X	X	X
Nicotinamide	Sigma (Munich, Germany)	10 mM	X		
N-acetylcysteine	Sigma	1 mM	X	X	X
A-83-01	Tocris (Bristol, UK)	500 nM	X		
SB202190	Sigma	500 nM	X		
Leu-Gastrin	Sigma	10 nM	X		
Mouse recombinant	Thermo	50 ng/mL	X	X	X
R-Spondin conditioned media	Made in the lab	10%		X	
Mouse recombinant Noggin	Peprotech (Rocky Hill, NJ, USA)	100 ng/mL		X	
Y-27632	Sigma	10 μM	X	X	X
Matrigel, Growth factor reduced (GFR), phenol free	Corning (Corning, NY, USA)	100%	X	X	X

2.4. Isolation of Feline Intestinal Cells

Ileum and colon sections (10 cm each) were harvested from 6 donors and stored in cold transport buffer (1× phosphate buffered saline (PBS), 50 ng/mL gentamicin (Thermo), 1% penicillin/streptomycin (Thermo), 1% fetal bovine serum (FBS) and 250 μg/mL Fungizone (Thermo)) until the time of isolation. The tissue was cut in half and washed 3 × 10 min with shaking in cold PBS to ensure that all fecal material had been removed. Crypts containing stem cells were isolated within 16 h of sacrificing the animals by first cutting the tissue into smaller 1 cm pieces and then either adding 2 mM EDTA to tissue sample for 1 h at 4 °C or 20 mL of Gentle Cell Dissociation Reagent (Stem Cell Technologies, Vancouver, BC, Canada) for 1 h at room temp. Tissue sections were transferred to a clean tube and 10 mL of cold PBS + 1% BSA was added to tube. Tubes were shaken to release the crypt fraction. Fractions enriched in crypts were filtered with 70 μm filters and the process was repeated to collect four to five fractions for each tissue. The fractions were observed under a light microscope and those containing the highest number of crypts were pooled and spun at 500× g for 5 min at 4 °C. The supernatant was removed, and crypts were washed 1× with cold DMEM/F12 (Thermo) and spun at 500× g for 5 min at 4 °C. The media was removed, and crypts were re-suspended in 100% Matrigel, plated in 50 µL drops in 24-well non-tissue culture treated plates (Corning) and following polymerization of the Matrigel, 500 µL organoid media was added to each well and was replaced every 48 h. Following isolation of the crypts and seeding into organoid media, the size of the organoids was monitored over time at day 3, 6, 9, 12, and 15 days by observing them under bright field microscopy with a Nikon Eclipse Ti-S microscope. Their size was measured with a 10× objective using the Nikon NIS software.

2.5. RNA Isolation, cDNA, and qPCR

RNA was harvested from cells using NucleoSpin RNA extraction kit (Macherey-Nagel, Dueren, Germany) as per manufacturer's instructions. cDNA was made using iSCRIPT reverse transcriptase (BioRad, Hercules, CA, USA) from 250 ng of total RNA as per manufacturer's instructions. Quantitative-PCR was performed using iTaq SYBR green (BioRad) as per manufacturer's instructions, GAPDH was used as normalizing gene. Pre-designed feline specific KiCqStart SYBR Green pre-designed primers were purchased from Sigma-Aldrich (Table 2).

Table 2. Feline specific primers used to control cell population in tissue and organoids.

Gene	Cell Type	Gene ID	Sequence ID
GAPDH	House keeping	493876	NM_001009307
LGR5	Stem cell	101080720	XM_003989046
SMOC2	Stem cell	101082409	XM_003986725
MUC2	Goblet cell	101096605	XM_003993797
SI	Enterocyte	100144605	NM_001123332
SYP	Enteroendocrine	101084343	XM_004000526
LYZ	Paneth cell	100127109	XM_003989032

2.6. Passaging of Feline Mini-Gut Organoids

Ileum and colon mini-gut organoids were monitored under a light microscope and were passaged when centers became dark and filled with dead cells. For passaging, media was removed and Matrigel was dissolved in cold PBS. Organoids were spun at 500× g for 5 min at 4 °C and PBS was removed. Subsequently three different methods were used for passaging:

- **Mechanical passaging:** 1 mL of cold PBS was used to re-suspend the organoids. Using a 27-gauge needle on a 1 mL syringe, organoids were broken down by passing the solution up and down 10 times through the needle. Organoids were then spun at 500× g for 5 min at 4 °C. The supernatant was removed, and the crypts were re-suspended in 100% Matrigel, plated in 50 μL drops in 24-well non-tissue culture treated plates (Corning) and following polymerization of the Matrigel, organoid media 500 μL was added to each well.
- **Trypsin-based passaging:** 1 mL of cold PBS was used to re-suspend the organoids. Organoids were then spun at 500× g for 5 min at 4 °C. Organoids were washed a second time in cold PBS by resuspending the pellet in 1 mL of PBS and spinning at 500× g for 5 min at 4 °C. The supernatant was removed and organoids were incubated in 0.05% Trypsin-EDTA (Gibco) for 5 min at 37 °C. Trypsin digestion was stopped with the addition of serum containing media and samples were spun at 500× g for 5 min at 4 °C. Organoids were washed a second time in cold PBS by resuspending the pellet in 1 mL of PBS and spinning at 500× g for 5 min at 4 °C. The supernatant was removed, and the crypts were re-suspended in 100% Matrigel, plated in 50 μL drops in 24-well non-tissue culture treated plates (Corning) and following polymerization of the Matrigel, 500 μL organoid media was added to each well.
- **Gentle Cell Dissociation Reagent method:** Media was removed and Gentle Cell Dissociation Reagent (Stem cell technologies) was

added to the organoid containing pellet and incubated for 10 min at room temp. Organoids were spun at 500× g for 5 min at 4 °C and the supernatant was removed. Organoids were washed in DMEM/F12 and then spun at 500× g for 5 min at 4 °C. The supernatant was removed and the crypts were re-suspended in 100% Matrigel, plated in 50 μL drops in 24-well non-tissue culture treated plates (Corning) and following polymerization of the Matrigel, 500 μL organoid media was added to each well.

2.7. Infection of Cell Culture with Recombinant Viruses

Approximately 90% confluent monolayers of FCWF cells were washed with serum free DMEM culture media. Cells were inoculated with recFECV-GFP and recFECV-S79-GFP at a multiplicity of infection (MOI) of 0.01 or serum free culture media for the mock control. The infection was incubated for one hour and subsequently the inoculum was replaced by culture media containing FBS. The formation of plaques and the GFP signal was monitored 48 hours post-infection (hpi).

2.8. Infection of Organoids

Organoids were removed from Matrigel by adding cold PBS for 5 min, liquefied Matrigel and organoids were separated by centrifugation (500× g, 5 min). Supernatant and Matrigel were removed and organoids were resuspended in media and gently disrupted with a 27 G needle to allow virus to access both the apical and basolateral sides of the organoids. Following disruption, organoid media containing 104 pfu of FCoV was added to the organoids and allowed to incubate for 6 h in suspension. Following the 6-h incubation, Matrigel was added back to the cultures and organoids were observed over a three-day period.

2.9. Statistical Methods

Statistics were calculated by Prizm using an unpaired t-test.

3. RESULTS AND DISCUSSION

Currently, infectious disease research of feline intestinal pathogens has been hampered by the lack of cell-based systems allowing for the screening of viruses and antiviral compounds. To fill this gap, we developed a method to isolate and propagate ileum and colon mini-gut organoids from felines. Six SPF cats were euthanized and the ileum and colon were collected. Tissue sections were washed thoroughly to remove all fecal material and were stored at 4 °C in a transport buffer containing PBS as well as antibiotics and antifungals. Organoids were prepared within 16 h of tissue harvesting. To determine the best method to isolate crypts containing stem cells from feline intestinal tissues, samples were split in two parts and two commonly used approaches (EDTA-based dissociation and Gentle Cell Dissociation Reagent™) for harvesting human and murine intestinal crypts [38,39] were compared side by side. Feline ileum and colon samples were washed thoroughly to remove all mucus or remaining contaminants and crypts containing stem cells were then isolated following incubation in EDTA or Gentle Cell Dissociation Reagent™. Microscopic evaluation of the crypts prior to seeding showed no large difference in the quantity or quality of crypts between the two isolation methods (data not shown). Equal numbers of crypts were seeded into Matrigel and the number and size of organoids was followed over 12 days. Unlike human and murine crypts, which form organoids with 16–24 h [38,39,40], the feline crypts took 72 h to seal and form cyst like organoids (Figure 1A). Following initial cyst formation, the ileum and colon organoids continued to grow over the twelve-day period (Figure 1A). To determine which isolation method produced the best organoids, the number and size of organoids was followed over time. Results show that organoids isolated using the EDTA approach produced a greater number of organoids and the organoids were larger in size compared to the Gentle Cell Dissociation Reagent™ method (Figure 1B,C). The increase in the number of organoids between day 3 and 6 does not illustrate a growth in organoid number but is due to the fact that at day 3 the crypts are too small to be effectively counted. This number of organoids then remained constant over the rest of the period as they continued to grow larger (Figure 1B,C), which was consistent with murine and human organoid generation [38,39].

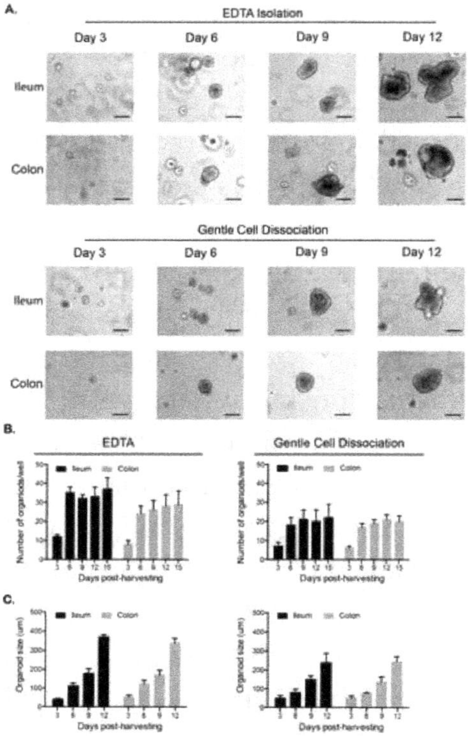

Figure 1. Ileum and colon organoids can be established from primary feline tissue. (A–C). The 1 cm2 sections of feline ileum and colon tissue were incubated with EDTA or Gentle Cell Dissociation Reagent to allow for the isolation of intestinal crypts. Isolated crypts were resuspended in Matrigel and followed over a 12-day period. (A). Bright field images of ileum and colon organoids. Representative images are shown (n = 6 donors). Scale bar = 100 µm. (B). The number of organoids from EDTA and Gentle cell Dissociation Reagent isolation were counted over the indicated time course. Error bars represent standard deviation (n = 6 donors). (C). Following EDTA and Gentle cell Dissociation Reagent isolation, organoids were imaged in three-day intervals using a Nikon Eclipse Ti-S. Their size was measured using the Nikon NIS software. Error bars represent standard deviation (n = 6 donors).

To determine if the organoids displayed similar cell types as the natural feline intestine, tissue and organoids samples from three donors were lysed and the relative expression of each cell type was quantified by q-RT-PCR. As the natural intestine is made of stem cells, absorptive enterocytes and secretory cells we evaluated known markers from each of these populations. As there are no markers to analyze feline organoids, we used the best-known markers for murine and human, hypothesizing that they will constitute a marker for feline tissue. As a marker of stem cells, we used LGR5 and SMOC2. LGR5 is the historical marker described by the Clevers lab, however while it constitutes an excellent marker in tissue it is known to be suboptimal for organoids [38,39,41]. On the contrary, SMOC2 represents an excellent marker for both tissue and tissue-derived organoids. As a marker for Goblet we used mucin 2 (Muc2), for enterocytes we used sucrose isomaltase (SI), for enteroendocrine cells we used synaptophysin (SYP) and for Paneth cells we used lysozyme (LYZ). Results showed that ileum from feline tissue samples displays markers for stem cells (LGR5, SMOC2), Goblet cells (Muc2), enterocytes (SI), enteroendocrine cells (SYP) and Paneth cells (LYZ) (Figure 2A). Similarly, feline colon tissues expressed similar amounts of markers for all cell types except Paneth cells which are not present in colon tissue. The ratios found in feline tissues are similar to those observed in humans [41].

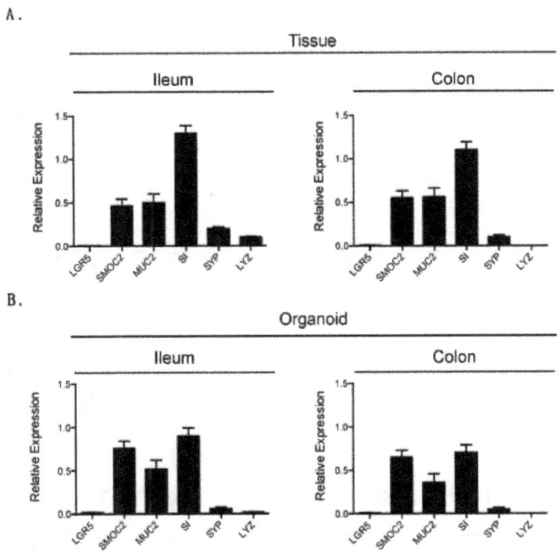

Figure 2. Feline organoids and originating tissue contain similar cell types. (A). The relative expression of intestinal cell type specific markers was evaluated in feline ileum and colon tissue by q-RT-PCR. (B). Same as A except using feline ileum- and colon-derived organoids. Results are mean +/- s.d. and are expressed as a relative expression to the house-keeping gene GAPDH. n = 3 donors and q-RT-PCR was done as technical triplicate.

Importantly, organoids derived from feline ileum expressed the same cell type-specific markers as their tissue counterpart (Figure 2B). The relative expressions of the different markers were slightly different compared to the expression in tissue (Figure 2A,B). Ileum organoids displayed a higher relative expression of the stem cells marker (SMOC2) and a lower relative expression of the enterocyte maker (SI) compared to normal feline tissue. This suggests that there are more stem cells than enterocytes. The increase in stem cell number is expected as organoids are cultured under high Wnt3a conditions which favor stem cell numbers. Similarly, colon organoids also displayed a high amount of stem cell markers and their cellular composition looked similar to those of ileum organoids (Figure 2B). These differences in expression of the different cell type specific markers are not specific to the feline ileum and colon organoids but this is also observed in both murine and human intestinal organoids [40,42]. This shows that although organoids are extremely close to the originating tissue, the fact that they are ex vivo mini-organs causes subtle differences in their expression profile and differentiation pathways. All together these data show that we have developed a novel protocol to isolate and generate intestinal organoids from feline intestinal tissue and that these organoids closely resemble their tissue counterpart.

3.1. Passaging and Maintenance of Feline Intestinal or Ganoids

As the ileum and colon organoids continued to grow and retain a similar identity to the natural tissue, we wanted to determine whether they could be passaged and maintained in culture. Trypsinization, mechanical disruption and passaging using Gentle Cell Dissociation Reagent™ were tested to establish the method that best supported the continued maintenance of feline ileal and colon organoids. For all methods, organoids were removed from the Matrigel and either incubated with low concentrations of trypsin, mechanical disrupted using a 27 G needle, or were incubated with Gentle Cell Dissociation Reagent™ (see materials and methods for full details). Following disruption, organoids were centrifuged to separate out organoid structures from dead cells. The organoids were then seeded into Matrigel and the formation of new organoids was followed for five days. One day post-passaging many small organoids could be seen in all conditions for both the ileum and colon organoids (Figure 3A). Organoids which were passaged by trypsinization were small and stressed, as shown by their loss of tight borders (dark rim at the periphery of the organoids under phase microscope) and the presence of many dissociated cells. On the contrary, organoids that were obtained using the mechanical disruption and Gentle Cell Dissociation Reagent™ methods were larger and their borders look more discrete (Figure 3A). Similar to the original isolation, many organoids were too small to be counted on the first day after passaging for all methods (Figure 3B). Observations five days post passaging showed that using mechanical disruption for passaging both ileum and colon organoids leads to a greater number of organoids that are larger in comparison to organoids that were passaged using trypsin (Figure 3B,C).

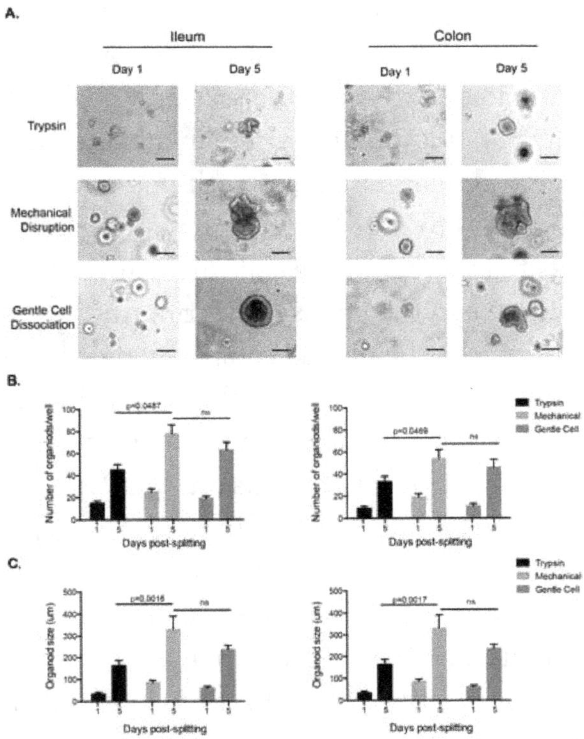

Figure 3. Mechanical disruption is the preferred method for passaging feline organoids. (A–C). Feline ileum- and colon-derived organoids were grown for 10 days post-harvesting prior to passaging. Three methods of

passaging were used, and new organoids were followed over five days. (A). Bright field images of ileum and colon organoids. Representative images are shown. N = 3 donors and each donor were followed overtime as a triplicate series. Scale bar = 100 µm (B). The number of organoids from each passing method was counted over the indicated time course. Results are mean +/- s.d. (n = 3 donors). (C). The size of the organoids was determined using a Nikon Eclipse Ti-S. Results are mean +/- s.d. (n = 3 donors).

Upon isolation of feline ileum and colon crypts, organoids were maintained in a media based upon one that is commonly used to support the growth of human organoids [39,43,44]. To determine if additional media compositions would support or enhance feline intestinal organoid growth, we tested three different formulations normally used to support human intestinal and colon organoids, murine small intestine or murine colon (see material and methods) [37,38,39,40,45]. Feline ileum and colon organoids were passaged using mechanical disruption and were maintained in each of the media conditions. Ten days post passaging, the number of organoids and the size of the organoids were counted for each condition. Results show that the human media condition supported a greater number and a larger size of both ileum and colon organoids (Table 3). Additionally, to determine how many passages each media type could support, organoids were split using mechanical dissociation and followed over time. The mouse intestine media did not support long term growth of either ileum or colon organoids and both types of organoids stopped growing and died within two passages in this media (Table 3). The mouse colon media supported longer passaging than its intestine counterpart, however this was still limited to a few passages. The human media was the only one that supported longer term passaging. However, unlike human and mouse organoids [38,39,40], both the feline ileum and colon organoids became arrested after 14–15 passages (Table 3). This was reproduced with several animals (n = 6) and was consistent between animals suggesting that an additional media component may be required to support culturing to the extent that human and murine organoids can reach. Importantly, it seems that Wnt3A is a critical component required for the maintenance of feline organoids as the mouse intestine media, which lacks Wnt3A, supported the fewest number of passages. Overall, we could see that feline intestinal and colon organoids could be maintained using mechanical disruption and Wnt3A containing media. In recent years, intestine- and colon-derived organoids have been generated from both large farm animals (bovine and porcine) [46] and domestic animals (canine) [47,48]. Bovine, porcine and canine intestinal and colon organoids have been shown to require Wnt3a for their generation and growth [46,47,48]. Additionally, while many of these studies did not compare different passaging techniques, they often used mechanical disruption as a common method to passage and maintain the organoids [46,47].

Table 3. Optimization of feline ileum and colon culturing conditions. The number and size of organoids were measured 10 days post-splitting for every other passage.

	Organoid	Avg # of Organoids/Well	Avg. Size of Organoids	No of Passages
Human Media	Colon	83 +/- 11	384.2 +/- 31.5	15
Mouse Intestine Media	Colon	54 +/- 6	312.5 +/- 23.6	2
Mouse Colon Media	Colon	68 +/- 9	359.4 +/- 29.1	6
Human Media	Ileum	68 +/- 9	346.2 +/- 24.8	14
Mouse Intestine Media	Ileum	49 +/- 6	297.6 +/- 21.3	2
Mouse Colon Media	Ileum	53 +/- 7	318.9 +/- 35.4	4

3.2. Infection of Feline Intestinal Organoids

Currently, the molecular mechanism by which FECVs can persist in the gut could not been studied due to the lack of cell culture models supporting the replication of FECV field viruses. To evaluate if the feline intestinal and colon organoid system could support FECV infection, two different FCoV viruses (recFECV-GFP and recFECV-S79-GFP) were constructed for this study (Figure 4A). recFECV-GFP is a recombinant FCoV with the entire genome sequence of a serotype FECV field strain while recFECV-S79-GFP contains the genomic backbone of the same serotype I FECV field strains with the spike protein gene of cell-culture-adapted serotype II laboratory strain 79–1146. To determine the ability of recFECV-GFP and recFECV-S79-GFP to replicate in cell culture, these recombinant viruses were first tested on FCWF cells. Inoculation of FCWF cells with recFECV-GFP, did not lead to any formation of cytopathic effect (CPE) or fluorescence as standard cell culture systems do not support the growth of serotype I FCoV field strains (Figure 4B), while as expected, recFECV-S79-GFP growth was supported in FCWF cells demonstrated by their ability to form a characteristic plaque phenotype accompanied by green fluorescence (Figure 4B).

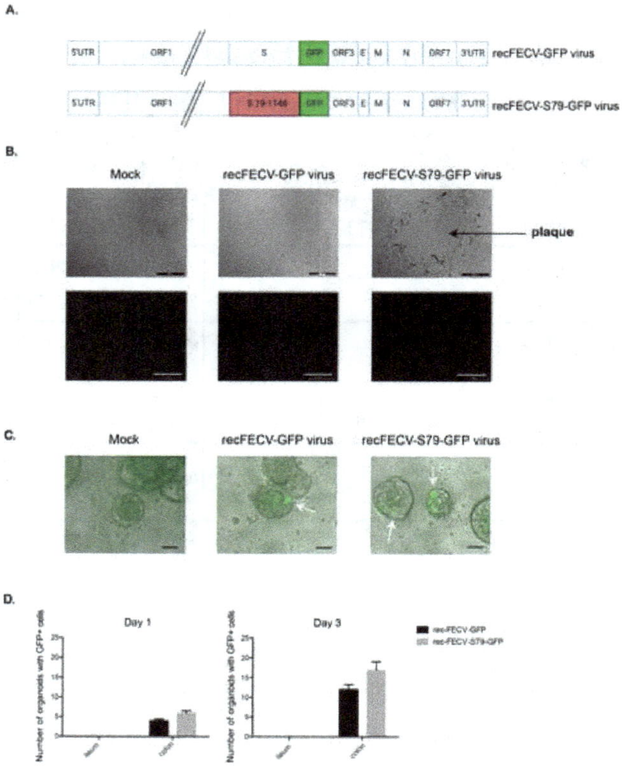

Figure 4. Feline colon organoids support infection of feline coronavirus (FCOV). (A). Diagram depicting the newly generated feline enteric coronavirus (FECV) recombinant viruses. (B). Felis catus whole fetus (FCWF) feline cells were infected with rec-FECV-GFP or recFECV-S79-GFP viruses and their ability to express GFP and produce plaques was monitored 48 h post-infection. Mock samples indicate media alone conditions. Representative images are shown. Scale bar is 200 μm. (C).

Feline colon organoids infected with rec-FECV-GFP or recFECV-S79-GFP for 72 h. Representative images shown. White arrow heads indicate GFP positive cells. Scale bar = 100 μm (D). Feline ileum and colon organoids (passage 3) were infected with rec-FECV-GFP or recFECV-S79-GFP viruses. Mock samples indicate media alone conditions. Infections were monitored 1 and 3 days post-infection by wide field microscopy. Results are mean +/- s.d. n = 3 donors for each infection.

Previously, Desmarets et al. (2013) [49] established permanent feline intestinal epithelial cell cultures of ileocyte and colonocyte origin that were shown to sustain propagation of FECV field strains in vitro. However, such valuable cell lines are not broadly available to the scientific community. To test if the established feline ileum and colon organoids could be used to support feline coronavirus infection, five days post-passaging, organoids were removed from the Matrigel and incubated with feline coronavirus recFECV-GFP and recFECV-S79-GFP and followed over 72 h. Unexpectedly, no fluorescence was obtained upon infection of ileum organoids with either recFECV-GFP or recFECV-S79-GFP (Figure 4D). The lack of fluorescence of ileum organoids upon infection with both viruses is somewhat surprising, since ileum has been described as a tissue supporting FCoV-growth in vivo [50]. Whether the lack of green fluorescence of ileum organoids is due to (i) a very low viral replication rate and GFP expression which is below the detection limit or (ii) the complete lack of virus infection, cannot be ruled out in the current experimental setup. Luciferase-expressing recombinant feline enteric coronaviruses (recFECVs) could be more appropriate instruments for conducting additional investigations into these theories. The limited or absence of infection in ileum organoids by the recombinant viruses may indicate that ileum organoids are not suitable for studying the biology of serotype I FECVs in a laboratory setting. However, it was shown that both genetically modified viruses were capable of infecting feline colon organoids, as shown in Figure 4C and 4D. Within 24 hours after infection, cells expressing green fluorescent protein (GFP) were observable, and the quantity of infected cells progressively rose over a span of three days for both viruses (Figure 4C,D). Unlike the FCWF cells, our feline colon organoids were capable of facilitating infection by recFECV-GFP. Given the observed progressive rise in the number of GFP positive cells, this data provides compelling evidence for the ability of colon organoids to facilitate initial infection, de novo virus production, and dissemination of the serotype I FECV field strain. Essentially, we suggest that organoids produced from the colon of cats serve as a fundamental model for studying the infection of field strain FECV of serotype I in intestinal cells. These cultures currently provide opportunities for researchers to investigate the molecular pathways that cause FECV to survive and perhaps convert to FIPV. This technique outlined here offers the community a systematic method to create organoids grown from the colon of felines. This method addresses the issue of limited access to readily available culture models. In addition, the organoid model developed here will enable the scientific community to explore host/pathogen interactions [37] and immune response [40,51] against FCoV more effectively, as primary intestinal epithelial cells are often more receptive to immunological stimuli compared to immortalized cells.

CONCISE SUMMARY

Feline coronaviruses (FCoVs) have a global presence and can infect both wild and domestic cat populations. FCoVs are classified into two primary biotypes: the mild feline enteric coronavirus (FECV) and the lethal feline infectious peritonitis virus (FIPV). Feline Infectious Peritonitis Virus (FIPV) arises as a result of genetic changes occurring in Feline Enteric Coronavirus (FECV) after a prolonged infection. Currently, the molecular mechanism behind FECV-persistence and the variables that contribute to FIPV growth cannot be examined due to the inability of field FECV isolates to grow in existing cell culture models. The objective of this study was to create feline ileum and colon organoids capable of supporting the growth of field FECVs. We have identified the most effective techniques for isolating, cultivating, and transferring feline ileum and colon organoids. Significantly, we have shown that colon organoids can facilitate the infection of GFP-expressing recombinant field FECV, whereas previous feline cell culture models were not susceptible to this infection. These organoids, when combined with recombinant FECVs, provide an opportunity to investigate the molecular pathways by which FECV can stay in the gut for an extended duration and how the change to FIPV occurs.

REFERENCES

1. Pedersen, N.C. A review of feline infectious peritonitis virus infection: 1963–2008. J. Feline Med. Surg. 2009, 11, 225–258.
2. Addie, D.D.; Schaap, I.A.T.; Nicolson, L.; Jarrett, O. Persistence and transmission of natural type I feline coronavirus infection. J. Gen. Virol. 2003, 84, 2735–2744.
3. Pedersen, N.C. Serologic studies of naturally occurring feline infectious peritonitis. Am. J. Vet. Res. 1976, 37, 1449–1453.
4. Sparkes, A.H.; Gruffydd-Jones, T.J.; Harbour, D.A. Feline coronavirus antibodies in UK cats. Vet. Rec. 1992, 131, 223–224.
5. Kipar, A.; May, H.; Menger, S.; Weber, M.; Leukert, W.; Reinacher, M. Morphologic features and development of granulomatous vasculitis in feline infectious peritonitis. Vet. Pathol. 2005, 42, 321–330.
6. Weiss, R.C.; Scott, F.W. Pathogenesis of feline infetious peritonitis: Pathologic changes and immunofluorescence. Am. J. Vet. Res. 1981, 42, 2036–2048.
7. Weiss, R.C.; Scott, F.W. Pathogenesis of feline infectious peritonitis: Nature and development of viremia. Am. J. Vet. Res. 1981, 42, 382–390.
8. Pedersen, N.C. Morphologic and physical characteristics of feline infectious peritonitis virus and its growth in autochthonous peritoneal cell cultures. Am. J. Vet. Res. 1976, 37, 567–572.
9. Tekes, G.; Thiel, H.J. Feline Coronaviruses: Pathogenesis of Feline Infectious Peritonitis. Adv. Virus Res. 2016, 96, 193–218.
10. Pedersen, N.C.; Liu, H.; Gandolfi, B.; Lyons, L.A. The influence of age and genetics on natural resistance to experimentally induced feline infectious peritonitis. Vet. Immunol. Immunopathol. 2014, 162, 33–40.

11. Pedersen, N.C. An update on feline infectious peritonitis: Diagnostics and therapeutics. Vet. J. 2014, 201, 133–141.

12. Pedersen, N.C. An update on feline infectious peritonitis: Virology and immunopathogenesis. Vet. J. 2014, 201, 123–132.

13. Kim, Y.; Liu, H.; Galasiti Kankanamalage, A.C.; Weerasekara, S.; Hua, D.H.; Groutas, W.C.; Chang, K.O.; Pedersen, N.C. Reversal of the Progression of Fatal Coronavirus Infection in Cats by a Broad-Spectrum Coronavirus Protease Inhibitor. PLoS Pathog. 2016, 12, e1005531.

14. Dickinson, P.J.; Bannasch, M.; Thomasy, S.M.; Murthy, V.D.; Vernau, K.M.; Liepnieks, M.; Montgomery, E.; Knickelbein, K.E.; Murphy, B.; Pedersen, N.C. Antiviral treatment using the adenosine nucleoside analogue GS-441524 in cats with clinically diagnosed neurological feline infectious peritonitis. J. Vet. Intern. Med. 2020, 34, 1587–1593.

15. Perera, K.D.; Rathnayake, A.D.; Liu, H.; Pedersen, N.C.; Groutas, W.C.; Chang, K.O.; Kim, Y. Characterization of amino acid substitutions in feline coronavirus 3C-like protease from a cat with feline infectious peritonitis treated with a protease inhibitor. Vet. Microbiol. 2019, 237, 108398.

16. Pedersen, N.C.; Perron, M.; Bannasch, M.; Montgomery, E.; Murakami, E.; Liepnieks, M.; Liu, H. Efficacy and safety of the nucleoside analog GS-441524 for treatment of cats with naturally occurring feline infectious peritonitis. J. Feline Med. Surg. 2019, 21, 271–281.

17. Murphy, B.G.; Perron, M.; Murakami, E.; Bauer, K.; Park, Y.; Eckstrand, C.; Liepnieks, M.; Pedersen, N.C. The nucleoside analog GS-441524 strongly inhibits feline infectious peritonitis (FIP) virus in tissue culture and experimental cat infection studies. Vet. Microbiol. 2018, 219, 226–233.

18. Pedersen, N.C.; Kim, Y.; Liu, H.; Galasiti Kankanamalage, A.C.; Eckstrand, C.; Groutas, W.C.; Bannasch, M.; Meadows, J.M.; Chang, K.O. Efficacy of a 3C-like protease inhibitor in treating various

forms of acquired feline infectious peritonitis. J. Feline Med. Surg. 2018, 20, 378–392.

19. Hohdatsu, T.; Sasamoto, T.; Okada, S.; Koyama, H. Antigenic analysis of feline coronaviruses with monoclonal antibodies (MAbs): Preparation of MAbs which discriminate between FIPV strain 79-1146 and FECV strain 79-1683. Vet. Microbiol. 1991, 28, 13–24.

20. Hohdatsu, T.; Okada, S.; Koyama, H. Characterization of monoclonal antibodies against feline infectious peritonitis virus type II and antigenic relationship between feline, porcine, and canine coronaviruses. Arch. Virol. 1991, 117, 85–95.

21. Pedersen, N.C.; Black, J.W.; Boyle, J.F.; Evermann, J.F.; McKeirnan, A.J.; Ott, R.L. Pathogenic differences between various feline coronavirus isolates. Adv. Exp. Med. Biol. 1984, 173, 365–380.

22. Decaro, N.; Buonavoglia, C. An update on canine coronaviruses: Viral evolution and pathobiology. Vet. Microbiol. 2008, 132, 221–234.

23. Herrewegh, A.A.; Smeenk, I.; Horzinek, M.C.; Rottier, P.J.; de Groot, R.J. Feline coronavirus type II strains 79-1683 and 79-1146 originate from a double recombination between feline coronavirus type I and canine coronavirus. J. Virol. 1998, 72, 4508–4514.

24. Lin, C.N.; Chang, R.Y.; Su, B.L.; Chueh, L.L. Full genome analysis of a novel type II feline coronavirus NTU156. Virus Genes 2013, 46, 316–322.

25. Terada, Y.; Matsui, N.; Noguchi, K.; Kuwata, R.; Shimoda, H.; Soma, T.; Mochizuki, M.; Maeda, K. Emergence of pathogenic coronaviruses in cats by homologous recombination between feline and canine coronaviruses. PLoS ONE 2014, 9, e106534.

26. Tekes, G.; Spies, D.; Bank-Wolf, B.; Thiel, V.; Thiel, H.J. A reverse genetics approach to study feline infectious peritonitis. J. Virol. 2012, 86, 6994–6998.

27. Tekes, G.; Hofmann-Lehmann, R.; Stallkamp, I.; Thiel, V.; Thiel, H.J. Genome organization and reverse genetic analysis of a type I feline coronavirus. J. Virol. 2008, 82, 1851–1859.

28. Tekes, G.; Hofmann-Lehmann, R.; Bank-Wolf, B.; Maier, R.; Thiel, H.J.; Thiel, V. Chimeric feline coronaviruses that encode type II spike protein on type I genetic background display accelerated viral growth and altered receptor usage. J. Virol. 2010, 84, 1326–1333.

29. Thiel, V.; Thiel, H.J.; Tekes, G. Tackling feline infectious peritonitis via reverse genetics. Bioengineered 2014, 5, 396–400.

30. Haijema, B.J.; Volders, H.; Rottier, P.J. Switching species tropism: An effective way to manipulate the feline coronavirus genome. J. Virol. 2003, 77, 4528–4538.

31. Ettayebi, K.; Crawford, S.E.; Murakami, K.; Broughman, J.R.; Karandikar, U.; Tenge, V.R.; Neill, F.H.; Blutt, S.E.; Zeng, X.L.; Qu, L.; et al. Replication of human noroviruses in stem cell-derived human enteroids. Science 2016, 353, 1387–1393.

32. Pyrc, K.; Sims, A.C.; Dijkman, R.; Jebbink, M.; Long, C.; Deming, D.; Donaldson, E.; Vabret, A.; Baric, R.; van der Hoek, L.; et al. Culturing the unculturable: Human coronavirus HKU1 infects, replicates, and produces progeny virions in human ciliated airway epithelial cell cultures. J. Virol. 2010, 84, 11255–11263.

33. Sato, T.; Clevers, H. Growing self-organizing mini-guts from a single intestinal stem cell: Mechanism and applications. Science 2013, 340, 1190–1194.

34. Ehmann, R.; Kristen-Burmann, C.; Bank-Wolf, B.; Konig, M.; Herden, C.; Hain, T.; Thiel, H.J.; Ziebuhr, J.; Tekes, G. Reverse Genetics for Type I Feline Coronavirus Field Isolate to Study the Molecular Pathogenesis of Feline Infectious Peritonitis. MBio 2018, 9.

35. Mettelman, R.C.; O'Brien, A.; Whittaker, G.R.; Baker, S.C. Generating and evaluating type I interferon receptor-deficient and feline TMPRSS2-expressing cells for propagating serotype I feline infectious peritonitis virus. Virology 2019, 537, 226–236.

36. O'Brien, A.; Mettelman, R.C.; Volk, A.; André, N.M.; Whittaker, G.R.; Baker, S.C. Characterizing replication kinetics and plaque production of type I feline infectious peritonitis virus in three feline cell lines. Virology 2018, 525, 1–9.

37. Stanifer, M.L.; Mukenhirn, M.; Muenchau, S.; Pervolaraki, K.; Kanaya, T.; Albrecht, D.; Odendall, C.; Hielscher, T.; Haucke, V.; Kagan, J.C.; et al. Asymmetric distribution of TLR3 leads to a polarized immune response in human intestinal epithelial cells. Nat. Microbiol. 2020, 5, 181–191.

38. Sato, T.; Vries, R.G.; Snippert, H.J.; van de Wetering, M.; Barker, N.; Stange, D.E.; van Es, J.H.; Abo, A.; Kujala, P.; Peters, P.J.; et al. Single Lgr5 stem cells build crypt-villus structures in vitro without a mesenchymal niche. Nature 2009, 459, 262–265.

39. Sato, T.; Stange, D.E.; Ferrante, M.; Vries, R.G.; Van Es, J.H.; Van den Brink, S.; Van Houdt, W.J.; Pronk, A.; Van Gorp, J.; Siersema, P.D.; et al. Long-term expansion of epithelial organoids from human colon, adenoma, adenocarcinoma, and Barrett's epithelium. Gastroenterology 2011, 141, 1762–1772.

40. Pervolaraki, K.; Stanifer, M.L.; Munchau, S.; Renn, L.A.; Albrecht, D.; Kurzhals, S.; Senis, E.; Grimm, D.; Schroder-Braunstein, J.; Rabin, R.L.; et al. Type I and Type III Interferons Display Different Dependency on Mitogen-Activated Protein Kinases to Mount an Antiviral State in the Human Gut. Front. Immunol. 2017, 8, 459.

41. Triana, S.; Stanifer, M.L.; Shahraz, M.; Mukenhirn, M.; Kee, C.; Ordoñez-Rueda, D.; Paulsen, M.; Benes, V.; Boulant, S.; Alexandrov, T. Single-cell transcriptomics reveals immune response of intestinal cell types to viral infection. bioRxiv Preprint Serv. Biol. 2020. Available online: https://doi.org/10.1101/2020.08.19.255893 (accessed on 19 August 2020).

42. Fujii, M.; Matano, M.; Toshimitsu, K.; Takano, A.; Mikami, Y.; Nishikori, S.; Sugimoto, S.; Sato, T. Human Intestinal Organoids Maintain Self-Renewal Capacity and Cellular Diversity in Niche-Inspired Culture Condition. Cell Stem Cell 2018, 23, 787–793.

43. Koo, B.K.; Stange, D.E.; Sato, T.; Karthaus, W.; Farin, H.F.; Huch, M.; van Es, J.H.; Clevers, H. Controlled gene expression in primary Lgr5 organoid cultures. Nat. Methods 2011, 9, 81–83.

44. Jung, P.; Sato, T.; Merlos-Suarez, A.; Barriga, F.M.; Iglesias, M.; Rossell, D.; Auer, H.; Gallardo, M.; Blasco, M.A.; Sancho, E.; et al. Isolation and in vitro expansion of human colonic stem cells. Nat. Med. 2011, 17, 1225–1227.

45. Stanifer, M.L.; Kee, C.; Cortese, M.; Zumaran, C.M.; Triana, S.; Mukenhirn, M.; Kraeusslich, H.G.; Alexandrov, T.; Bartenschlager, R.; Boulant, S. Critical Role of Type III Interferon in Controlling SARS-CoV-2 Infection in Human Intestinal Epithelial Cells. Cell Rep. 2020, 32, 107863.
46. Derricott, H.; Luu, L.; Fong, W.Y.; Hartley, C.S.; Johnston, L.J.; Armstrong, S.D.; Randle, N.; Duckworth, C.A.; Campbell, B.J.; Wastling, J.M.; et al. Developing a 3D intestinal epithelium model for livestock species. Cell Tissue Res. 2019, 375, 409–424.
47. Kramer, N.; Pratscher, B.; Meneses, A.M.C.; Tschulenk, W.; Walter, I.; Swoboda, A.; Kruitwagen, H.S.; Schneeberger, K.; Penning, L.C.; Spee, B.; et al. Generation of Differentiating and Long-Living Intestinal Organoids Reflecting the Cellular Diversity of Canine Intestine. Cells 2020, 9, 822.
48. Chandra, L.; Borcherding, D.C.; Kingsbury, D.; Atherly, T.; Ambrosini, Y.M.; Bourgois-Mochel, A.; Yuan, W.; Kimber, M.; Qi, Y.; Wang, Q.; et al. Derivation of adult canine intestinal organoids for translational research in gastroenterology. BMC Biol. 2019, 17, 33.
49. Desmarets, L.M.; Theuns, S.; Olyslaegers, D.A.; Dedeurwaerder, A.; Vermeulen, B.L.; Roukaerts, I.D.; Nauwynck, H.J. Establishment of feline intestinal epithelial cell cultures for the propagation and study of feline enteric coronaviruses. Vet. Res. 2013, 44, 71.
50. Kipar, A.; Meli, M.L.; Baptiste, K.E.; Bowker, L.J.; Lutz, H. Sites of feline coronavirus persistence in healthy cats. J. Gen. Virol. 2010, 91, 1698–1707.
51. Pervolaraki, K.; Rastgou Talemi, S.; Albrecht, D.; Bormann, F.; Bamford, C.; Mendoza, J.L.; Garcia, K.C.; McLauchlan, J.; Hofer, T.; Stanifer, M.L.; et al. Differential induction of interferon stimulated genes between type I and type III interferons is independent of interferon receptor abundance. PLoS Pathog. 2018, 14, e1007420.

22

THE INFECTION OF ASTROCYTES DURING RABIES ENCEPHALITIS IS INFLUENCED BY THE SPECIFIC STRAIN OF THE VIRUS AND THE ROUTE OF INFECTION, AS SHOWN BY A NEW QUANTITATIVE 3D ANALYSIS OF CELL TROPISM

1. INTRODUCTION: AN OVERVIEW

The rabies virus (RABV) is a virus that primarily affects the nervous system and invariably leads to fatal disease in mammals once neurological symptoms appear [1]. RABV is a member of the Rhabdoviridae family in the Mononegavirales order. It is a non-segmented, single-stranded RNA virus with negative RNA polarity. The 12 kb genome of RABV contains five important virus proteins: nucleoprotein N, phosphoprotein P, matrix protein M, glycoprotein G, and the large polymerase L [3]. Aside from their crucial roles in genome replication and viral assembly, the RABV proteins have also been found to have several other functions. The pathogenicity of RABV is mostly related to its strong interference with the innate immune system through the actions of N, P, and M [4,5,6,7,8,9,10]. Additionally, RABV regulates neuronal survival through the actions of G [11,12,13,14]. The majority of pathogenicity experiments, however, were conducted using virus backbones that had previously been weakened. The mechanisms underlying the variations in disease-causing potential between highly virulent wild virus isolates and lab-adapted, less pathogenic RABV strains remain unclear. Furthermore, the impact of molecular disparities between virulent and attenuated viruses on virus multiplication and transmission in the infected animal, as well as the intricate interaction between the virus and host that ultimately leads to either sickness or an incomplete infection, remains unclear. Within a living organism, once neurons are infected, the RABV virus spreads from the infected neurons to the neurons that are connected to them through synapses. This spread occurs by retrograde axonal transport of the virus along microtubules, which are long, tube-like structures. This transport is crucial for the RABV virus to invade the nervous system and infect the central nervous system (CNS) through the peripheral nervous system. The process of co-internalization, along with the neuronal p75NTR receptor, followed by the retrograde axonal transport of RABV particles in endocytic vesicles, and the subsequent anterograde axonal transport of newly formed RABV, has been observed using live virus particle tracking in sensory neurons [20,21]. This highlights the ability of the virus to exploit neuron-specific mechanisms for long-distance transport to synaptic membranes. Nevertheless, the process of internalization and movement of lab-adapted viruses within axons, as well as the utilization of vaccine virus vectors for tracing the spread across synapses, indicate that the overall ability of axonal transport and trans-synaptic transmission is insufficient to account for the contrasting mechanisms observed between highly virulent RABV and less potent laboratory strains. The variations observed in the efficiency of trans-synaptic dissemination across different laboratory strains of RABV [24] suggest that the effectiveness of the underlying processes, rather than the capacity itself, may play a role in the spread of RABV in living organisms. Several receptors for RABV (rabies virus) have been discussed, including nAChR (nicotinic acetylcholine receptor), NCAM (neuronal cell adhesion molecule), p75NTR, and mGluR2 (metabotropic glutamate receptor subtype 2), which support RABV entry [25,26,27,28]. Nevertheless, the presence of these receptors is not crucial for central nervous system (CNS) infection by RABV. under fact, a diverse range of non-neuronal cell types can be infected under laboratory conditions, suggesting that RABV is not limited to infecting neurons based on receptor specificity. The majority of in vivo studies indicate a highly specific infection targeting the nervous system. Immediately following a bite, muscle cells get infected, as discussed in [30]. Additionally, infection of non-neuronal cells in the central nervous system (CNS) is possible, as indicated by references [29,31,32]. The utilization of recombinant Cre recombinase-expressing RABV allowed for the detection of abortively infected glial cells in mouse brains that were infected. This finding strongly indicates that these cells were infected and then eliminated by a robust type I interferon response [33]. Hence, the failure of infection in non-neuronal cells and the activation of innate immune responses may have a significant impact on both the infection process and the regulation of subsequent adaptive immunity pathways. A model using astrocytes infected in vitro indicates that attenuated RABV, unlike wild-type RABV, triggers inflammatory reactions in astrocytes by increasing the synthesis of double-strand RNA (dsRNA) and its recognition by retinoic acid-inducible gene I (RIG-I)/ melanoma differentiation-associated protein 5 (MDA5) [34]. Field isolates of RABV with high virulence have the ability to avoid or postpone the immune responses of the host, which can lead to the reproduction of the virus reaching harmful levels. In contrast, attenuated viruses do not induce early innate immune responses through astrocytes or other glial cells. However, every productive infection of the Rabies virus (RABV) ultimately leads to the development of rabies, a condition that is consistently linked to encephalitis. Nevertheless, there are discernible disparities between field viruses and attenuated lab RABV, since the former induces less tissue damage and neuroinflammation [34, 36, 37]. In order to assess the cell tropism of highly virulent field and lab-adapted RABVs in the central nervous system (CNS), we utilized a new technique called uDISCO (ultimate 3D imaging of solvent-cleared organs) [38,39,40]. This technique allows us to detect and measure RABV infection in neurons and astrocytes within brain tissue that has been cleared using solvents, while also being compatible with immunostaining. Following the capture of large confocal z-stacks in thick tissue slices using a confocal laser scan, three-dimensional (3D) reconstructions were conducted to study the cellular context of RABV infection. The study investigated the occurrence of RABV infection in neurons and astrocytes following intramuscular (i.m.) and intracerebral (i.c.) inoculation. The findings provided new knowledge about the capacity of non-neuronal astrocytes to be infected by RABV, the relationship between RABV virulence and infection in astrocytes, and the influence of the infection route on its establishment. The data provide significant evidence for a concept that suggests distinctions between field and lab-adapted viruses in their ability to reproduce in astrocytes in vivo, unlike in vitro circumstances where all viruses can infect non-neuronal astrocytes. Astrocytes infected with highly pathogenic RABV and the buildup of virus proteins that counteract interferon may hinder the prompt activation of the innate immune response, thereby aiding in immune evasion, even when glial cells are infected.

2. MATERIALS AND METHODS

2.1. Cell Lines, Primary Brain Cell Preparation, and Cultivation

Mouse neuroblastoma cells (Na42/13) were used for virus amplification. All cells were provided by the Collection of Cell Lines in Veterinary Medicine (CCLV), Friedrich-Loeffler-Institut, Insel Riems, Germany. Primary neurons and astrocytes were prepared from 1-day-old neonatal Sprague Dawley rats (P0–P1) [41]. The rats were decapitated, the heads were disinfected in 70% ethanol, the brains were removed, and the hippocampi were separated and mechanically minced. The hippocampal cells were transferred into ice-cold Hank´s Balanced Salt Solution (HBSS) and stored on ice. After preparation, the brain cells were dissociated by adding 0.25% trypsin + EDTA (Gibco; Thermo Fisher Scientific, Darmstadt, Germany) and DNase I (Applichem, Darmstadt, Germany) and incubated for 15 min at 37 °C. Afterwards, further dissociation in Neurobasal-A Medium (NB-A, Gibco; Thermo Fisher Scientific) was performed by using a glass Pasteur pipette with a fire-polished tip. The cells were purified with an OptiprepTM gradient (with concentrations from bottom to the top: 17.3%, 12.4%, 9.9%, and 7.4%)

and centrifuged for 15 min at 800× g without brake. Next, the cells were washed with NB-A medium and counted with the LUNA-II Automated Cell Counter (Logos Biosystems, Villeneuve d'Ascq, France). The neuronal cells were seeded on coverslips in a 24-well plate and cultured for two weeks at 37 °C and 5% CO2 in serum-free NB-A supplemented with 2% B27 (Gibco; Thermo Fisher Scientific), 1% GlutaMAX (Gibco; Thermo Fisher Scientific), 1% Penicillin-Streptomycin (stock: 10,000 U/mL penicillin and 10 mg/mL streptomycin; Sigma-Aldrich, Taufkirchen, Germany), and 0.2% Gentamicin (stock: 50 mg/mL; Gibco; Thermo Fisher Scientific) until they were used for infection experiments.

2.2. Viruses

RABV isolates used in this study comprised five recombinant virus clones and one non-recombinant lab virus. The recombinant field virus clones rRABV Dog and rRABV Fox have been described before [42]. SAD L16 is a recombinant virus clone of attenuated vaccine virus SAD B19 live vaccine virus [43]. The Evelyn Rokitnicki Abelseth (ERA) strain [44] is a progenitor virus strain of live vaccine virus SAD B19 [45] and was obtained from the FLI virus archive (FLI ID N 12829).

A cDNA full length clone (pRABV Rac) from a raccoon RABV isolate (Alabama, USA 1991; FLI archive ID N 13205) [46] was generated by full length RT-PCR amplification of the 12 kb cDNA genome with the primer pair 5′-TCGATCCCGGGTCACGCTTAACAACAAAA-3′/5′-TAATA CACCTGCCCATGCCGACCCACGCTTAACAAAAAAACAA-3′. After PCR amplification of a 2.7 kb vector DNA fragment from pCMV HaHd ampR Br 322 ori with the primers 5′-TCTGTTTGCTTGATGGTTTTTTTT GTCTTTGTTG TTTTTTTGTTAAGCGTGGG TCGGCATGG CATCTCCAC-3′and 5′-TTTTTGTAGATGATACTGTCTA CTTCTTCT CTGATTT TGTTGTTAAG CGTGACCCGGGACTCCGGGTTTCGTC-3′,the fragments were combined by linear to linear recombination, and the resultant recombinant rRABV Rac virus was rescued and amplified according to previously described protocols [42]. The sequence of the full-length cDNA clone was deposited at GenBank (accession no. MN862283).

A CVS-11 (Challenge Virus Street-11; sequence accession no. LT839616) cDNA full length clone (pCVS-11) was generated by RT-PCR amplification of 7.5kb and 3.7 kb cDNA fragments from CVS-11 RNA with the primers pairs 5′-AGTTTCAGACGTCTCAGTC-3′/5′-CTAGTAGGGATGATCTAGATC-3′ and 5′c-GACTGAGACGTCTGAAACT-3′/5′ CATTGCAGATAGGATAGAG-3′, respectively. The resultant fragments were assembled in combination with EcoRI-linearized 3.4 kb vector plasmid pCVS-11-termini by Hot Fusion reaction [47]. Plasmid pCVS-11-termini was generated by PCR amplification of a 2.7 kb DNA fragment from pCMV HaHd ampR Br 322 ori [42] with the primers 5′-GACCCGGGACTCCGGGTTTC-3′ and 5′-GGGTCGGCATGGCATCTCCA-3′, and Hot Fusion assembly with a synthetic DNA fragment containing the CVS-11 (accession no. LT839616) regions from nucleotide positions 1–580 (3′-genome end) and 11759–11927 (5′-genome end) separated by a non-viral EcoRI restriction site. The full-length cDNA clone was 99.99% identical to CVS-11 GenBank sequence no. LT839616 (1 nt mismatch in the N gene at nucleotide position 1263, which was already present in #LT839616 as a minor SNP (single nucleotide polymorphism), resulting in the amino acid exchange E398V). Recombinant rCVS-11 was rescued in Na42/13 neuroblastoma cells by co-transfection of pCVS-11, pCAGGS-based expression plasmids for RABV N, P, and L [48], and a pCAGGS-T7Pol vector comprising a codon-optimized bacteriophage T7 RNA polymerase gene. Three to six days after transfection, the supernatants were transferred to new Na42/13 cells. Two days after the transfer, infectious virus was identified by N and G protein-specific indirect immunofluorescence.

rRABV Dog, rRABV Fox, rRABV Rac, and rCVS-11 were amplified on Na42/13 cells. BSR T7/5 cells [49] were used for amplification of SAD L16 and ERA viruses. Infectious virus titers in cell culture supernatants were determined by end point dilution and titration on Na42/13 cells.

2.3. Antibodies

To verify the presence of RABV, a polyclonal rabbit serum against recombinant RABV P protein (P160-5, immunofluorescence 1:5000, uDISCO 1:3000) and a polyclonal goat serum against RABV N (goat anti-RV N, immunofluorescence 1:4000), which have been described previously [40,50], were used. The polyclonal chicken anti-glial fibrillary acidic protein (GFAP) antibody (Thermo Fisher Scientific, Darmstadt, Germany; #PA1-10004, uDISCO 1:1500), the polyclonal rabbit anti-GFAP (Dako, #Z0334, immunofluorescence 1:500), the polyclonal guinea pig anti-NeuN antibody (Synaptic Systems, Goettingen, Germany; #266004, uDISCO 1:800), and the rabbit anti-MAP2 antibody (Abcam, Cambridge, UK; #ab32454, immunofluorescence 1:250) were purchased from their respective suppliers.

2.4. Infection of in Vitro Cell Cultures and Immunofluorescence Staining

Two-week-old primary rat hippocampal cell cultures were infected with 1 × 103 infectious units of rRABV Dog, rRABV Fox, rCVS-11, and SAD L16, and cultivated for 24 h at 37 °C and 5% CO_2. Indirect immunofluorescence was performed by standard techniques after fixation with 4% paraformaldehyde (PFA) in phosphate-buffered saline (PBS) for 30 min and 15 min permeabilization with 0.5% Triton X-100 in PBS. Afterwards, samples were blocked with 0.025% skim milk powder in PBS for 15 min. Immunostainings were executed by 1.5 h incubation with primary antibodies, three wash steps with PBS, followed by 1 h incubation with secondary antibodies and additional Hoechst33342 (1 μg/mL) for staining of the nuclear chromatin. Specimens were mounted on coverslips and were analyzed by confocal laser scanning microscopy.

2.5. Brain Samples and Mouse Infections

Three- to four-week-old BALB/c mice (Charles-River, Germany) were infected with rRABV Rac, rCVS-11, ERA, and SAD L16 viruses using two different inoculation routes and two different viral doses, essentially as described before [51]. Two groups of six animals each were anesthetized and infected i.m. with 102 or 105 TCID50/30 μL, and an additional group of three mice was infected i.c. with 102 TCID50/30 μL. The weight and the clinical score, ranging from zero to four, of all mice were observed for 21 days post infection (dpi). When reaching a clinical score of two or three (ruffled fur, slowed movement, weight loss >15%), the animals were anaesthetized with isoflurane and euthanized through cervical dislocation. Samples were taken, fixed with 4% paraformaldehyde (PFA) for one week, and stored for further processing. All remaining animals were euthanized at 21 dpi. Mouse experimental studies on the characterization of lyssaviruses were evaluated by the responsible animal care, use, and ethics committee of the State Office for Agriculture, Food Safety, and Fishery in Mecklenburg-Western Pomerania (LALFF M-V) and gained approval with permissions 7221.3–2–001/18.

2.6. Archived Mouse Brains Infected with rRABV Dog and rRABV Fox

To minimize animal experiments, archived PFA-fixed brains from previous pathogenicity trials [42] were used for the analysis of rRABV Dog and rRABV Fox virus infections in mice. Similar to the mouse experiments with rRABV Rac, rCVS-11, ERA, and SAD L16 described above, mice were inoculated via the i.c. or the i.m. route.

2.7. Ultimate 3D Imaging of Solvent-Cleared Organs (uDISCO)

The clearing of brain tissue slices was performed as described previously [40] in modification of earlier publications [38,39]. Briefly, the PFA-fixed tissues were sectioned into 1 mm-thick slices using a vibratome (Leica VT1200S; Leica Biosystems, Wetzlar, Germany). All subsequent incubations steps were performed with gentle oscillation. To increase antibody diffusion and reduce tissue autofluorescence [38], the sections were pretreated with increasing concentrations of methanol (20%, 40%, 60%, 80%, and twice in 100%; dilutions with distilled water, incubation for 1 h each) and bleached by overnight incubation at 4 °C with 5% H2O2 in 100% methanol. After removal of the bleaching solution, the samples were rehydrated with decreasing concentrations of methanol (80%, 60%, 40%, and 20%; dilutions with distilled water, incubation for 1 h each) and a subsequent wash with PBS for 1 h. To permeabilize the samples, they were washed twice for 1 h each with 0.2% Triton X-100 in PBS and subsequently incubated for 48 h at 37 °C in 0.2% Triton X-100/20% DMSO/0.3 M glycine in PBS. The samples were then blocked by incubation with 0.2% Triton X-100/10% DMSO/6% donkey serum in PBS for 48 h at 37 °C. Primary antibodies were diluted in 3% donkey serum/5% DMSO in PTwH (0.2% Tween-20 in PBS with 10 µg/mL heparin), and incubation of samples was performed at 37 °C for 5 days. The antibody solution was refreshed after 2.5 days. The samples were washed with PTwH by exchanging the solution four times during the course of the day and subsequently incubated overnight in PTwH. Incubation with secondary antibodies was performed in 3% donkey serum in PTwH for 5 days at 37 °C, refreshing the secondary antibody solution once after 2.5 days. Subsequent washing was performed as described above following primary antibody incubation.

For tissue clearing, the samples were dehydrated with a series of tert-butanol (TBA) solutions (30%, 50%, 70%, 80%, 90%, and 96%; dilutions with distilled water, incubation for 2 h each), leaving 96% TBA on overnight. Following further dehydration in 100% TBA for 2 h, the samples were cleared in BABB-D15 [39] [1:2 mixture of benzyl alcohol (BA) and benzyl benzoate (BB), which is mixed with diphenyl ether (DPE) at a ratio of 15:1 and supplemented with 0.4 vol% DL-α-tocopherol] until they were optically transparent (2–6 h).

For confocal laser scanning microscopy, the samples were mounted in 3D-printed imaging chambers (printer: Ultimaker 2 + [Ultimaker, Utrecht, Netherlands], material: co-polyester, nozzle: 0.25 mm, layer height: 0.06 mm, wall thickness: 0.88 mm, wall count: 4, infill: 100%, no support structure; the corresponding [40]).

2.8. Confocal Laser Scanning Microscopy and Image Processing

The immunofluorescent staining of primary brain cells and infected tissues was visualized with a confocal laser scanning microscope (Leica DMI 6000 TCS SP5; Leica Microsystems, Wetzlar, Germany) equipped with a long free working distance 40× water immersion objective (NA = 1.1; Leica, #15506360). For image processing, a Dell Precision 7920 workstation was used (CPU: Intel Xeon Gold 5118, GPU: Nvidia Quadro P5000, RAM: 128 GB 2666 MHz DDR4, SSD: 2 TB; Dell, Frankfurt am Main, Germany).

For quantification of infected cells in thick tissue sections, the image was split into individual channels using Fiji, an ImageJ (v1.52h) distribution package [52]. A bleach correction was performed (simple ratio; background intensity: 5.0), and brightness and contrast were adjusted for each channel. Objects were identified and counted with the 3D Objects Counter plugin [53]. The resulting objects map was overlaid with the RABV P channel to quantify infected objects. For each sample, at least six regions were imaged and analyzed[54]. All other maximum z- and 3D projections.

2.9. Statistical Analysis

Statistical significance was determined using two-way ANOVA followed by Tukey's multiple comparison test using GraphPad Prism 7.05.

3. RESULTS

3.1. Astrocytes Are Readily Infected by both Field Viruses and Lab Strains in Mixed Primary Brain Cell Cultures

To investigate whether field and lab-adapted RABVs differ in their ability to infect primary neurons and non-neuronal astrocytes, hippocampal brain cells were prepared from neonatal rats and were cultivated as mixed cultures containing neurons and glial cells. After 13 days of cultivation, the cultures were infected with 103 infectious units of the recombinant field viruses rRABV Dog and rRABV Fox, and the lab-adapted strains rCVS-11 and SAD L16. After 24 h of infection, the cells were fixed and analyzed by confocal laser scanning microscopy using indirect immunofluorescence stainings against RABV nucleoprotein N, neuron marker MAP2 (microtubule-associated protein 2) and astrocyte marker GFAP (glial fibrillary acidic protein). All viruses infected both MAP2-positive neurons and GFAP-positive astrocytes, as demonstrated by the formation of RABV-positive cytoplasmic inclusion bodies (Figure 1). These data indicated that there was no obvious difference between the tested field and the lab-adapted viruses in their ability to infect neurons and non-neuronal astrocytes.

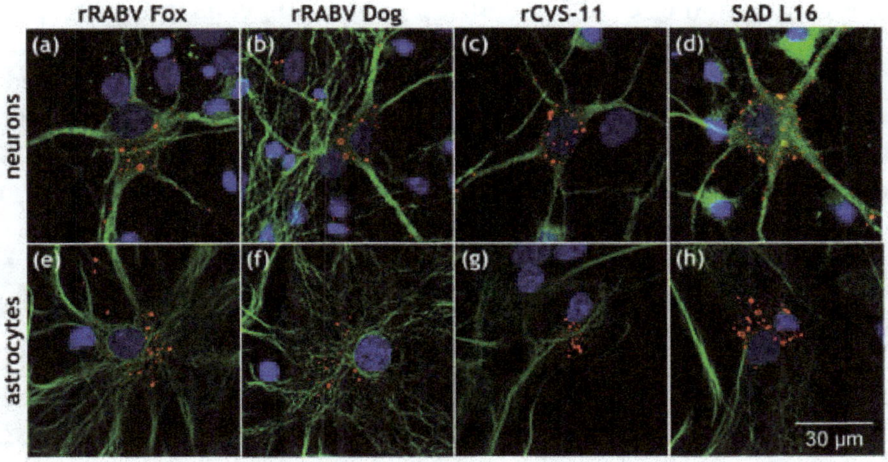

Figure 1. In vitro infection of primary hippocampus neurons and astrocytes by field and lab rabies virus (RABV). (a–d) Indirect immunofluorescence detection of RABV nucleoprotein N (red) and microtubule-associated protein 2 (MAP2) (green). Nuclei were counterstained with Hoechst33342 (blue). (e–f) Indirect immunofluorescence detection of RABV nucleoprotein N (red) and glial fibrillary acidic protein (GFAP) (green). Nuclei were counterstained with Hoechst33342 (blue).

Although all four viruses were able to infect GFAP-positive cells in the hippocampal cell cultures, the majority of the infected cells were MAP2-positive neurons. To quantitatively compare astrocyte infection between viruses, the number of RABV N-positive cells was counted for an area of 25 mm2, and the percentage of GFAP-positive and -negative RABV-infected cells was determined (Table 1). All four viruses resulted in infection of GFAP-positive cells at levels ranging from 4.9% to 6.7%. Although a higher percentage of GFAP-positive cells could not be excluded because a high number of cells could not be classified into either positive or negative for GFAP, these results indicated that all four viruses could readily establish an infection in cultivated astrocytes and did not substantially differ in their ability to do so.

Table 1. Infection of GFAP- and MAP2-expressing cells in hippocampus cell cultures. RABV-infected cells were counted after indirect immunofluorescence detection of RABV N protein and subdivided in cells that were either negative (-) or positive (+) for GFAP/MAP2. Because of overlaying positive and negative cells in GFAP and MAP2 stainings, a clear assignment was not always possible. Such cells were classified as uncertain. CVS: Challenge Virus Street.

Virus	GFAP (-)		GFAP (+)		Uncertain	
	Count	%	Count	%	Count	%
SAD L16	256	48.3	30	5.7	244	46
rCVS-11	58	64.4	6	6.7	26	28.9
rRABV Dog	252	46.8	29	5.4	258	47.9
rRABV Fox	1002	55.2	89	4.9	712	39.5
Virus	MAP2 (-)		MAP2 (+)		Uncertain	
	Count	%	Count	%	Count	%
SAD L16	18	3.5	387	74.2	116	22.3
rCVS-11	1	1.0	44	43.2	57	55.9
rRABV Dog	27	5.3	347	73.3	109	21.4
rRABV Fox	190	7.3	2096	80.3	324	12.4

3.2. Field RABV Infects Neurons and Astrocytes In Vivo as Demonstrated by High-Resolution 3D Analysis of Infected Brain Tissue

In order to test whether the primary cell culture infection experiments are comparable to the more complex in vivo situation, a sample preparation and imaging pipeline was established, which allows three-dimensional, high-resolution confocal laser scanning image acquisition from RABV-infected tissue for 3D reconstruction and quantification of infected cells. To this end, 1 mm-thick brain slices (Figure 2a) from clinically diseased rRABV Fox field virus-infected mice (i.m. infection route) were immunostained for RABV P, GFAP, and NeuN, optically cleared, and imaged. Acquisition of z-stacks by confocal laser scanning microscopy resulted in z-volumes of 400 μm × 400 μm × 50–100 μm (x, y, and z) from different areas of the infected brain. RABV-infected neurons of variable morphologies were detected by presence of NeuN and RABV P protein accumulation in neuronal cell bodies and neurites (Figure 2b–e). Astrocytes were detected by filamentous GFAP-positive structures (Figure 2b–g).

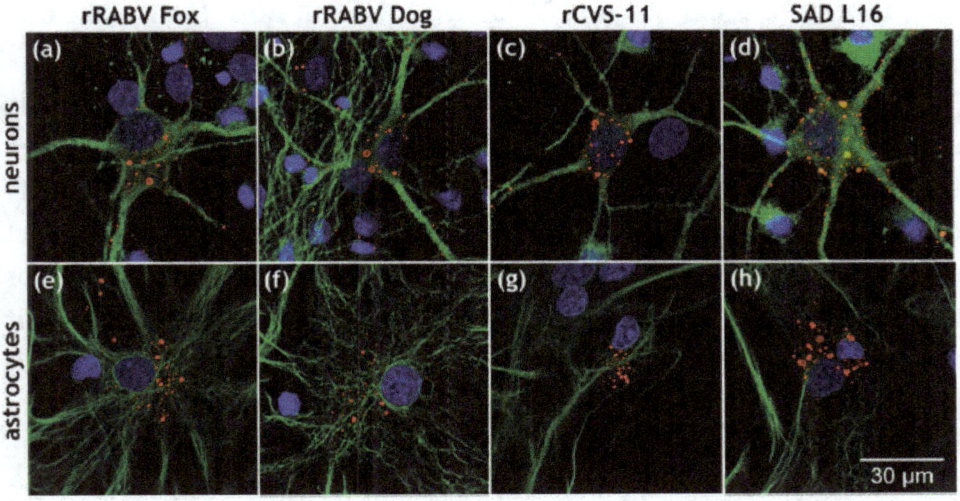

Figure 2. 3D immunofluorescence imaging of field RABV-infected neurons and astrocytes in a mouse brain. (a) Workflow for 3D immunofluorescence imaging, including vibratome sectioning into 1 mm slices, pretreatment, immunostaining, and subsequent optical clearing with organic solvents. Confocal imaging and acquisition of cleared tissue slices was done in custom-made imaging containers (see lower image). (b,c) Maximum z- and (d–e) 3D projections of z-stack (x, y, z = 400 μm, 400 μm, 59 μm for D and 400 μm, 400 μm, 103 μm for E) after indirect immunofluorescence for RABV phosphoprotein P (red), GFAP (green), and NeuN (blue). White arrows in Figure 2b indicate RABV P and GFAP-positive astrocytes. (f) Maximum projection of detail from Figure 2b (see white box) with GFAP-positive cell (green) and associated RABV P fluorescence (red). (g) 3D projection of detail view from Figure 2f.

Notably, besides accumulation of RABV P in NeuN-positive neurons, P also accumulated at GFAP-positive filaments (Figure 2b,c,f,g; white arrows), indicating a robust infection of astrocytes by field virus rRABV Fox. Evaluation of brain areas with differing localization patterns of the highlighted cellular subpopulations (Figure 2b,c) revealed that astrocyte infections were not restricted to particular regions of the brain but appeared in all imaged neuron layers or brain areas.

Because of the complex 3D morphology of both neurons (long axons in Figure 2b,c) and astrocytes (filamentous structure of GFAP signals in Figure 2f,g), 3D reconstruction of the imaged tissues (Figure 2d) was performed in order to achieve reliable visualization and quantification of cells in downstream analyses.

3.3. Similar Levels of Field Virus Infection in both Neurons and Astrocytes in the Infected Mouse Brain

Whereas immunofluorescence analysis in Figure 2 clearly demonstrated infection of astrocytes and neurons, they also revealed that only a fraction of both cell types was positive for RABV P. To quantify the ratio of infected and non-infected neurons and astrocytes, 3D object segmentation and counting was performed for NeuN- (Figure 3a,b) or GFAP (Figure 3d,e)-positive cells. The object map was merged with their respective RABV fluorescence (Figure 3c,f), and the number of RABV-positive neurons and astrocytes was determined by manual counting of RABV P- and cell marker-positive cells. For the z-stack shown in Figure 2b and Figure 3, total numbers of 762 neurons and 272 astrocytes were counted (Table 2, region 2), of which 16 and 18 were RABV positive, respectively.

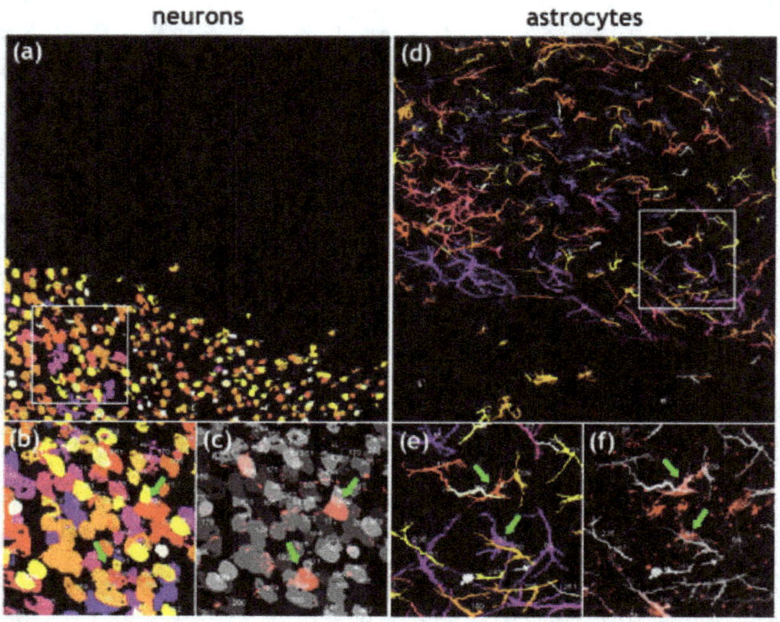

Figure 3. Quantification of RABV-infected neurons and astrocytes. (a) Maximum z-projection of objects map for NeuN-positive neurons generated from a confocal z-stack (see Figure 2b). Individual neurons in the z-stack were identified by NeuN-specific fluorescence and converted to objects. Numbers indicate individual cell counts. n = 762 neurons in a volume of 400 µm × 400 µm × 59 µm. The object colors indicate different

z-positions (darker colors in the back and brighter colors in the front). (b) Detail of area indicated by white box in Figure 3a. Green arrows indicate RABV-positive neuron cell bodies. (c) Overlay of objects map (greyscale) with RABV P signals allows identification and counting of rRABV Fox-infected neurons. (d) Maximum z-projection of objects map for GFAP-positive astrocytes generated from a confocal z-stack (see Figure 2b). Individual astrocytes in the z-stack were identified by GFAP-specific fluorescence and converted to objects. Numbers indicate individual cell counts. n = 272 astrocytes in a volume of 400 μm × 400 μm × 59 μm. The object colors indicate different z-positions (darker colors in the back and brighter colors in the front). (e) Detail of area indicated by white box in Figure 3d. Green arrows indicate RABV-positive astrocytes. (f) Overlay of objects map (greyscale) with RABV P signals allows identification and counting of rRABV Fox-infected astrocytes.

Table 2. Quantification of rRABV Fox-infected neurons and astrocytes. Results of the analysis of six different areas of one infected mouse brain. The quantification pipeline of region 2 is depicted in Figure 3. Neurons and astrocytes were segmented, automatically counted, and overlaid with the respective RABV P signals. Infected cells were counted manually.

	Neurons					Astrocytes				
Region	Counted	Infected	%	Mean	SD	Counted	Infected	%	Mean	SD
1	485	21	4.3	3.9	1.4	362	11	3.0	7.0	4.9
2	762	16	2.1			272	18	6.6		
3	448	13	2.9			106	18	17.0		
4	2518	132	5.2			488	9	1.8		
5	3029	91	3.0			273	20	7.3		
6	3847	234	6.1			541	32	5.9		
Σ	11098	507				2042	108			

Analysis of six z-stacks from different RABV-infected brain areas of the same sample led to the detection of 11,089 neurons and 2042 astrocytes, of which 3.9% and 7.0% were RABV positive, respectively (Table 2). The fraction of RABV-positive cells ranged from 2.1% to 6.1% (SD: ± 1.4) for neurons and 1.8% to 17.0% for astrocytes (SD: ± 4.9) (Table 2). These data indicated that rRABV Fox infects astrocytes and neurons in the mouse brain to comparable levels.

Observation of three different rRABV Fox-infected mice revealed that the infection level for astrocytes differed between single animals from 1.2% to 15.6% (SD: ± 6.7) . Nevertheless, in spite of some variance between individual mice and/or between different analyzed brain regions, detection of infected non-neuronal astrocytes in all animals at surprisingly high levels indicates that astrocyte infection by RABV has been underestimated thus far.

3.4. Astrocyte Infection by RABV Depends on the Type of Virus (Field vs. Lab-Adapted).

To compare the astrocyte tropism of field and lab-adapted viruses after i.m. inoculation, brains of mice infected with three different field viruses (rRABV Fox, rRABV Dog, and rRABV Rac) and two lab-adapted viruses (rCVS-11 and ERA) were analyzed. SAD L16 was excluded from these analyses since it is not able to induce clinical signs after i.m. inoculation.

Whereas rRABV Fox and rRABV Dog caused disease even after infection with a very low virus dose [42], infections with rRABV Rac, rCVS-11, and ERA did not show any clinical signs at a dose of 102 TCID50. High dose infection with the ERA strain (105 TCID50) led to 100% disease development, whereas high dose infections with rRABV Rac and rCVS-11 only caused disease in 50% and 16.7% of the infected mice, respectively. Accordingly, available tissue samples for the latter two viruses were limited to two and one infected brain. Imaging and quantification of at least six confocal z-stacks from different RABV-infected regions per infected brain was performed.

RABV infections were readily detectable in the brain for all five viruses (Figure 4). Whereas RABV P antigen accumulated at GFAP-positive structures in rRABV Fox, rRABV Dog, and rRABV Rac-infected brains (Figure 4a,c,e), similar accumulations were not observed in rCVS-11 or ERA-infected brains, in which only infection of NeuN-positive neurons was detected (Figure 4b,d).

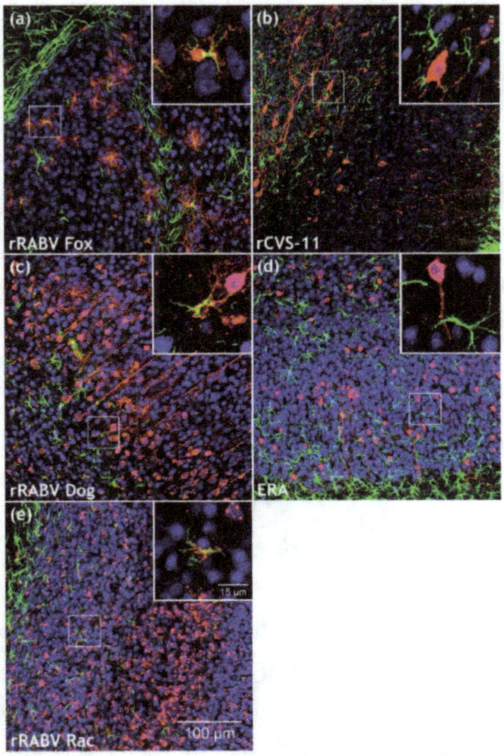

Figure 4. Comparison of field and lab RABV-infected brains after intramuscular (i.m.) infection with rRABV Fox, rRABV Dog, rRABV Rac, rCVS-11, and Evelyn Rokitnicki Abelseth (ERA). (a–e) Maximum z-projections of z-stacks [x, y = 400 µm, 400 µm (a–e); z = 59 µm (a, rRABV Fox), 66 µm (b, rCVS-11), 49 µm (c, rRABV Dog), 100 µm (d, ERA) and 89 µm (e, rRABV Rac)] after indirect immunofluorescence for RABV phosphoprotein P (red), GFAP (green), and NeuN (blue). To improve visualization of the maximum z-projections, some z-stacks were reduced in thickness. Insets in Figure 4a,c,e show RABV P accumulation (red) at GFAP-positive cells (green). Insets in Figure 4b,d show NeuN- (blue) and RABV P (red)-positive neurons. For the individual channels of the detail images.

Infections with rRABV Fox, rRABV Dog, rRABV Rac, rCVS-11, and ERA led to a mean of 8.3%, 7.3%, 18.9%, 6.9%, and 15.1% RABV-positive neurons (Figure 5), respectively, demonstrating a comparable level of neuron infection by all five viruses in clinically diseased mice. However, significant differences were observed for the astrocyte infections. Whereas rRABV Fox, rRABV Dog, and rRABV Rac infected 7.6% (SD:

± 6.7), 10.1% (SD: ± 7.7), and 16.5% (SD: ± 15.0) of the astrocytes, no RABV-positive astrocytes were detected in rCVS-11 and ERA-infected samples (Figure 5). These data indicated that, in contrast to the tested field viruses, the lab-adapted RABVs rCVS-11 and ERA did not infect astrocytes to detectable levels in vivo after i.m. inoculation.

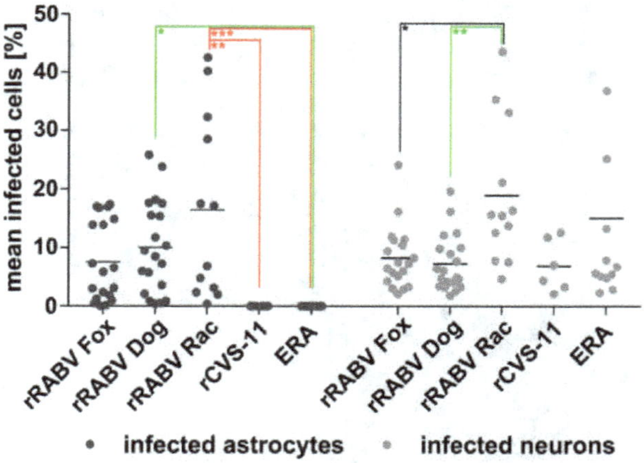

Figure 5. Percentage of field and lab RABV-infected neurons and astrocytes after i.m. inoculation. Per virus, 3 to 9 × 103 astrocytes and 1.5 to 4 × 104 neurons were counted in 19 (rRABV Fox), 20 (rRABV Dog), 12 (rRABV Rac and ERA), and six (rCVS-11) independent confocal z-stacks in three (rRABV Fox and rRABV Dog), two (rRABV RAC and ERA), and one (rCVS-11) animal. Each dot represents the frequency of infected astrocytes or neurons in an analyzed z-stack. Mean values are provided as horizontal lines. * p ≤ .05; ** p ≤ .01; *** p ≤. 001 (two-way ANOVA with Tukey's multiple comparison test).

3.5. Confirmation of the Specific Astrocyte Tropism of Field RABV, ERA, and SAD L16 after i.c. Inoculation and Route-Dependent Astrocyte Infection by rCVS-11

To test whether the inoculation route affects astrocyte infection and whether the highly attenuated SAD L16 virus is comparable to the SAD vaccine progenitor strain ERA, brain samples from two animals i.c.-infected with rRABV Fox, rRABV Dog, rCVS-11, or SAD L16 and one infected with the ERA strain were analyzed.

With astrocyte infections at frequencies of 10.9% (SD: ± 10.3), 11.6% (SD: ± 5.6), and 27.2% (SD: ± 12.8) (Figure 6), rRABV Fox, rRABV Dog, and rRABV Rac led to robust astrocyte infection via the i.c. route and thus confirmed their ability to establish astrocyte infection in vivo (Figure 7a,c,e). Lack of detectable astrocyte infection for the ERA and SAD L16 further confirmed that these viruses are not able to infect astrocytes to a detectable level (Figure 7d,f and Figure 6), even after direct virus administration into the brain.

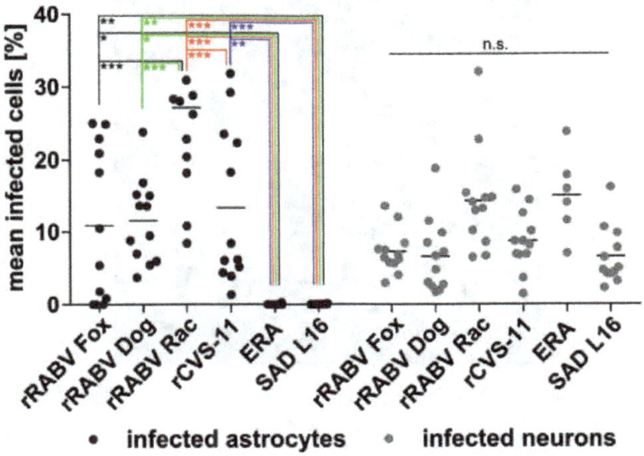

Figure 6. Percentage of field and lab RABV-infected neurons and astrocytes after i.c. inoculation. Per virus, 5 to 9 × 103 astrocytes and 1.3 to 3 × 104 neurons were counted in 12 (rRABV Dog, rRABV Fox, rRABV Rac, rCVS-11, and SAD L16) and six (ERA) independent confocal z-stacks in two (rRABV Fox, rRABV Dog, rRABV Rac, rCVS-11, and SAD L16) and one (ERA) animal. Each dot represents the frequency of infected astrocytes or neurons in an analyzed z-stack. Mean values are provided as horizontal lines. * p ≤ .05; ** p ≤. 01; *** p ≤. 001; n.s. = not significant (two-way ANOVA with Tukey's multiple comparison test).

Notably, in contrast to the i.m. inoculation route and to the other i.c.-inoculated lab-adapted RABVs SAD L16 and ERA, 13.4% (SD: ± 10.4) of the astrocytes were RABV-positive in rCVS-11-infected animals, indicating that, in the case of rCVS-11, the infection route has a substantial influence on astrocyte infection in the brain. However, the ratio of rCVS-11-infected neurons (8.7%; SD: ± 4.0) remained comparable to i.m. infections.

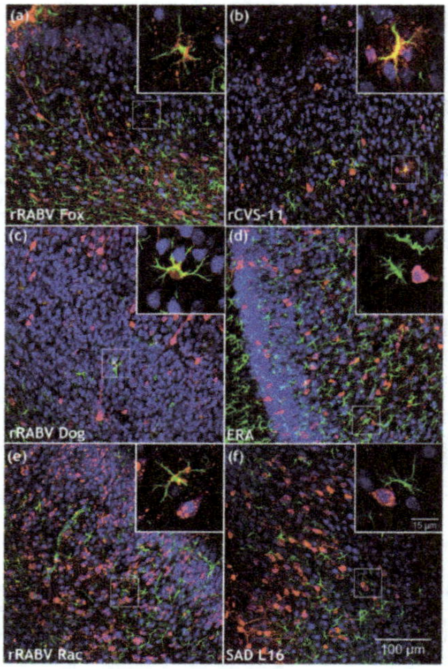

Figure 7. Comparison of field and lab RABV-infected brains after i.c. infection. (a–f) Maximum z-projections of z-stacks [x, y = 400 µm, 400 µm (a–f); z = 21 µm (a, rRABV Fox), 48 µm (b, rCVS-11), 74 µm (c, rRABV Dog), 98 µm (d, ERA), 75 µm (e, rRABV Rac) and 67 µm (f, SAD L16)] after indirect immunofluorescence for RABV phosphoprotein P (red), GFAP (green), and NeuN (blue). Insets in Figure 7a,b,c,e show RABV P accumulation (red) at GFAP-positive cells (green). To improve visualization of the maximum z-projections, some z-stacks were reduced in thickness. Insets in Figure 7d,f show NeuN- (blue) and RABV P (red)-positive neurons. For the individual channels of the detail images.

4. DISCUSSION

By use of a novel immunofluorescence-compatible technique for 3D immunofluorescence imaging of solvent-cleared brain tissue slices [38,39] adapted to the imaging of RABV infections in brain tissue [40], we investigated the cell tropism of six different RABV isolates and lab-adapted strains. Both neurons and astrocytes display heterogeneous morphologies with pronounced three-dimensional projections. Compared to conventional thin sections or even in silico 3D reconstructions of serial thin sections, optical slicing of 1 mm-thick tissue samples during image acquisition allowed fast and seamless 3D reconstruction of immunostained tissues and high-resolution dissection of the detected antigens [40]. For the first time, this allowed systematic analysis and comparison of the tropism of the different viruses in large 3D volumes of infected mouse brains. One example for the superiority of the employed technique is the unambiguous detection of rCVS-11-infected astrocytes after i.c. inoculation (Figure 7 and Figure 6), whereas conventional thin layer immunohistochemistry analyses led to the assumption that infection of glial cells by CVS does not occur, even after i.c. inoculation [55].

Importantly, more than 104 neurons and 103 astrocytes were quantitatively investigated for RABV infection in their authentic environment per virus and inoculation route. This provided reliable datasets about the frequencies of RABV detection in the two cellular subpopulations in vivo and statistically significant differences in astrocyte infections, which depended on the degree of virulence and the inoculation route of the tested RABVs (Figure 5 and Figure 6).

Because of the clonal and the defined nucleotide sequences, the recombinant viruses rRABV Fox and rRABV Dog [42], rRABV Rac (this work), SAD L16 [43], and rCVS-11 (this work) were selected for the analyses. Since SAD L16 was directly derived from the live vaccine SAD B19 vaccine strain [56], which is equally apathogenic after intramuscular inoculation, the non-recombinant SAD B19 progenitor vaccine virus strain ERA was included to allow for the investigation of attenuated live vaccine virus astrocyte tropism after peripheral i.m. inoculation. Compared to SAD B19, the ERA strain has a less intense cell culture passage history [45] and is still pathogenic in mice after peripheral inoculation [57]. Indeed, whereas SAD L16 was not able to cause disease after i.m. inoculation, ERA was even more pathogenic than rCVS-11 at the high i.m. inoculation dose of 105 TCID50 with 100% and 16.7% of mice developing clinical signs, respectively.

Both the original field virus isolates and the corresponding genetically modified virus clones, rRABV Fox and rRABV Dog, have been demonstrated to have similar ability to cause disease and efficiently develop the disease after being injected into mice intramuscularly [42]. Virulence in raccoons has been demonstrated for the two previously stated field virus isolates, as well as for the raccoon RABV isolate used in the creation of rRABV Rac [46]. It is worth mentioning that whereas the raccoon virus isolate was extremely harmful in its original host, the genetically modified clone rRABV Rac produced in our study was less harmful in mice. It only caused illness in 50% of the mice when given a high dosage of the virus through intramuscular injection. Unlike the rRABV Fox and rRABV Dog, the lab viral clone rCVS-11 was less effective when administered through the i.m. route. However, it still managed to cause sickness in one out of six mice when a high virus dose of 105 infectious units per mouse was used. This finding is consistent with prior research, which showed that the non-recombinant progenitor CVS-11 strain produced disease in only 16.7% of the mice when given a high intramuscular dose, and, was not pathogenic when given a low dose intramuscular infection [51]. Unlike the results observed in living organisms, when rat hippocampal neuron/astrocyte cultures were infected with rRABV Fox, rRABV Dog, SAD L16, and rCVS-11 in a laboratory setting, similar levels of astrocyte infections were observed. This was confirmed by the presence of GFAP-positive, RABV-infected cells, which ranged from 4.9% to 6.7% for all viruses, as shown in Figure 1 and Table 1. The data presented here align with previous findings that demonstrate the sensitivity of grown astrocytes to RABV infection, as observed with both field and lab-adapted viruses [29,34]. Furthermore, the data indicate a preference for viral replication in neurons [32]. While Tsiang et al. found that 90-99% of the primary astrocytes did not show any RABV antigen, our findings suggest that there is minimal variation among the tested viruses. We observed virus antigen detection in approximately 5% of the cultivated astrocytes, regardless of whether they originated from field viruses or lab strains (Table 1). Overall, the comparison of the in vitro and in vivo data indicates that the vulnerability of a specific group of cells in the central nervous system (CNS) observed under laboratory conditions (Figure 1) may not accurately represent the actual situation in a living organism (Figure 2, Figure 4, and Figure 6).

Possible explanations for this phenomenon include changes in the antiviral response patterns under less controlled cell culture conditions, variations in the replication and spread rates of the viruses in living organisms, or disparities in the release of infected neurons that are not connected through synapses, thereby enabling infection of non-synaptically connected cells in the central nervous system. While there is currently no available information about the latter, it is established that astrocytes in mouse brains are incompletely infected by a chimeric SAD L16 virus that expresses CVS-11 glycoprotein [33]. This indicates that the replication of rCVS-11, ERA, and SAD L16 in astrocytes was hindered by strong innate immune responses after infection through the i.m. or i.c. inoculation methods, respectively. Nevertheless, given the measurement of viral genome copies and virus mRNA levels was not conducted in this study, it is still uncertain whether the replication kinetics of the laboratory strains truly vary from those of the field strains in the infected brains. However, it is possible that a more effective reproduction of rRABV Fox, rRABV Dog, and rRABV Rac, and subsequently the buildup of the primary interferon antagonist phosphoprotein P [8,9] in astrocytes, resulted in the suppression of antiviral reactions. Lab-attenuated RABV has been found to differ from wild-type RABV by causing an enhanced generation of type I interferon (IFN) and expression of inflammatory cytokines through the mitochondrial antiviral-signaling protein (MAVS) pathway [34]. The immunofluorescence detection of RABV P in Figure 2, Figure 4, and Figure 6 revealed the presence of high amounts of the primary interferon antagonist in astrocytes infected with the field virus. The absence of P detection in the astrocytes of lab RABV-infected animals is unlikely to be caused by a significant differential in P gene expression of field viruses, given P was easily detectable in lab RABV-infected neurons. In addition to reducing the release of inflammatory cytokines, promoting virus replication in astrocytes can lead to a less noticeable or delayed overall antiviral response to field viruses. This phenomenon aligns with the documented ability of field viruses to evade the immune response in infected animals [35] and could potentially signify a significant immunological distinction from laboratory-adapted strains. Nevertheless, additional investigation will be required to validate this idea. Interestingly, while the ERA strain was not found in astrocytes following i.c. inoculation, the infection of astrocytes by rCVS-11 occurred at a rate of 13.4% after i.c. inoculation (Figure 6). Furthermore, the levels of astrocyte infection by rCVS-11 were higher than those observed for the field viruses rRABV Fox and rRABV Dog following both i.c. and i.m. infection. This demonstrated that the capacity of rCVS-11 to initiate an infection in these types of cells was contingent upon the route of infection. Given that viral infection from the outer regions of the body depends on the spread of viruses through synapses, it is possible that the virus may not be detectable by immune and non-neuronal cells. In this case, infection of astrocytes could be considered a later stage of brain infection, where there is a high level of infection in neurons and the release of virus particles outside of synapses, which may make it easier for astrocytes to become infected. As a result, the immunological reactions facilitated by astrocytes would be postponed, and the removal of the virus before the development of the disease would not be feasible. When compared to other viruses in the field, the slower replication and spreading kinetics of rCVS-11 could result in lower levels or the complete absence of detectable astrocyte infection after i.m. inoculation (Figure 5), even if the virus is primarily capable of infecting these types of cells (Figure 7b). The effectiveness of trans-synaptic retrograde propagation can vary between a highly neurotropic CVS-24 viral variant and a SAD L16-like vaccination virus [24]. At a more advanced level, this can also differentiate highly infectious field viruses like rRABV Fox and rRABV Dog from rCVS-11. Moreover, it is possible to consider that intracerebral (i.c.) inoculations serve as a direct route to infect the brain, causing simultaneous and multiple infections of neurons and astrocytes in a manner that does not involve transmission via synapses. The initial presence of a greater number of infected cells during central nervous system (CNS) infection, as opposed to trans-synaptic invasion following intramuscular (i.m.) infection, can result in a more rapid spread of the virus and the infection of multiple brain regions. This can lead to a higher release of the virus during the later stages of infection and the infection of astrocytes. This pattern is similar to what has been hypothesized for i.m. infections with highly virulent field viruses. Indeed, previous studies have demonstrated that astrocyte activation occurs more rapidly following infections within the central nervous system (i.c.) compared to infections in peripheral areas [58]. The kinetics of both virus and cell response may vary between the two routes of infection. The study found that there was a higher occurrence of neuronal cell death following intracranial (i.c.) infection with RABV compared to no detectable apoptosis in rats infected intramuscularly (i.m.) [59]. This suggests that there are qualitative changes in the way the host reacts to the virus. Astrocytes play a crucial role in the central nervous system by being the primary source of IFN-β expression and controlling viruses through the activation of Toll-like receptor (TLR) and RIG-I-like receptor (RLR) pathways. These pathways, along with the observed differences in infection routes, such as in the case of rCVS-11, may contribute to variations in apoptosis and other host reactions. However, additional research is necessary to discover whether the kinetics of the virus, which may vary depending on the route of infection, can impact the preference of astrocytes in living organisms, and how this preference influences subsequent host responses. Given the previous evidence of variations in the activation of the innate immune response by dsRNA between natural and weakened viruses [34], it is necessary to examine the potential of rRABV Fox, rRABV Dog, rRABV Rac, rCVS-11, ERA, and SAD L16 to activate the innate immune response in astrocytes. This investigation aims to determine if the route dependency of rCVS-11 is influenced by varying levels of innate immune activation. The probable reason for the absence of astrocyte infection by SAD L16 and ERA is the result of robust virus inhibition and clearance, as earlier studies have indicated that a similar virus may only lead to incomplete astrocyte infection. Future studies will investigate if a reduced induction of antiviral response may enable the more virulent rCVS-11 or extremely virulent field viruses to surpass a threshold of virus replication and antagonist expression, hence promoting continued replication. Although the specific mechanisms are not explained here, the similar outcomes seen for SAD L16 and the still harmful ERA strain strongly indicate that a significant distinction from the other examined viruses is the overall inhibition of observable astrocyte multiplication by these vaccine strains. We utilized innovative 3D immunofluorescence techniques to conduct thorough investigations of the cell tropism of both highly virulent field RABVs and less virulent or attenuated lab strains. Specifically, the use of high-resolution 3D pictures and quantitative downstream analysis has yielded new and valuable information about the infection processes during the clinical phase of this fatal illness. Despite the absence of observable astrocyte infection, the illness symptoms and fatal progression occur when the virus efficiently spreads in the brain following infection with any of the six tested viruses (Figure 7). Nevertheless, variations in virus kinetics and the innate immunological responses of astrocytes can impact the rate of development, the immune-related harm caused, and the subsequent dissemination of the virus to peripheral salivary glands. The latter may pose a significant challenge in terms of the transmission and persistence of viruses in host populations. While these factors need to be examined in distinct experiments, this study offers an innovative and quantitative foundation for a fresh and dynamic perspective on the interactions between RABV and its host in living organisms at the cellular level. Furthermore, this method highlights the capacity of immunostaining-compatible tissue clearing/3D imaging tools to precisely examine virus-host interactions at a detailed level, both at the cellular and subcellular levels. Although this study primarily examined typical cellular markers for neurons and astrocytes, future methods that incorporate markers related to the immune system and host pathways will enable the direct examination of infection processes in complex and morphologically intact tissues using high-resolution imaging.

CONCISE SUMMARY

While standard immunohistochemistry for neurotropic rabies virus (RABV) typically exhibits a strong affinity for neurons, it is worth noting that non-neuronal cells can also be affected. Specifically, astrocyte infection that does not result in full viral replication is believed to be a primary cause of innate immune responses in the central nervous system. Although in vitro studies have shown variations between field and less virulent lab-adapted RABVs, there is a lack of a systematic and quantitative evaluation of astrocyte tropism in vivo. In this study, a solvent-based technique called tissue clearing was employed to assess the cell tropism of RABV (rabies virus) in infected brains. The utilization of immunofluorescence analysis on tissue slices that were 1 mm in thickness allowed for the accurate identification and measurement of the frequencies of astrocyte and neuron infections in three dimensions. When comparing three extremely infectious viral clones obtained from fox, dog, and raccoon with three strains that were modified to the laboratory, significant variations were seen in their capacity to infect astrocytes in a living organism. Although all viruses and infection techniques resulted in neuron infection frequencies ranging from 7% to 19%, significant variations were seen for astrocytes. While astrocyte infection by field viruses was observed regardless of the method of inoculation (8-27% infection rate), only one strain that had been modified in the laboratory infected astrocytes in a manner that depended on the route of inoculation (0% infection rate after intramuscular (i.m.) inoculation and 13% infection rate after intracerebral (i.c.) inoculation). Both lab-adapted vaccine viruses did not infect astrocytes at all, with a 0% infection rate in both intracerebral and intramuscular routes of administration. This implies that the capacity to effectively infect astrocytes in a living organism is what sets apart field and weakened laboratory strains of the rabies virus.

REFERENCES

1. Fooks, A.R.; Cliquet, F.; Finke, S.; Freuling, C.; Hemachudha, T.; Mani, R.S.; Müller, T.; Nadin-Davis, S.; Picard-Meyer, E.; Wilde, H.; et al. Rabies. Nat. Rev. Dis. Primers 2017, 3, 17091.

2. Walker, P.J.; Blasdell, K.R.; Calisher, C.H.; Dietzgen, R.G.; Kondo, H.; Kurath, G.; Longdon, B.; Stone, D.M.; Tesh, R.B.; Tordo, N.; et al.

ICTV Virus Taxonomy Profile: Rhabdoviridae. J. Gen. Virol. 2018, 99, 447–448.

3. Finke, S.; Conzelmann, K.-K. Replication strategies of rabies virus. Virus Res. 2005, 111, 120–131.
4. Zhang, G.; Wang, H.; Mahmood, F.; Fu, Z.F. Rabies virus glycoprotein is an important determinant for the induction of innate immune responses and the pathogenic mechanisms. Vet. Microbiol. 2013, 162, 601–613.
5. Besson, B.; Sonthonnax, F.; Duchateau, M.; Ben Khalifa, Y.; Larrous, F.; Eun, H.; Hourdel, V.; Matondo, M.; Chamot-Rooke, J.; Grailhe, R.; et al. Regulation of NF-κB by the p105-ABIN2-TPL2 complex and RelAp43 during rabies virus infection. PLoS Pathog. 2017, 13, e1006697.
6. Masatani, T.; Ito, N.; Shimizu, K.; Ito, Y.; Nakagawa, K.; Sawaki, Y.; Koyama, H.; Sugiyama, M. Rabies virus nucleoprotein functions to evade activation of the RIG-I-mediated antiviral response. J. Virol. 2010, 84, 4002–4012.
7. Ben Khalifa, Y.; Luco, S.; Besson, B.; Sonthonnax, F.; Archambaud, M.; Grimes, J.M.; Larrous, F.; Bourhy, H. The matrix protein of rabies virus binds to RelAp43 to modulate NF-κB-dependent gene expression related to innate immunity. Sci. Rep. 2016, 6, 39420.
8. Brzózka, K.; Finke, S.; Conzelmann, K.-K. Identification of the rabies virus alpha/beta interferon antagonist: Phosphoprotein P interferes with phosphorylation of interferon regulatory factor 3. J. Virol. 2005, 79, 7673–7681.
9. Brzózka, K.; Finke, S.; Conzelmann, K.-K. Inhibition of interferon signaling by rabies virus phosphoprotein P: Activation-dependent binding of STAT1 and STAT2. J. Virol. 2006, 80, 2675–2683.
10. Ito, N.; Moseley, G.W.; Blondel, D.; Shimizu, K.; Rowe, C.L.; Ito, Y.; Masatani, T.; Nakagawa, K.; Jans, D.A.; Sugiyama, M. Role of interferon antagonist activity of rabies virus phosphoprotein in viral pathogenicity. J. Virol. 2010, 84, 6699–6710.

11. Faber, M.; Pulmanausahakul, R.; Hodawadekar, S.S.; Spitsin, S.; McGettigan, J.P.; Schnell, M.J.; Dietzschold, B. Overexpression of the rabies virus glycoprotein results in enhancement of apoptosis and antiviral immune response. J. Virol. 2002, 76, 3374–3381.

12. Jackson, A.C.; Rasalingam, P.; Weli, S.C. Comparative pathogenesis of recombinant rabies vaccine strain SAD-L16 and SAD-D29 with replacement of Arg333 in the glycoprotein after peripheral inoculation of neonatal mice: Less neurovirulent strain is a stronger inducer of neuronal apoptosis. Acta Neuropathol. 2006, 111, 372–378.

13. Morimoto, K.; Hooper, D.C.; Spitsin, S.; Koprowski, H.; Dietzschold, B. Pathogenicity of different rabies virus variants inversely correlates with apoptosis and rabies virus glycoprotein expression in infected primary neuron cultures. J. Virol. 1999, 73, 510–518.

14. Sarmento, L.; Li, X.-q.; Howerth, E.; Jackson, A.C.; Fu, Z.F. Glycoprotein-mediated induction of apoptosis limits the spread of attenuated rabies viruses in the central nervous system of mice. J. Neurovirol. 2005, 11, 571–581.

15. Dietzschold, B.; Li, J.; Faber, M.; Schnell, M. Concepts in the pathogenesis of rabies. Future Virol. 2008, 3, 481–490.

16. Tsiang, H.; Lycke, E.; Ceccaldi, P.E.; Ermine, A.; Hirardot, X. The anterograde transport of rabies virus in rat sensory dorsal root ganglia neurons. J. Gen. Virol. 1989, 70, 2075–2085.

17. Astic, L.; Saucier, D.; Coulon, P.; Lafay, F.; Flamand, A. The CVS strain of rabies virus as transneuronal tracer in the olfactory system of mice. Brain Res. 1993, 619, 146–156.

18. Lycke, E.; Tsiang, H. Rabies virus infection of cultured rat sensory neurons. J. Virol. 1987, 61, 2733–2741.

19. Ceccaldi, P.E.; Gillet, J.P.; Tsiang, H. Inhibition of the transport of rabies virus in the central nervous system. J. Neuropathol. Exp. Neurol. 1989, 48, 620–630.

20. Gluska, S.; Zahavi, E.E.; Chein, M.; Gradus, T.; Bauer, A.; Finke, S.; Perlson, E. Rabies Virus Hijacks and accelerates the p75NTR retrograde axonal transport machinery. PLoS Pathog. 2014, 10, e1004348.

21. Bauer, A.; Nolden, T.; Schröter, J.; Römer-Oberdörfer, A.; Gluska, S.; Perlson, E.; Finke, S. Anterograde glycoprotein-dependent transport of newly generated rabies virus in dorsal root ganglion neurons. J. Virol. 2014, 88, 14172–14183.

22. Wickersham, I.R.; Finke, S.; Conzelmann, K.-K.; Callaway, E.M. Retrograde neuronal tracing with a deletion-mutant rabies virus. Nat. Methods 2007, 4, 47–49.

23. Wickersham, I.R.; Lyon, D.C.; Barnard, R.J.O.; Mori, T.; Finke, S.; Conzelmann, K.-K.; Young, J.A.T.; Callaway, E.M. Monosynaptic restriction of transsynaptic tracing from single, genetically targeted neurons. Neuron 2007, 53, 639–647.

24. Reardon, T.R.; Murray, A.J.; Turi, G.F.; Wirblich, C.; Croce, K.R.; Schnell, M.J.; Jessell, T.M.; Losonczy, A. Rabies Virus CVS-N2c(ΔG) Strain Enhances Retrograde Synaptic Transfer and Neuronal Viability. Neuron 2016, 89, 711–724.

25. Thoulouze, M.I.; Lafage, M.; Schachner, M.; Hartmann, U.; Cremer, H.; Lafon, M. The neural cell adhesion molecule is a receptor for rabies virus. J. Virol. 1998, 72, 7181–7190.

26. Tuffereau, C.; Bénéjean, J.; Blondel, D.; Kieffer, B.; Flamand, A. Low-affinity nerve-growth factor receptor (P75NTR) can serve as a receptor for rabies virus. EMBO J. 1998, 17, 7250–7259.

27. Lentz, T.; Burrage, T.; Smith, A.; Crick, J.; Tignor, G. Is the acetylcholine receptor a rabies virus receptor? Science 1982, 215, 182–184.

28. Wang, J.; Wang, Z.; Liu, R.; Shuai, L.; Wang, X.; Luo, J.; Wang, C.; Chen, W.; Wang, X.; Ge, J.; et al. Metabotropic glutamate receptor subtype 2 is a cellular receptor for rabies virus. PLoS Pathog. 2018, 14, e1007189.

29. Ray, N.B.; Power, C.; Lynch, W.P.; Ewalt, L.C.; Lodmell, D.L. Rabies viruses infect primary cultures of murine, feline, and human microglia and astrocytes. Arch. Virol. 1997, 142, 1011–1019.

30. Davis, B.M.; Rall, G.F.; Schnell, M.J. Everything You Always Wanted to Know About Rabies Virus (But Were Afraid to Ask). Annu. Rev. Virol. 2015, 2, 451–471.

31. Jackson, A.C.; Phelan, C.C.; Rossiter, J.P. Infection of Bergmann glia in the cerebellum of a skunk experimentally infected with street rabies virus. Can. J. Vet. Res. 2000, 64, 226–228.

32. Tsiang, H.; Koulakoff, A.; Bizzini, B.; Berwald-Netter, Y. Neurotropism of rabies virus. An in vitro study. J. Neuropathol. Exp. Neurol. 1983, 42, 439–452.

33. Pfefferkorn, C.; Kallfass, C.; Lienenklaus, S.; Spanier, J.; Kalinke, U.; Rieder, M.; Conzelmann, K.-K.; Michiels, T.; Staeheli, P. Abortively Infected Astrocytes Appear to Represent the Main Source of Interferon Beta in the Virus-Infected Brain. J. Virol. 2016, 90, 2031–2038.

34. Tian, B.; Zhou, M.; Yang, Y.; Yu, L.; Luo, Z.; Tian, D.; Wang, K.; Cui, M.; Chen, H.; Fu, Z.F.; et al. Lab-Attenuated Rabies Virus Causes Abortive Infection and Induces Cytokine Expression in Astrocytes by Activating Mitochondrial Antiviral-Signaling Protein Signaling Pathway. Front. Immunol. 2018, 8, 2011.

35. Ito, N.; Moseley, G.W.; Sugiyama, M. The importance of immune evasion in the pathogenesis of rabies virus. J. Vet. Med. Sci. 2016, 78, 1089–1098.

36. Suja, M.S.; Mahadevan, A.; Madhusudana, S.N.; Shankar, S.K. Role of apoptosis in rabies viral encephalitis: A comparative study in mice, canine, and human brain with a review of literature. Patholog. Res. Int. 2011, 2011, 374286.

37. Wang, Z.W.; Sarmento, L.; Wang, Y.; Li, X.-q.; Dhingra, V.; Tseggai, T.; Jiang, B.; Fu, Z.F. Attenuated Rabies Virus Activates, while Pathogenic Rabies Virus Evades, the Host Innate Immune Responses in the Central Nervous System. J. Virol. 2005, 79, 12554–12565.

38. Renier, N.; Wu, Z.; Simon, D.J.; Yang, J.; Ariel, P.; Tessier-Lavigne, M. iDISCO: A simple, rapid method to immunolabel large tissue samples for volume imaging. Cell 2014, 159, 896–910.

39. Pan, C.; Cai, R.; Quacquarelli, F.P.; Ghasemigharagoz, A.; Lourbopoulos, A.; Matryba, P.; Plesnila, N.; Dichgans, M.; Hellal, F.; Ertürk, A. Shrinkage-mediated imaging of entire organs and organisms using uDISCO. Nat. Methods 2016, 13, 859–867.

40. Zaeck, L.; Potratz, M.; Freuling, C.M.; Müller, T.; Finke, S. High-Resolution 3D Imaging of Rabies Virus Infection in Solvent-Cleared Brain Tissue. J. Vis. Exp. 2019, 30.

41. Brewer, G.J.; Torricelli, J.R. Isolation and culture of adult neurons and neurospheres. Nat. Protoc. 2007, 2, 1490–1498.

42. Nolden, T.; Pfaff, F.; Nemitz, S.; Freuling, C.M.; Höper, D.; Müller, T.; Finke, S. Reverse genetics in high throughput: Rapid generation of complete negative strand RNA virus cDNA clones and recombinant viruses thereof. Sci. Rep. 2016, 6, 23887.

43. Schnell, M.J.; Mebatsion, T.; Conzelmann, K.K. Infectious rabies viruses from cloned cDNA. EMBO J. 1994, 4195–4203.

44. Abelseth, M.K. An attenuated rabies vaccine for domestic animals produced in tissue culture. Can. Vet. J. 1964, 5, 279–286.

45. Höper, D.; Freuling, C.M.; Müller, T.; Hanke, D.; von Messling, V.; Duchow, K.; Beer, M.; Mettenleiter, T.C. High definition viral vaccine strain identity and stability testing using full-genome population data--The next generation of vaccine quality control. Vaccine 2015, 33, 5829–5837.

46. Vos, A.; Nolden, T.; Habla, C.; Finke, S.; Freuling, C.M.; Teifke, J.; Müller, T. Raccoons (Procyon lotor) in Germany as potential reservoir species for Lyssaviruses. Eur J. Wildl Res. 2013, 59, 637–643.

47. Fu, C.; Donovan, W.P.; Shikapwashya-Hasser, O.; Ye, X.; Cole, R.H. Hot Fusion: An efficient method to clone multiple DNA fragments as well as inverted repeats without ligase. PLoS ONE 2014, 9, e115318.

48. Finke, S.; Granzow, H.; Hurst, J.; Pollin, R.; Mettenleiter, T.C. Intergenotypic replacement of lyssavirus matrix proteins demonstrates the role of lyssavirus M proteins in intracellular virus accumulation. J. Virol. 2010, 84, 1816–1827.

49. Buchholz, U.J.; Finke, S.; Conzelmann, K.K. Generation of bovine respiratory syncytial virus (BRSV) from cDNA: BRSV NS2 is not essential for virus replication in tissue culture, and the human RSV leader region acts as a functional BRSV genome promoter. J. Virol. 1999, 73, 251–259.

50. Orbanz, J.; Finke, S. Generation of recombinant European bat lyssavirus type 1 and inter-genotypic compatibility of lyssavirus genotype 1 and 5 antigenome promoters. Arch. Virol. 2010, 155, 1631–1641.

51. Eggerbauer, E.; Pfaff, F.; Finke, S.; Höper, D.; Beer, M.; Mettenleiter, T.C.; Nolden, T.; Teifke, J.-P.; Müller, T.; Freuling, C.M. Comparative analysis of European bat lyssavirus 1 pathogenicity in the mouse model. PLoS Negl. Trop. Dis. 2017, 11, e0005668.

52. Schindelin, J.; Arganda-Carreras, I.; Frise, E.; Kaynig, V.; Longair, M.; Pietzsch, T.; Preibisch, S.; Rueden, C.; Saalfeld, S.; Schmid, B.; et al. Fiji: An open-source platform for biological-image analysis. Nat. Methods 2012, 9, 676–682.

53. Bolte, S.; Cordelières, F.P. A guided tour into subcellular colocalization analysis in light microscopy. J. Microsc. 2006, 224, 213–232.

54. de Chaumont, F.; Dallongeville, S.; Chenouard, N.; Hervé, N.; Pop, S.; Provoost, T.; Meas-Yedid, V.; Pankajakshan, P.; Lecomte, T.; Le Montagner, Y.; et al. Icy: An open bioimage informatics platform for extended reproducible research. Nat. Methods 2012, 9, 690–696.

55. Jackson, A.C.; Reimer, D.L. Pathogenesis of experimental rabies in mice: An immunohistochemical study. Acta Neuropathol. 1989, 78, 159–165.

56. Conzelmann, K.K.; Cox, J.H.; Schneider, L.G.; Thiel, H.J. Molecular cloning and complete nucleotide sequence of the attenuated rabies virus SAD B19. Virology 1990, 175, 485–499.

57. Weiland, F.; Cox, J.H.; Meyer, S.; Dahme, E.; Reddehase, M.J. Rabies Virus Neuritic Paralysis: Immunopathogenesis of Nonfatal Paralytic Rabies. J. Virol. 1992, 5096–5099.

58. Zlotnik, I. The reaction of astrocytes to acute virus infections of the central nervous system. Br. J. Exp. Pathol 1968, 49, 555–564.

59. Jackson, A.C. Rabies. Scientific Basis of the Disease and Its Management, 3rd ed.; Elsevier Science: San Diego, CA, USA, 2013; ISBN 9780123965479.

60. Detje, C.N.; Lienenklaus, S.; Chhatbar, C.; Spanier, J.; Prajeeth, C.K.; Soldner, C.; Tovey, M.G.; Schlüter, D.; Weiss, S.; Stangel, M.; et al. Upon intranasal vesicular stomatitis virus infection, astrocytes in the olfactory bulb are important interferon Beta producers that protect from lethal encephalitis. J. Virol. 2015, 89, 2731–2738.

61. Kallfass, C.; Ackerman, A.; Lienenklaus, S.; Weiss, S.; Heimrich, B.; Staeheli, P. Visualizing production of beta interferon by astrocytes and microglia in brain of La Crosse virus-infected mice. J. Virol. 2012, 86, 11223–11230.

INDEX

www.ingramcontent.com/pod-product-compliance
Lightning Source LLC
LaVergne TN
LVHW011839200926
848484LV00002B/379
* 9 7 8 1 7 8 7 1 5 4 3 5 3 *